Gregory Levitin (Ed.)

Computational Intelligence in Reliability Engineering

T0137861

Studies in Computational Intelligence, Volume 40

Editor-in-chief
Prof. Janusz Kacprzyk
Systems Research Institute
Polish Academy of Sciences
ul. Newelska 6
01-447 Warsaw
Poland
E-mail: kacprzyk@ibspan.waw.pl

Further volumes of this series
can be found on our homepage:
springer.com

Vol. 23. M. Last, Z. Volkovich, A. Kandel (Eds.)
Algorithmic Techniques for Data Mining, 2006
ISBN 3-540-33880-2

Vol. 24. Alakananda Bhattacharya, Amit Konar,
Ajit K. Mandal
Parallel and Distributed Logic Programming,
2006
ISBN 3-540-33458-0

Vol. 25. Zoltán Ésik, Carlos Martín-Vide,
Victor Mitrana (Eds.)
*Recent Advances in Formal Languages
and Applications,* 2006
ISBN 3-540-33460-2

Vol. 26. Nadia Nedjah, Luiza de Macedo Mourelle
(Eds.)
Swarm Intelligent Systems, 2006
ISBN 3-540-33868-3

Vol. 27. Vassilis G. Kaburlasos
*Towards a Unified Modeling and Knowledge-
Representation based on Lattice Theory,* 2006
ISBN 3-540-34169-2

Vol. 28. Brahim Chaib-draa, Jörg P. Müller (Eds.)
Multiagent based Supply Chain Management, 2006
ISBN 3-540-33875-6

Vol. 29. Sai Sumathi, S.N. Sivanandam
*Introduction to Data Mining and its
Applications,* 2006
ISBN 3-540-34689-9

Vol. 30. Yukio Ohsawa, Shusaku Tsumoto (Eds.)
*Chance Discoveries in Real World Decision
Making,* 2006
ISBN 3-540-34352-0

Vol. 31. Ajith Abraham, Crina Grosan, Vitorino
Ramos (Eds.)
Stigmergic Optimization, 2006
ISBN 3-540-34689-9

Vol. 32. Akira Hirose
Complex-Valued Neural Networks, 2006
ISBN 3-540-33456-4

Vol. 33. Martin Pelikan, Kumara Sastry, Erick
Cantú-Paz (Eds.)
*Scalable Optimization via Probabilistic
Modeling,* 2006
ISBN 3-540-34953-7

Vol. 34. Ajith Abraham, Crina Grosan, Vitorino
Ramos (Eds.)
Swarm Intelligence in Data Mining, 2006
ISBN 3-540-34955-3

Vol. 35. Ke Chen, Lipo Wang (Eds.)
Trends in Neural Computation, 2007
ISBN 3-540-36121-9

Vol. 36. Ildar Batyrshin, Janusz Kacprzyk, Leonid
Sheremetor, Lotfi A. Zadeh (Eds.)
*Perception-based Data Mining and Decision
Making in Economics and Finance,* 2006
ISBN 3-540-36244-4

Vol. 37. Jie Lu, Da Ruan, Guangquan Zhang (Eds.)
E-Service Intelligence, 2007
ISBN 3-540-37015-3

Vol. 38. Art Lew, Holger Mauch
Dynamic Programming, 2007
ISBN 3-540-37013-7

Vol. 39. Gregory Levitin (Ed.)
*Computational Intelligence in Reliability
Engineering,* 2007
ISBN 3-540-37367-5

Vol. 40. Gregory Levitin (Ed.)
*Computational Intelligence in Reliability
Engineering,* 2007
ISBN 3-540-37371-3

Gregory Levitin (Ed.)

Computational Intelligence in Reliability Engineering

New Metaheuristics, Neural and Fuzzy Techniques in Reliability

With 90 Figures and 53 Tables

 Springer

Dr. Gregory Levitin
Research & Development Division
The Israel Electronic Corporation Ltd.
PO Box 10
31000 Haifa
Israel
E-mail: levitin@iec.co.il

ISSN print edition: 1860-949X
ISSN electronic edition: 1860-9503
ISBN 978-3-642-07219-2   e-ISBN 978-3-540-37372-8

Springer is a part of Springer Science+Business Media
springer.com
© Springer-Verlag Berlin Heidelberg 2007
Softcover reprint of the hardcover 1st edition 2007

Cover design: deblik, Berlin

Preface

This two-volume book covers the recent applications of computational intelligence techniques in reliability engineering. Research in the area of computational intelligence is growing rapidly due to the many successful applications of these new techniques in very diverse problems. "Computational Intelligence" covers many fields such as neural networks, fuzzy logic, evolutionary computing, and their hybrids and derivatives. Many industries have benefited from adopting this technology. The increased number of patents and diverse range of products developed using computational intelligence methods is evidence of this fact.

These techniques have attracted increasing attention in recent years for solving many complex problems. They are inspired by nature, biology, statistical techniques, physics and neuroscience. They have been successfully applied in solving many complex problems where traditional problem-solving methods have failed.

The book aims to be a repository for the current and cutting-edge applications of computational intelligent techniques in reliability analysis and optimization.

In recent years, many studies on reliability optimization use a universal optimization approach based on metaheuristics. These metaheuristics hardly depend on the specific nature of the problem that is solved and, therefore, can be easily applied to solve a wide range of optimization problems. The metaheuristics are based on artificial reasoning rather than on classical mathematical programming. Their important advantage is that they do not require any information about the objective function besides its values corresponding to the points visited in the solution space. All metaheuristics use the idea of randomness when performing a search, but they also use past knowledge in order to direct the search. Such algorithms are known as randomized search techniques.

Genetic algorithms are one of the most widely used metaheuristics. They were inspired by the optimization procedure that exists in nature, the biological phenomenon of evolution. The first volume of this book starts with a survey of the contributions made to the optimal reliability design literature in the resent years. The next chapter is devoted to using the metaheuristics in multiobjective reliability optimization. The volume also contains chapters devoted to different applications of the genetic algorithms in reliability engineering and to combinations of this algorithm with other computational intelligence techniques.

The second volume contains chapters presenting applications of other metaheuristics such as ant colony optimization, great deluge algorithm, cross-entropy method and particle swarm optimization. It also includes chapters devoted to such novel methods as cellular automata and support vector machines. Several chapters present different applications of artificial neural networks, a powerful adaptive technique that can be used for learning, prediction and optimization. The volume also contains several chapters describing different aspects of imprecise reliability and applications of fuzzy and vague set theory.

All of the chapters are written by leading researchers applying the computational intelligence methods in reliability engineering.

This two-volume book will be useful to postgraduate students, researchers, doctoral students, instructors, reliability practitioners and engineers, computer scientists and mathematicians with interest in reliability.

I would like to express my sincere appreciation to Professor Janusz Kacprzyk from the Systems Research Institute, Polish Academy of Sciences, Editor-in-Chief of the Springer series "Studies in Computational Intelligence", for providing me with the chance to include this book in the series.

I wish to thank all the authors for their insights and excellent contributions to this book. I would like to acknowledge the assistance of all involved in the review process of the book, without whose support this book could not have been successfully completed. I want to thank the authors of the book who participated in the reviewing process and also Prof. F. Belli, University of Paderborn, Germany, Prof. Kai-Yuan Cai, Beijing University of Aeronautics and Astronautics, Dr. M. Cepin, Jozef Stefan Institute, Ljubljana , Slovenia, Prof. M. Finkelstein, University of the Free State, South Africa, Prof. A. M. Leite da Silva, Federal University of Itajuba, Brazil, Prof. Baoding Liu, Tsinghua University, Beijing, China, Dr. M. Muselli, Institute of Electronics, Computer and Telecommunication Engineering, Genoa, Italy, Prof. M. Nourelfath, Université Laval, Quebec, Canada, Prof. W. Pedrycz, University of Alberta, Edmonton, Canada, Dr. S. Porotsky, FavoWeb, Israel, Prof. D. Torres, Universidad Central de Venezuela, Dr. Xuemei Zhang, Lucent Technologies, USA for their insightful comments on the book chapters.

I would like to thank the Springer editor Dr. Thomas Ditzinger for his professional and technical assistance during the preparation of this book.

Haifa, Israel, 2006   Gregory Levitin

Contents

1 The Ant Colony Paradigm for Reliable Systems Design
Yun-Chia Liang, Alice E. Smith... 1
  1.1 Introduction... 1
  1.2 Problem Definition 5
    1.2.1 Notation... 5
    1.2.2 Redundancy Allocation Problem... 6
  1.3 Ant Colony Optimization Approach... 7
    1.3.1 Solution Encoding... 7
    1.3.2 Solution Construction.. 8
    1.3.3 Objective Function 9
    1.3.4 Improving Constructed Solutions Through Local Search 10
    1.3.5 Pheromone Trail Intensity Update... 10
    1.3.6 Overall Ant Colony Algorithm.. 11
  1.4 Computational Experience.. 11
  1.5 Conclusions.. 16
  References... 18

2 Modified Great Deluge Algorithm versus Other Metaheuristics in Reliability Optimization
Vadlamani Ravi 21
  2.1 Introduction.. 21
  2.2 Problem Description 23
  2.3 Description of Various Metaheuristics 25
    2.3.1 Simulated Annealing (SA) 25
    2.3.2 Improved Non-equilibrium Simulated Annealing (INESA)........... 26
    2.3.3 Modified Great Deluge Algorithm (MGDA) 26
      2.3.3.1 Great Deluge Algorithm 27
      2.3.3.2 The MGDA.. 27
  2.4 Discussion of Results.. 30
  2.5 Conclusions.. 33
  References... 33
  Appendix 34

3 Applications of the Cross-Entropy Method in Reliability
Dirk P. Kroese, Kin-Ping Hui 37
   3.1 Introduction 37
      3.1.1 Network Reliability Estimation.. 37
      3.1.2 Network Design 38
   3.2 Reliability 39
      3.2.1 Reliability Function.. 42
      3.2.2 Network Reliability 44
   3.3 Monte Carlo Simulation 45
      3.3.1 Permutation Monte Carlo and the Construction Process.................. 46
      3.3.2 Merge Process 48
   3.4 Reliability Estimation using the CE Method 50
      3.4.1 CE Method 52
      3.4.2 Tail Probability Estimation 53
      3.4.3 CMC and CE (CMC-CE) 54
      3.4.4 CP and CE (CP-CE) 56
      3.4.5 MP and CE (MP-CE) 57
      3.4.6 Numerical Experiments... 59
      3.4.7 Summary of Results 62
   3.5 Network Design and Planning 62
      3.5.1 Problem Description... 63
      3.5.2 The CE Method for Combinatorial Optimization.......................... 64
         3.5.2.1 Random Network Generation.. 64
         3.5.2.2 Updating Generation Parameters.. 65
         3.5.2.3 Noisy Optimization 66
      3.5.3 Numerical Experiment 66
   3.6 Network Recovery and Expansion 68
      3.6.1 Problem Description... 68
      3.6.2 Reliability Ranking 69
         3.6.2.1 Edge Relocated Networks 69
         3.6.2.2 Coupled Sampling 70
         3.6.2.3 Synchronous Construction Ranking (SCR) 71
      3.6.3 CE Method 74
         3.6.3.1 Random Network Generation.. 74
         3.6.3.2 Updating Generation Parameters.. 74
      3.6.4 Hybrid Optimization Method... 77
         3.6.4.1 Multi-optima Termination 77
         3.6.4.2 Mode Switching.. 78
      3.6.5 Comparison Between the Methods... 79
   References 80

4 Particle Swarm Optimization in Reliability Engineering
Gregory Levitin, Xiaohui Hu, Yuan-Shun Dai 83
   4.1 Introduction 83
   4.2 Description of PSO and MO-PSO 84
      4.2.1 Basic Algorithm 85

4.2.2 Parameter Selection in PSO...86
    4.2.2.1 Learning Factors...86
    4.2.2.2 Inertia Weight..87
    4.2.2.3 Maximum Velocity...87
    4.2.2.4 Neighborhood Size..87
    4.2.2.5 Termination Criteria..88
4.2.3 Handling Constraints in PSO...88
4.2.4 Handling Multi-objective Problems with PSO...............................89
4.3 Single-Objective Reliability Allocation...91
    4.3.1 Background..91
    4.3.2 Problem Formulation...92
        4.3.2.1 Assumptions..92
        4.3.2.2 Decision variables..92
        4.3.2.3 Objective Function...93
        4.3.2.4 The Problem..94
    4.3.3 Numerical Comparison...95
4.4 Single-Objective Redundancy Allocation..96
    4.4.1 Problem Formulation...96
        4.4.1.1 Assumptions..96
        4.4.1.2 Decision Variable...96
        4.4.1.3 Objective Function...97
    4.4.2 Numerical Comparison...98
4.5 Single Objective Weighted Voting System Optimization......................99
    4.5.1 Problem Formulation...99
    4.5.2 Numerical Comparison...101
4.6 Multi-Objective Reliability Allocation...105
    4.6.1 Problem Formulation...105
    4.6.2 Numerical Comparison...106
4.7 PSO Applicability and Efficiency..108
References...109

5 Cellular Automata and Monte Carlo Simulation for Network Reliability
and Availability Assessment
Claudio M. Rocco S., Enrico Zio...113
    5.1 Introduction...113
    5.2 Basics of CA Computing..115
        5.2.1 One-dimensional CA...116
        5.2.2 Two-dimensional CA..118
        5.2.3 CA Behavioral Classes...118
    5.3 Fundamentals of Monte Carlo Sampling and Simulation...............119
        5.3.1 The System Transport Model..119
        5.3.2 Monte Carlo Simulation for Reliability Modeling......................120
    5.4 Application of CA for the Reliability Assessment of Network Systems .122
        5.4.1 S-T Connectivity Evaluation Problem.......................................123
        5.4.2 S-T Network Steady-state Reliability Assessment......................124
            5.4.2.1 Example...125

5.4.2.2 Connectivity Changes...125
5.4.2.3 Steady-state Reliability Assessment....................126
5.4.3 The All-Terminal Evaluation Problem 127
5.4.3.1 The CA Model...127
5.4.3.2 Example...128
5.4.3.3 All-terminal Reliability Assessment: Application.................128
5.4.4 The *k*-Terminal Evaluation Problem 130
5.4.5 Maximum Unsplittable Flow Problem 130
5.4.5.1 The CA Model...130
5.4.5.2 Example...132
5.4.6 Maximum Reliability Path 134
5.4.6.1 Shortest Path..134
5.4.6.2 Example...135
5.4.6.3 Example...136
5.4.6.4 Maximum Reliability Path Determination................136
5.5 MC-CA network availability assessment............................138
5.5.1 Introduction...138
5.5.2 A Case Study of Literature.....................................140
5.6 Conclusions 141
References 142
Appendix 143

6 Network Reliability Assessment through Empirical Models Using a Machine Learning Approach
Claudio M. Rocco S., Marco Muselli 145
6.1 Introduction: Machine Learning (ML) Approach to Reliability
Assessment 145
6.2 Definitions 147
6.3 Machine Learning Predictive Methods...............................149
6.3.1 Support Vector Machines..149
6.3.2 Decision Trees..154
6.3.2.1 Building the Tree...156
6.3.2.2 Splitting Rules...157
6.3.2.3 Shrinking the Tree...159
6.3.3 Shadow Clustering (SC)...159
6.3.3.1 Building Clusters...162
6.3.3.2 Simplifying the Collection of Clusters 164
6.4 Example...164
6.4.1 Performance Results..166
6.4.2 Rule Extraction Evaluation 169
6.5 Conclusions 171
References 172

7 Neural Networks for Reliability-Based Optimal Design
Ming J Zuo, Zhigang Tian, Hong-Zhong Huang............................175
7.1 Introduction 175

7.1.1 Reliability-based Optimal Design 175
7.1.2 Challenges in Reliability-based Optimal Design 177
7.1.3 Neural Networks 177
7.1.4 Content of this Chapter 178
7.2 Feed-forward Neural Networks as a Function Approximator 179
7.2.1 Feed-forward Neural Networks 179
7.2.2 Evaluation of System Utility of a Continuous-state Series-parallel
System 182
7.2.3 Other Applications of Neural Networks as a Function
Approximator 186
  7.2.3.1 Reliability Evaluation of a k-out-of-n System Structure 186
  7.2.3.2 Performance Evaluation of a Series-parallel System Under
  Fuzzy Environment 187
  7.2.3.3 Evaluation of All-terminal Reliability in Network Design 187
  7.2.3.4 Evaluation of Stress and Failure Probability in Large-scale
  Structural Design 188
7.3 Hopfield Networks as an Optimizer 189
7.3.1 Hopfield Networks 189
7.3.2 Network Design with Hopfield ANN 190
7.3.3 Series System Design with Quantized Hopfield ANN 192
7.4 Conclusions 194
References 195

8 Software Reliability Predictions using Artificial Neural Networks
Q.P. Hu, M. Xie and S.H. Ng 197
8.1 Introduction 197
8.2 Overview of Software Reliability Models 200
8.2.1 Traditional Models for Fault Detection Process 200
  8.2.1.1 NHPP Models 200
  8.2.1.2 Markov Models 201
  8.2.1.3 Bayesian Models 201
  8.2.1.4 ANN Models 201
8.2.2 Models for Fault Detection and Correction Processes 202
  8.2.2.1 Extensions on Analytical Models 202
  8.2.2.2 Extensions on ANN Models 203
8.3 Combined ANN Models 204
8.3.1 Problem Formulation 205
8.3.2 General Prediction Procedure 205
  8.3.2.1 Data Normalization 206
  8.3.2.2 Network Training 206
  8.3.2.3 Fault Prediction 207
8.3.3 Combined Feedforward ANN Model 207
  8.3.3.1 ANN Framework 207
  8.3.3.2 Performance Evaluation 208
  8.3.3.3 Network Configuration 209
8.3.4 Combined Recurrent ANN Model 209

8.3.4.1 ANN Framework 209
8.3.4.2 Robust Configuration Evaluation 210
8.3.4.3 Network Configuration through Evolution 211
8.4 Numerical Analysis 212
8.4.1 Feedforward ANN Application 213
8.4.2 Recurrent ANN Application... 215
8.4.3 Comparison of Combined Feedforward & Recurrent Model 216
8.5 Comparisons with Separate Models... 216
8.5.1 Combined ANN Models vs Separate ANN Model 217
8.5.2 Combined ANN Models vs Paired Analytical Model 218
8.6 Conclusions and Discussions... 219
References 220

9 Computation Intelligence in Online Reliability Monitoring
Ratna Babu Chinnam, Bharatendra Rai 223
9.1 Introduction 223
9.1.1 Individual Component versus Population Characteristics.............. 223
9.1.2 Diagnostics and Prognostics for Condition-Based Maintenance...... 225
9.2 Performance Reliability Theory.. 228
9.3 Feature Extraction from Degradation Signals................................. 230
9.3.1 Time, Frequency, and Mixed-Domain Analysis 231
9.3.2 Wavelet Preprocessing of Degradation Signals.......................... 233
9.3.3 Multivariate Methods for Feature Extraction 236
9.4 Fuzzy Inference Models for Failure Definition 237
9.5 Online Reliability Monitoring with Neural Networks 239
9.5.1 Motivation for Using FFNs for Degradation Signal Modeling 240
9.5.2 Finite-Duration Impulse Response Multi-layer Perceptron
Networks 241
9.5.3 Self-Organizing Maps 242
9.5.4 Modeling Dispersion Characteristics of Degradation Signals.......... 243
9.6 Drilling Process Case Study 246
9.6.1 Experimental Setup 247
9.6.2 Actual Experimentation.. 247
9.6.3 Sugeno FIS for Failure Definition... 248
9.6.4 Online Reliability Estimation using Neural Networks 251
9.7 Summary, Conclusions and Future Research 253
References 254

10 Imprecise reliability: An introductory overview
Lev V. Utkin, Frank P.A. Coolen.. 261
10.1 Introduction 261
10.2 System Reliability Analysis... 266
10.3 Judgements in Imprecise Reliability.. 272
10.4 Imprecise Probability Models for Inference 274
10.5 Second-order Reliability Models.. 278
10.6 Reliability of Monotone Systems 281

10.7 Multi-state and Continuum-state Systems 283
10.8 Fault Tree Analysis...284
10.9 Repairable Systems..285
10.10 Structural Reliability..287
10.11 Software Reliability 288
10.12 Human Reliability...291
10.13 Risk Analysis..292
10.14 Security Engineering...293
10.15 Concluding Remarks and Open Problems 295
References..297

11 Posbist Reliability Theory for Coherent Systems
Hong-Zhong Huang, Xin Tong, Ming J Zuo 307
11.1 Introduction...307
11.2 Basic Concepts in the Possibility Context 310
   11.2.1 Lifetime of the System 311
   11.2.2 State of the System..312
11.3 Posbist Reliability Analysis of Typical Systems 313
   11.3.1 Posbist Reliability of a Series System 313
   11.3.2 Posbist Reliability of a Parallel System...315
   11.3.3 Posbist Reliability of a Series-parallel Systems 316
   11.3.4 Posbist Reliability of a Parallel-series System 317
   11.3.5 Posbist Reliability of a Cold Standby System 317
11.4 Posbist Fault Tree Analysis of Coherent Systems 319
   11.4.1 Posbist Fault Tree Analysis of Coherent Systems...........................321
      11.4.1.1 Basic Definitions of Coherent Systems 321
      11.4.1.2 Basic Assumptions..322
   11.4.2 Construction of the Model of Posbist Fault Tree Analysis..............322
      11.4.2.1 The Structure Function of Posbist Fault Tree 323
      11.4.2.2 Quantitative Analysis..324
11.5 The Methods for Developing Possibility Distributions..........................326
   11.5.1 Possibility Distributions Based on Membership Functions.............326
      11.5.1.1 Fuzzy Statistics 327
      11.5.1.2 Transformation of Probability Distributions to Possibility
      Distributions 327
      11.5.1.3 Heuristic Methods...328
      11.5.1.4 Expert Opinions..330
   11.5.2 Transformation of Probability Distributions to Possibility
   Distributions...330
      11.5.2.1 The Bijective Transformation Method......................................330
      11.5.2.2 The Conservation of Uncertainty Method 331
   11.5.3 Subjective Manipulations of Fatigue Data 333
11.6 Examples..335
   11.6.1 Example 1..335
      11.6.1.1 The Series System 336
      11.6.1.2 The Parallel System 336

11.6.1.3 The Cold Standby System 337
11.6.2 Example 2 337
11.6.3 Example 3 339
11.7 Conclusions 342
References 344

12 Analyzing Fuzzy System Reliability Based on the Vague Set Theory
Shyi-Ming Chen 347
12.1 Introduction 347
12.2 A Review of Chen and Jong's Fuzzy System Reliability Analysis
Method 348
12.3 Basic Concepts of Vague Sets 353
12.4 Analyzing Fuzzy System Reliability Based on Vague Sets 358
12.4.1 Example 359
12.5 Conclusions 361
References 361

13 Fuzzy Sets in the Evaluation of Reliability
Olgierd Hryniewicz 363
13.1 Introduction 363
13.2 Evaluation of Reliability in Case of Imprecise Probabilities 365
13.3 Possibilistic Approach to the Evaluation of Reliability 371
13.4 Statistical Inference with Imprecise Reliability Data 374
13.4.1 Fuzzy Estimation of Reliability Characteristics 374
13.4.2 Fuzzy Bayes Estimation of Reliability Characteristics 381
13.5 Conclusions 383
References 384

**14 Grey Differential Equation GM(1,1) Modeling In Repairable System
Modeling**
Renkuan Guo 387
14.1 Introduction 387
14.1.1 Small Sample Difficulties and Grey Thinking 387
14.1.2 Repair Effect Models and Grey Approximation 389
14.2 The Foundation of GM(1,1) Model 391
14.2.1 Equal-Spaced GM(1,1) Model 391
14.2.2 The Unequal-Spaced GM(1,1) Model 394
14.2.3 A two-stage GM(1,1) Model for Continuous Data 396
14.2.4 The Weight Factor in GM(1,1) Model 397
14.3 A Grey Analysis on Repairable System Data 399
14.3.1 Cement Roller Data 399
14.3.2 An Interpolation-least-square Modeling 400
14.3.3 A two-stage Least-square Modeling Approach 404
14.3.4 Prediction of Next Failure Time 407
14.4 Concluding Remarks 408
References 409

The Ant Colony Paradigm for Reliable Systems Design

Yun-Chia Liang

Department of Industrial Engineering and Management,
Yuan Ze University

Alice E. Smith

Department of Industrial and Systems Engineering, Auburn University

1.1 Introduction

This chapter introduces a relatively new meta-heuristic for combinatorial optimization, the ant colony. The ant colony algorithm is a multiple solution global optimizer that iterates to find optimal or near optimal solutions. Like its siblings genetic algorithms and simulated annealing, it is inspired by observation of natural systems, in this case, the behavior of ants in foraging for food. Since there are many difficult combinatorial problems in the design of reliable systems, applying new meta-heuristics to this field makes sense. The ant colony approach with its flexibility and exploitation of solution structure is a promising alternative to exact methods, rules of thumb and other meta-heuristics.

The most studied design configuration of the reliability systems is a series system of s independent k-out-of- :G subsystems as illustrated in Figure 1. A subsystem i is functioning properly if at least k_i of its n _i components are operational and a series-parallel system is where _i = one for all subsystems. In this problem, multiple component choices are used in parallel in each subsystem. Thus, the problem is to select the optimal combination of components and redundancy levels to meet system level constraints while maximizing system reliability. Such a redundancy allocation problem (RAP) is NP-hard [6] and has been well studied (see Tillman, et al. [45] and Kuo & Prasad [25]).

Y.-C. Liang and A.E. Smith: *The Ant Colony Paradigm for Reliable Systems Design*, Computational Intelligence in Reliability Engineering (SCI) **40**, 1–20 (2007)
www.springerlink.com   © Springer-Verlag Berlin Heidelberg 2007

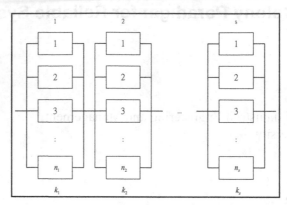

Fig. 1. Typical series-parallel system configuration.

Exact optimization approaches to the RAP include dynamic programming [2, 20, 35], integer programming [3, 22, 23, 33], and mixed-integer and nonlinear programming [31, 46]. Because of the exponential increase in search space with problem size, heuristics have become a common alternative to exact methods. Meta-heuristics, in particular, are global optimizers that offer flexibility while not being confined to specific problem types or instances. Genetic algorithms (GA) have been applied by Painton & Campbell [37], Levitin et al. [26], and Coit & Smith [7, 8]. Ravi et al. propose simulated annealing (SA) [39], fuzzy logic [40], and a modified great deluge algorithm [38] to optimize the complex system reliability. Kulturel-Konak et al. [24] use a Tabu search (TS) algorithm embedded with an adaptive version of the penalty function in [7] to solve RAPs. Three types of benchmark problems which consider the objectives of system cost minimization and system reliability maximization respectively were used to evaluate the algorithm performance. Liang and Wu [27] employ a variable neighborhood descent (VND) algorithm for the RAP. Four neighborhood search methods are defined to explore both the feasible and infeasible solution space.

Ant Colony Optimization (ACO) is one of the adaptive meta-heuristic optimization methods inspired by nature which include simulated annealing (SA), particle swarm optimization (PSO), GA and TS. ACO is distinct from other meta-heuristic methods in that it *constructs* a new solution set (colony) in each generation (iteration), while others focus on *improving* the set of solutions or a single solution from previous iterations. ACO was inspired by the behavior of physical ants. Ethologists have studied how blind animals, such as ants, could establish shortest paths from their nest to food sources and found that the medium used to communicate information among individual ants regarding paths is a chemical substance called pheromone. A moving ant lays some pheromone on the ground, thus marking the path. The pheromone, while dissipating over time, is reinforced if other ants use the same trail. Therefore, superior trails increase their pheromone level over time while inferior ones reduce to nil. Inspired by the behavior of ants, Marco Dorigo introduced the ant colony optimization approach in his Ph.D. thesis in 1992 [13] and expanded it in his further work including [14, 15, 18, 19]. The primary characteristics of ant colony optimization are:

1. a method to construct solutions that balances pheromone trails (characteristics of past solutions) with a problem-specific heuristic (normally, a simple greedy rule),
2. a method to both reinforce and dissipate pheromone,
3. a method capable of including local (neighborhood) search to improve solutions.

ACO methods have been successfully applied to common combinatorial optimization problems including traveling salesman [16, 17], quadratic assignment [32, 44], vehicle routing [4, 5, 21], telecommunication networks [12], graph coloring [10], constraint satisfaction [38], Hamiltonian graphs [47] and scheduling [1, 9, 11]. A comprehensive survey of ACO algorithms and applications can be found in [19].

The application of ACO algorithms to reliability system problems was first proposed by Liang and Smith [28, 29], and then enhanced by the same authors in [30]. Liang and Smith employ ACO variations to solve a system reliability maximization RAP. Section III uses the ACO algorithm in [30] as a paradigm to demonstrate the application of ACO to RAP.

Thus far, the applications of ACO to reliability system are still very limited. Shelokar et al. [43] propose ant algorithms to solve three types of system reliability models: complex (neither series nor parallel), N-stage mixed series-parallel, and a complex bridge network system. In order to solve problems with different number of objectives and different types of decision variables, the authors develop three ant algorithms for single objective combinatorial problem, single objective continuous problem, and bi-objective continuous problem, respectively. The ant algorithm of single objective combinatorial version use the pheromone information only to construct the solutions, and no online pheromone updating rule is applied. Two local search methods, swap and random exchange, are performed to the best ant. For continuous problems, the authors divided the colony into two groups – global ants and local ants. The global ant concept can be considered as a pure GA mechanism since these ants apply crossover and mutation and no pheromone is deposited. Local ants are improved by a stochastic hill-climbing technique, and an improving ant can deposit the improvement magnitude of the objective on the trails. Lastly, a clustering technique and Pareto concept are combined with the continuous version of ant algorithms to solve bi-objective problems. The authors compared their algorithms with methods in the literature such as SA, a generalized Lagrange function approach, and a random search method. The results on four sets of test problems show the superiority of ACO algorithms.

Ouiddir et al. [36] develop an ACO algorithm for multi-state electrical power system problems. In this system redesign problem, the objective is to minimize the investment over the study period while satisfying availability or performance criteria. The proposed ant algorithm is based on the Ant Colony System (ACS) of [17] and [30]. A universal moment generating function is used to calculate the availability of the repairable multistate system. The algorithm is tested on a small problem with five subsystems, each with four to six component options. Samrout et al. [41] apply ACO to determine the component replacement conditions in series-parallel systems minimizing the preventive maintenance cost. Three algorithms are proposed – two based on Ant System (AS) [18] and one based on ACS [17]. Different transition rules and pheromone updating rules are employed in each algorithm. Local search is not used. Given different mission times and availability constraints, the performance of the ACO algorithms is compared with a GA from the literature. In this paper, results are mixed: one of the AS based methods and the ACS based method outperform the GA while the other AS algorithm is dominated by the GA. Nahas and Nourelfath [34] use an AS algorithm to optimize the reliability of a series system with multiple choices and budget constraints. Online pheromone updating and local search are not used. The authors apply a penalty function to determine the magnitude of pheromone deposition. Four examples with up to 25 component options are tested to verify the performance of the proposed algorithm. The computational results show that the AS algorithm is effective with respect to solution quality and computational expense.

The remaining chapter is organized as follows. Section II offers the notation list and defines the system reliability maximization RAP. A detailed introduction of an ant colony paradigm on solving RAP is provided in Section III using the work of Liang and Smith as a basis. Computational results on a set of benchmark problems are discussed in Section IV. Finally, concluding remarks are summarized in Section V.

1.2 Problem Definition

1.2.1 Notation

Redundancy Allocation Problem (RAP)

k minimum number of components required to function a pure parallel system

n total number of components used in a pure parallel system

k-out-of-n: G a system that functions when at least of its components function

R overall reliability of the series-parallel system

C cost constraint

W weight constraint

s number of subsystems

a_i number of available component choices for subsystem i

r_{ij} reliability of component j available for subsystem i

c_{ij} cost of component j available for subsystem i

w_{ij} weight of component j available for subsystem i

y_{ij} quantity of component j used in subsystem i

$\mathbf{y}_i$ $(y_{i1},...,y_{ia_i})$

n_i $= \sum_{j=1}^{a_i} y_{ij}$, total number of components used in subsystem i

n_{max} maximum number of components that can be in parallel (user assigned)

k_i minimum number of components in parallel required for subsystem i to function

$R_i(y_i \mid k_i)$ reliability of subsystem i, given k_i

$C_i(y_i)$ total cost of subsystem i

$W_i(y_i)$ total weight of subsystem i

R_u unpenalized system reliability of solution u

R_{up} penalized system reliability of solution u

R_{mp} penalized system reliability of the rank m^{th} solution

C_u total system cost of solution u

W_u	total system weight of solution u
AC	set of available component choices

Ant Colony Optimization (ACO)

i	index for subsystem, $i = 1,...,s$
j	index for components in a subsystem
τ_{ij}	pheromone trail intensity of combination (i, j)
τ_{ij}^{old}	pheromone trail intensity of combination (i, j) before update
τ_{ij}^{new}	pheromone trail intensity of combination (i, j) after update
τ_{i0}	$=1/a_i$, initial pheromone trail intensity of subsystem i
P_{ij}	transition probability of combination (i, j)
η_{ij}	problem-specific heuristic of combination (i, j)
α	relative importance of the pheromone trail intensity
β	relative importance of the problem-specific heuristic
l	index for component choices from set AC
ρ	$\in [0,1]$, trail persistence
q	$\in [0,1]$, a uniformly generated random number
q_0	$\in [0,1]$, a parameter which determines the relative importance of exploitation versus exploration
E	number of best solutions chosen for offline pheromone update
m	index (rank, best to worst) for solutions in a given iteration
γ	amplification parameter in the penalty function

1.2.2 Redundancy Allocation Problem

The RAP can be formulated to maximize system reliability given restrictions on system cost of C and system weight of W. It is assumed that system weight and system cost are linear combinations of component weight and cost, although this is a restriction that can be relaxed using heuristics.

$$\max \quad R = \prod_{i=1}^{s} R_i(\mathbf{y}_i \mid k_i) \qquad (1)$$

Subject to the constraints

$$\sum_{i=1}^{s} C_i(\mathbf{y}_i) \le C, \tag{2}$$

$$\sum_{i=1}^{s} W_i(\mathbf{y}_i) \le W, \tag{3}$$

If there is a known maximum number of components allowed in parallel, the following constraint is added:

$$k_i \le \sum_{j=1}^{a_i} y_{ij} \le n_{max} \quad \forall i = 1, 2, \dots, s \tag{4}$$

Typical assumptions are:
- The states of components and the system are either operating or failed.
- Failed components do not damage the system and are not repaired.
- The failure rates of components when not in use are the same as when in use (i.e., active redundancy is assumed).
- Component attributes (reliability, weight and cost) are known and deterministic.
- The supply of any component is unconstrained.

1.3 Ant Colony Optimization Approach

This section is taken from the authors' earlier work in using the ant colony approach for reliable systems optimization [28, 29, 30]. The generic components of ant colony are each defined and the overall flow of the method is defined. These should be applicable, with minor changes, to many problems in reliable systems combinatorial design.

1.3.1 Solution Encoding

As with other meta-heuristics, it is important to devise a solution encoding that provides (ideally) a one to one relationship with the solutions to be considered during search. For combinatorial problems this generally takes the form of a binary or k-nery string although occasionally other representations such as real numbers can be used. For the RAP, each ant represents a design of an entire system, a collection of n_i components in parallel $(k_i \le n_i \le n_{max})$ for s different subsystems. The n_i components are chosen from a_i available types of components. The a_i types are sorted in descending order of reliability; i.e., 1 represents the most reliable component type, etc. An index of $a_i + 1$ is assigned to a position where an additional component was not used (that is, was left blank) with attributes of zero. Each of the s subsystems is represented by $n_{\max}$ positions with each component listed according to its reliability index, as in [7], therefore a complete system design (that is, an ant) is an integer vector of length $n_{\max} \times s$.

1.3.2 Solution Construction

Also, as with other meta-heuristics, an initial solution set must be generated. For global optimizers the solution quality in this set is not usually important and that is true for the ant approach as well. In the ACO-RAP algorithm, ants use problem-specific heuristic information, denoted by η_{ij}, along with pheromone trail intensity, denoted by τ_{ij}, to construct a solution. n_i components ($k_i + 1 \le n_i \le n_{\max} - 4$) are selected for each subsystem using the probabilities calculated by equations 5 and 6, below. This range of components encourages the construction of a solution that is likely to be feasible, that is, be reliable enough (satisfying the $k_i + 1$ lower bound) but not violate the weight and cost constraints (satisfying the $n_{\max} -$ 4 upper bound). Solutions which contain more or less components per subsystem than these bounds are examined during the local search phase of the algorithm (described in Section III D).

The ACO problem specific heuristic chosen is $\eta_{ij} = \dfrac{r_{ij}}{c_{ij} + w_{ij}}$ where r_{ij}, c_{ij}, and w_{ij} represent the associated reliability, cost and weight of component j for subsystem i. This favors components with higher reliability and smaller cost and weight. Adhering to the ACO meta-heuristic concept, this is a simple and obvious rule. Uniform pheromone trail intensities for the initial iteration (colony of ants) are set over the component choices, that is, $\tau_{i0} = 1/a_i$. The pheromone trail intensities are subsequently changed as described in Section III E.

A solution is constructed by selecting component j for subsystem i according to:

$$j = \begin{cases} \arg \max_{l \in AC}[(\tau_{il})^{\alpha}(\eta_{il})^{\beta}] & q \le q_0 \\ \\ J & q > q_0 \end{cases} \tag{5}$$

and J is chosen according to the transition probability mass function given by

$$
P_{ij} = \begin{cases} \dfrac{(\tau_{ij})^{\alpha} (\eta_{ij})^{\beta}}{\displaystyle\sum_{l=1}^{a_i} (\tau_{il})^{\alpha} (\eta_{il})^{\beta}} & j \in AC \\ \\ 0 & \text{Otherwise} \end{cases} \tag{6}
$$

where α and β control the relative weight of the pheromone and the local heuristic, respectively, AC is the set of available component choices for subsystem i, q is a uniform random number, and q_0 determines the relative importance of the exploitation of superior solutions versus the diversification of search spaces. When $q \le q_0$ exploitation of known good solutions occurs. The component selected is the best for that particular subsystem, that is, has the highest product of pheromone intensity and ratio of reliability to cost and weight. When $q > q_0$, the search favors more exploration as all components are considered for selection with some probability.

1.3.3 Objective Function

Fitness (the common term for the analogy to objective function value for nature inspired heuristics) plays an important role in the ant colony approach as it determines the construction probabilities for the subsequent generation. After solution u is constructed, the unpenalized reliability R_u is calculated using equation (1). For solutions with cost that exceeds C and / or weight that exceeds W, the penalized reliability R_{up} is calculated:

$$
R_{up} = R_u \cdot \left(\frac{W}{W_u}\right)^{\gamma} \cdot \left(\frac{C}{C_u}\right)^{\gamma} \tag{7}
$$

where the exponent γ is an amplification parameter and W_u and C_u are the system weight and cost of solution u, respectively. This penalty function encourages the ACO-RAP algorithm to explore the feasible region and infeasible region that is near the border of the feasible area, and discourages, but allows, search further into the infeasible region.

1.3.4 Improving Constructed Solutions Through Local Search

After an ant colony is generated, each ant is improved using local search. Local search is an optional, but usually beneficial, aspect of the ant colony approach that allows a systematic enhancement of the constructed ants. For the RAP, starting with the first subsystem, a chosen component type is deleted and a different component type is added. All possibilities are enumerated. For example, if a subsystem has one of component 1, two of component 2 and one of component 3, then one alternative is to delete a component 1 and to add a component 2. Another possibility is to delete a component 3 and to add a component 1. Whenever an improvement of the objective function is achieved, the new solution replaces the old one and the process continues until all subsystems have been searched. This local search does not require recalculating the system reliability each time, only the reliability of the subsystem under consideration needs to be recalculated.

1.3.5 Pheromone Trail Intensity Update

The pheromone trail is a unique concept to the ant approach. Naturally, this idea is taken directly from studying physical ants and their deposits of the pheromone chemical. For the RAP, the pheromone trail update consists of two phases – online (ant-by-ant) updating and offline (colony) updating. Online updating is done after each solution is constructed and its purpose is to lessen the pheromone intensity of the components of the solution just constructed to encourage exploration of other component choices in the later solutions to be constructed. Online updating is by

$$\tau_{ij}^{new} = \rho \cdot \tau_{ij}^{old} + (1-\rho) \cdot \tau_{io} \qquad (8)$$

where $\rho \in [0,1]$ controls the pheromone persistence; i.e., $1-\rho$ represents the proportion of the pheromone evaporated. After all solutions in a colony have been constructed and subject to local search, pheromone trails are updated offline. Offline updating is to reflect the discoveries of this iteration. The offline intensity update is:

$$\tau_{ij}^{new} = \rho \cdot \tau_{ij}^{old} + (1-\rho) \cdot \sum_{m=1}^{E}(E-m+1) \cdot R_{mp} \qquad (9)$$

where $m = 1$ is the best feasible solution yet found (which may or may not be in the current colony) and the remaining E-1 solutions are the best ones in the current colony. In other words, only the best E ants are allowed to contribute pheromone to the trail intensity and the magnitudes of contributions are weighted by their ranks in the colony.

1.3.6 Overall Ant Colony Algorithm

Generally, ant colony algorithms are similar to other meta-heuristics in that they iterate over generations (termed colonies for ACO) until some termination criteria are met. If an algorithm is elitist (as most genetic algorithms and ant colonies are) the best solution found is also contained in the final iteration (colony). The termination criteria are usually a combination of total solutions considered (or total computational time) and lack of best solution improvement over some number iterations. These are experimentally determined. Of course, there is no downside to running the ACO overly long except waste of computer time.

The flow of the ACO-RAP is as follows:
Set all parameter values and initialize the pheromone trail intensities
Loop
    Sub-Loop
            Construct an ant using the pheromone trail intensity and the problem-specific heuristic (eqs. 5, 6)
            Apply the online pheromone intensity update rule (eq. 8)
    Continue until all ants in the colony have been generated
    Apply local search to each ant in the colony
    Evaluate all ants in the colony (eqs. 1, 7), rank them and record the best feasible one
    Apply the offline pheromone intensity update rule (eq. 9)
Continue until a stopping criterion is reached

1.4 Computational Experience

To show the effectiveness of the ant colony approach for reliable systems design results from [30] are given here. The ACO is coded in Borland C++ and run using an Intel Pentium III 800 MHz PC with 256 MB RAM. All computations use real float point precision without rounding or truncating values. The system reliability of the final solution is rounded to four digits behind the decimal point in order to compare with results in the literature.

The parameters of the ACO algorithm are set to the following values: $\alpha = 1$, $\beta = 0.5$, $q_0 = 0.9$, $\rho = 0.9$ and $E = 5$. This gives relatively more weight to the pheromone trail intensity than the problem-specific heuristic and greater emphasis on exploitation rather than exploration. The ACO is not very sensitive these values and tested well for quite a range of them. For the penalty function, $\gamma = 0.1$ except when the previous iteration has 90% or more infeasible solutions, then $\gamma = 0.3$. This increases the penalty temporarily to move the search back into the feasible region when all or nearly all solutions in the current colony are infeasible. This bi-level penalty improved performance on the most constrained instances of the test problems. Because of varying magnitudes of R, C and W, all η_{ij} and τ_{ij} are normalized between (0,1) before solution construction. 100 ants are used in each colony. The stopping criterion is when the number of colonies reaches 1000 or the best feasible solution has not changed for 500 consecutive colonies. This results in a maximum of 100,000 ants.

The 33 variations of the Fyffe et al. problem [20] as devised by Nakagawa and Miyazaki [35] were used to test the performance of ACO. In this problem set $C = 130$ and W is decreased incrementally from 191 to 159. In [20] and [35], the optimization approaches required that identical components be placed in redundancy, however for the ACO approach, as in Coit and Smith [7], different component types are allowed to reside in parallel (assuming that a value of $n_{max} = 8$ for all subsystems). This makes the search space size larger than 7.6×10^{33}. Since the heuristic benchmark for the RAP with component mixing is the GA of [7], it is chosen for comparison. Ten runs of each algorithm (GA and ACO) were made using different random number seeds for each problem instance.

The results are summarized in Table 1 where the comparisons between the GA and ACO results over 10 runs are divided into three categories: maximum, mean and minimum system reliability (denoted by Max R, Mean R and Min R, respectively). The shaded box shows the maximum reliability solution to an instance while considering all GA and ACO results. The ACO solutions are equivalent to or superior to the GA over all categories and all problem instances. When the problem instances are less constrained (the first 18), the ACO performs much better than the GA. When the problems become more constrained (the last 15), ACO is equal to GA for 12 instances and better than GA for three instances in terms of the Max R measure (best over ten runs). However, for Min R (worst over 10 runs) and Mean R (of 10 runs), ACO dominates GA.

Table 1. Comparison of the GA [7] and the ACO over 10 random number seeds each for the test problems from [35]. These results are from [30].

No	C	W	C&S [7] GA - 10 runs			ACO-RAP - 10 runs		
			Max R	Mean R	Min R	Max R	Mean R	Min R
1	130	191	0.9867	0.9862	0.9854	0.9868	0.9862	0.9860
2	130	190	0.9857	0.9855	0.9852	0.9859	0.9858	0.9857
3	130	189	0.9856	0.9850	0.9838	0.9858	0.9853	0.9852
4	130	188	0.9850	0.9848	0.9842	0.9853	0.9849	0.9848
5	130	187	0.9844	0.9841	0.9835	0.9847	0.9841	0.9837
6	130	186	0.9836	0.9833	0.9827	0.9838	0.9836	0.9835
7	130	185	0.9831	0.9826	0.9822	0.9835	0.9830	0.9828
8	130	184	0.9823	0.9819	0.9812	0.9830	0.9824	0.9820
9	130	183	0.9819	0.9814	0.9812	0.9822	0.9818	0.9817
10	130	182	0.9811	0.9806	0.9803	0.9815	0.9812	0.9806
11	130	181	0.9802	0.9801	0.9800	0.9807	0.9806	0.9804
12	130	180	0.9797	0.9793	0.9782	0.9803	0.9798	0.9796
13	130	179	0.9791	0.9786	0.9780	0.9795	0.9795	0.9795
14	130	178	0.9783	0.9780	0.9764	0.9784	0.9784	0.9783
15	130	177	0.9772	0.9771	0.9770	0.9776	0.9776	0.9776
16	130	176	0.9764	0.9760	0.9751	0.9765	0.9765	0.9765
17	130	175	0.9753	0.9753	0.9753	0.9757	0.9754	0.9753
18	130	174	0.9744	0.9732	0.9716	0.9749	0.9747	0.9741
19	130	173	0.9738	0.9732	0.9719	0.9738	0.9735	0.9731
20	130	172	0.9727	0.9725	0.9712	0.9730	0.9726	0.9714
21	130	171	0.9719	0.9712	0.9701	0.9719	0.9717	0.9710
22	130	170	0.9708	0.9705	0.9695	0.9708	0.9708	0.9708
23	130	169	0.9692	0.9689	0.9684	0.9693	0.9693	0.9693
24	130	168	0.9681	0.9674	0.9662	0.9681	0.9681	0.9681
25	130	167	0.9663	0.9661	0.9657	0.9663	0.9663	0.9663
26	130	166	0.9650	0.9647	0.9636	0.9650	0.9650	0.9650
27	130	165	0.9637	0.9632	0.9627	0.9637	0.9637	0.9637
28	130	164	0.9624	0.9620	0.9609	0.9624	0.9624	0.9624
29	130	163	0.9606	0.9602	0.9592	0.9606	0.9606	0.9606
30	130	162	0.9591	0.9587	0.9579	0.9592	0.9592	0.9592
31	130	161	0.9580	0.9572	0.9561	0.9580	0.9580	0.9580
32	130	160	0.9557	0.9556	0.9554	0.9557	0.9557	0.9557
33	130	159	0.9546	0.9538	0.9531	0.9546	0.9546	0.9546

Thus, the ACO tends to find better solutions than the GA, is significantly less sensitive to random number seed, and for the 12 most constrained instances, finds the best solution each and every run. While these differences in system reliability are not large, it is beneficial to use a search method that performs well over different problem sizes and parameters. Moreover, any system reliability improvement while adhering to the design constraints is of some value, even if the reliability improvement realized is relatively small.

The best design and its system reliability, cost and weight for each of the 33 instances are shown in Table 2. For instances 6 and 11, two designs with different system costs but with the same reliability and weight are found. All but instance 33 involve mixing of components within a subsystem which is an indication that superior designs can be identified by not restricting the search space to a single component type per subsystem.

It is difficult to make a precise computational comparison. CPU seconds vary according to hardware, software and coding. Both the ACO and the GA generate multiple solutions during each iteration, therefore the computational effort changes in direct proportion to number of solutions considered. The number of solutions generated in [7] (a population size of 40 with 1200 iterations) is about half of the ACO (a colony size of 100 with up to 1000 iterations). However, given the improved performance per seed of the ACO, a direct comparison per run is not meaningful. If the average solution of the ACO over ten seeds is compared to the best performance of GA over ten seeds, in 13 instances ACO is better, in 9 instances GA is better and in the remaining instances (11) they are equal, as shown in Figure 2. Since this is a comparison of average performance (ACO) versus best performance (GA), the additional computational effort of the ACO is more than compensated for. In summary, an average run of ACO is likely to be as good or better than the best of ten runs of GA. The difference in variability over all 33 test problems between ACO and the GA is clearly shown in Figure 3.

Given the well-structured neighborhood of the RAP, a meta-heuristic that exploits it is likely to be more effective and more efficient than one that does not. While the GA certainly performs well relative to previous approaches, the largely random mechanisms of crossover and mutation result in greater run to run variability than the ACO. Since the ACO shares the GA's attributes of flexibility, robustness and implementation ease and improves on its random behavior, it seems a very promising general method for other NP-hard reliability design problems such as those found in networks and complex structures.

Table 2. Configuration, reliability, cost and weight of the best solution to each problem. These results are from [30].

No.	W	R	Cost	Weight	Solution
1	191	0.9868	130	191	333,11,111,2222,333,22,333,3333,23,122,333,4444,12,12
2	190	0.9859	129	190	333,11,111,2222,333,22,333,3333,22,112,333,4444,11,22
3	189	0.9858	130	189	333,11,111,2222,333,22,333,3333,22,122,11,4444,11,12
4	188	0.9853	130	188	333,11,111,2222,333,22,333,3333,23,112,13,4444,12,12
5	187	0.9847	130	187	333,11,111,2222,333,22,333,3333,23,122,13,4444,11,12
6	186	0.9838	129	186	333,11,111,2222,333,22,333,3333,22,122,11,4444,11,22
			130	186	333,11,111,2222,333,24,333,3333,33,122,13,4444,12,12
7	185	0.9835	130	185	333,11,111,2222,333,22,333,3333,13,122,13,4444,11,22
8	184	0.9830	130	184	333,11,111,222,333,22,333,3333,33,112,11,4444,11,12
9	183	0.9822	128	183	333,11,111,222,333,22,333,3333,33,112,13,4444,11,12
10	182	0.9815	127	182	333,11,111,222,333,22,333,3333,33,122,13,4444,11,12
11	181	0.9807	125	181	333,11,111,222,333,22,333,3333,13,122,13,4444,11,22
			126	181	333,11,111,222,333,22,333,3333,23,122,11,4444,11,22
12	180	0.9803	128	180	333,11,111,222,333,22,333,3333,33,122,11,4444,11,22
13	179	0.9795	126	179	333,11,111,222,333,22,333,3333,33,122,13,4444,11,22
14	178	0.9784	125	178	333,11,111,222,333,22,333,3333,33,222,13,4444,11,22
15	177	0.9776	126	177	333,11,111,222,333,22,333,133,33,122,13,4444,11,22
16	176	0.9765	125	176	333,11,111,222,333,22,333,133,33,222,13,4444,11,22
17	175	0.9757	125	175	333,11,111,222,333,22,13,3333,33,122,11,4444,11,22
18	174	0.9749	123	174	333,11,111,222,333,22,13,3333,33,122,13,4444,11,22
19	173	0.9738	122	173	333,11,111,222,333,22,13,3333,33,222,13,4444,11,22
20	172	0.9730	123	172	333,11,111,222,333,22,13,133,33,122,13,4444,11,22
21	171	0.9719	122	171	333,11,111,222,333,22,13,133,33,222,13,4444,11,22
22	170	0.9708	120	170	333,11,111,222,333,22,13,133,33,222,33,4444,11,22
23	169	0.9693	121	169	333,11,111,222,333,22,33,133,33,222,13,4444,11,22
24	168	0.9681	119	168	333,11,111,222,333,22,33,133,33,222,33,4444,11,22
25	167	0.9663	118	167	333,11,111,222,33,22,13,133,33,222,33,4444,11,22
26	166	0.9650	116	166	333,11,11,222,333,22,13,133,33,222,33,4444,11,22
27	165	0.9637	117	165	333,11,111,222,33,22,33,133,33,222,33,4444,11,22
28	164	0.9624	115	164	333,11,11,222,333,22,33,133,33,222,33,4444,11,22
29	163	0.9606	114	163	333,11,11,222,33,22,13,133,33,222,33,4444,11,22
30	162	0.9592	115	162	333,11,11,222,33,22,33,133,33,222,13,4444,11,22
31	161	0.9580	113	161	333,11,11,222,33,22,33,133,33,222,33,4444,11,22
32	160	0.9557	112	160	333,11,11,222,33,22,33,333,33,222,13,4444,11,22
33	159	0.9546	110	159	333,11,11,222,33,22,33,333,33,222,33,4444,11,22

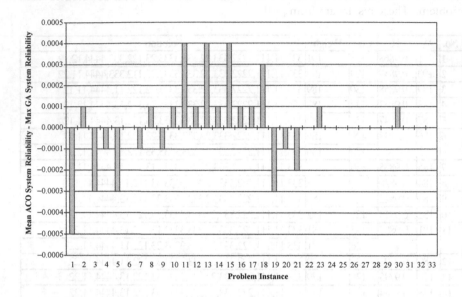

Fig. 2. Comparison of mean ACO with best GA performance over 10 seeds. These results are from [30].

1.5 Conclusions

This chapter cites the latest developments of ACO algorithms to reliability system problems. The main part of the chapter gives details of a general ant colony meta-heuristic to solve the redundancy allocation problem (RAP) which was devised over the past several years by the authors and published in [28, 29, 30]. The RAP is a well known NP-hard problem that has been the subject of much prior work, generally in a restricted form where each subsystem must consist of identical components in parallel to make computations tractable. Heuristic methods can overcome this limitation and offer a practical way to solve large instances of a relaxed RAP where different components can be placed in parallel. The ant colony algorithm for the RAP is shown to perform well with little variability over problem instance or random number seed. It is competitive with the best-known heuristics for redundancy allocation. Undoubtedly there will be much more work forthcoming in the literature that uses the ant colony paradigm to solve the many difficult combinatorial problems in the field of reliable system design.

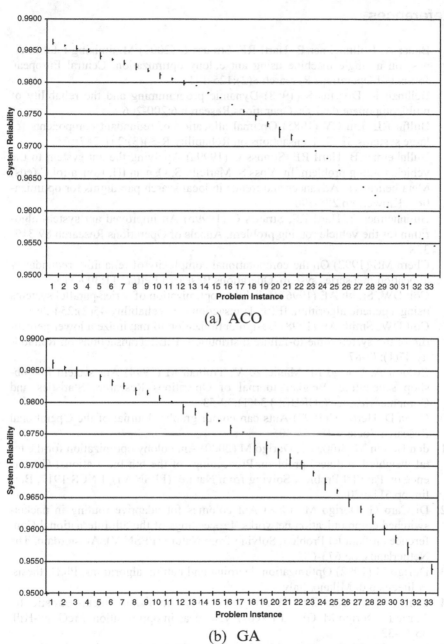

(a) ACO

(b) GA

Fig. 3. Range of performance over 10 seeds with mean shown as horizontal dash. These results are from [30].

References

1. Bauer A, Bullnheimer B, Hartl RF, Strauss C (2000) Minimizing total tardiness on a single machine using ant colony optimization. Central European Journal of Operations Research 8(2):125-141
2. Bellman R, Dreyfus S (1958) Dynamic programming and the reliability of multicomponent devices. Operations Research 6:200-206
3. Bulfin RL, Liu CY (1985) Optimal allocation of redundant components for large systems. IEEE Transactions on Reliability R-34(3):241-247
4. Bullnheimer B, Hartl RF, Strauss C (1999a) Applying the ant system to the vehicle routing problem. In: Voss S, Martello S, Osman IH, Roucairol C (eds) Meta-heuristics: Advances and trends in local search paradigms for optimization. Kluwer, pp 285-296
5. Bullnheimer B, Hartl RF, Strauss C (1999b) An improved ant system algorithm for the vehicle routing problem. Annals of Operations Research 89:319-328
6. Chern MS (1992) On the computational complexity of reliability redundancy allocation in a series system. Operations Research Letters 11:309-315
7. Coit DW, Smith AE (1996) Reliability optimization of series-parallel systems using a genetic algorithm. IEEE Transactions on Reliability 45(2):254-260
8. Coit DW, Smith AE (1998) Design optimization to maximize a lower percentile of the system-time-to-failure distribution. IEEE Transactions on Reliability 47(1):79-87
9. Colorni A, Dorigo M, Maniezzo V, Trubian M (1994) Ant system for job-shop scheduling. Belgian Journal of Operations Research, Statistics and Computer Science (JORBEL) 34(1):39-53
10. Costa D, Hertz A (1997) Ants can colour graphs. Journal of the Operational Research Society 48:295-305
11. den Besten M, Stützle T, Dorigo M (2000) Ant colony optimization for the total weighted tardiness problem. Proceedings of the 6th International Conference on Parallel Problem Solving from Nature (PPSN VI), LNCS 1917, Berlin, pp 611-620
12. Di Caro G, Dorigo M (1998) Ant colonies for adaptive routing in packet-switched communication networks. Proceedings of the 5th International Conference on Parallel Problem Solving from Nature (PPSN V), Amsterdam, The Netherlands, pp 673-682
13. Dorigo M (1992) Optimization, learning and natural algorithms. Ph.D. thesis, Politecnico di Milano, Italy
14. Dorigo M, Di Caro G (1999) The ant colony optimization meta-heuristic. In: Corne D, Dorigo M, Glover F (eds) New ideas in optimization. McGraw-Hill, pp 11-32
15. Dorigo M, Di Caro G, Gambardella LM (1999) Ant algorithms for discrete optimization. Artificial Life 5(2):137-172
16. Dorigo M, Gambardella LM (1997) Ant colonies for the travelling salesman problem. BioSystems 43:73-81

17. Dorigo M, Gambardella LM (1997) Ant colony system: A cooperative learning approach to the travelling salesman problem. IEEE Transactions on Evolutionary Computation 1(1):53-66
18. Dorigo M, Maniezzo V, Colorni A (1996) Ant system: Optimization by a colony of cooperating agents. IEEE Transactions on Systems, Man, and Cybernetics-Part B: Cybernetics 26(1):29-41
19. Dorigo M, Stützle T (2004) Ant colony optimization. The MIT Press, Cambridge
20. Fyffe DE, Hines WW, Lee NK (1968) System reliability allocation and a computational algorithm. IEEE Transactions on Reliability R-17(2):64-69
21. Gambardella LM, Taillard E, Agazzi G (1999) MACS-VRPTW: A multiple ant colony system for vehicle routing problems with time windows. In: Corne D, Dorigo M, Glover F (eds) New Ideas in Optimization. McGraw-Hill, pp 63-76
22. Gen M, Ida K, Tsujimura Y, Kim CE (1993) Large-scale 0-1 fuzzy goal programming and its application to reliability optimization problem. Computers and Industrial Engineering 24(4):539-549
23. Ghare PM, Taylor RE (1969) Optimal redundancy for reliability in series systems. Operations Research 17:838-847
24. Kulturel-Konak S, Coit DW, Smith AE (2003) Efficiently solving the redundancy allocation problem using tabu search. IIE Transactions 35(6):515-526
25. Kuo W, Prasad VR (2000) An annotated overview of system-reliability optimization. IEEE Transactions on Reliability 49(2):176-187
26. Levitin G, Lisnianski A, Ben-Haim H, Elmakis D (1998) Redundancy optimization for series-parallel multi-state systems. IEEE Transactions on Reliability 47(2):165-172
27. Liang YC, Wu CC (2005) A variable neighborhood descent algorithm for the redundancy allocation problem. Industrial Engineering and Management Systems 4(1):109-116
28. Liang YC, Smith AE (1999) An ant system approach to redundancy allocation. Proceedings of the 1999 Congress on Evolutionary Computation, Washington, D.C., pp 1478-1484
29. Liang YC, Smith AE (2000) Ant colony optimization for constrained combinatorial problems. Proceedings of the 5th Annual International Conference on Industrial Engineering – Theory, Applications and Practice, Hsinchu, Taiwan, ID 296
30. Liang YC, Smith AE (2004) An ant colony optimization algorithm for the redundancy allocation problem (RAP). IEEE Transactions on Reliability 53(3):417-23
31. Luus R (1975) Optimization of system reliability by a new nonlinear integer programming procedure. IEEE Transactions on Reliability R-24(1):14-16
32. Maniezzo V, Colorni A (1999) The ant system applied to the quadratic assignment problem. IEEE Transactions on Knowledge and Data Engineering 11(5):769-778
33. Misra KB, Sharma U (1991) An efficient algorithm to solve integer-programming problems arising in system-reliability design. IEEE Transac-

tions on Reliability 40(1):81-91
34. Nahas N, Nourelfath M (2005) Any system for reliability optimization of a series system with multiple-choice and budget constraints. Reliability Engineering and System Safety 87:1-12
35. Nakagawa Y, Miyazaki S (1981) Surrogate constraints algorithm for reliability optimization problems with two constraints. IEEE Transactions on Reliability R-30(2):175-180
36. Ouiddir R, Rahli M, Meziane R, Zeblah A (2004) Ant colony optimization or new redesign problem of multi-state electrical power systems. Journal of Electrical Engineering 55(3-4):57-63
37. Painton L, Campbell J (1995) Genetic algorithms in optimization of system reliability. IEEE Transactions on Reliability 44(2):172-178
38. Ravi V (2004) Optimization of complex system reliability by a modified great deluge algorithm. Asia-Pacific Journal of Operational Research 21(4):487-497
39. Ravi V, Murty BSN, Reddy PJ (1997) Nonequilibrium simulated annealing algorithm applied to reliability optimization of complex systems. IEEE Transactions on Reliability 46(2):233-239
40. Ravi V, Reddy PJ, Zimmermann H-J (2000) Fuzzy global optimization of complex system reliability. IEEE Transactions on Fuzzy Systems 8(3):241-248
41. Samrout M, Yalaoui F, Châtelet E, Chebbo N (2005) New methods to minimize the preventive maintenance cost of series-parallel systems using ant colony optimization. Reliability Engineering and System Safety 89:346-354
42. Schoofs L, Naudts B (2000) Ant colonies are good at solving constraint satisfaction problems. Proceedings of the 2000 Congress on Evolutionary Computation, San Diego, CA, pp 1190-1195
43. Shelokar P, Jayaraman VK, Kulkarni BD (2002) Ant algorithm for single and multiobjective reliability optimization problems. Quality and Reliability Engineering International 18:497-514
44. Stützle T, Dorigo M (1999) ACO algorithms for the quadratic assignment problem. In: Corne D, Dorigo M, Glover F (eds) New ideas in optimization. McGraw-Hill, pp 33-50
45. Tillman FA, Hwang CL, Kuo W (1977a) Optimization techniques for system reliability with redundancy - A review. IEEE Transactions on Reliability R-26(3):148-155
46. Tillman FA, Hwang CL, Kuo W (1977b) Determining component reliability and redundancy for optimum system reliability. IEEE Transactions on Reliability R-26(3):162-165
47. Wagner IA, Bruckstein AM (1999) Hamiltonian(t) - An ant inspired heuristic for recognizing Hamiltonian graphs. Proceedings of the 1999 Congress on Evolutionary Computation, Washington, D.C., pp 1465-1469

Modified Great Deluge Algorithm versus Other Metaheuristics in Reliability Optimization

Vadlamani Ravi

Institute for Development and Research in Banking Technology, Hyderabad, India

2.1 Introduction

Optimization of reliability of complex systems is an extremely important issue in the field of reliability engineering. Over the past three decades, reliability optimization problems have been formulated as non-linear programming problems within either single-objective or multi-objective environment. Tillman *et al.*. (1980) provides an excellent overview of a variety of optimization techniques applied to solve these problems. However, he reviewed the application of only derivative-based optimization techniques, as metaheuristics were not applied to the reliability optimization problems by that time.

Over the last decade, metaheuristics have also been applied to solve the reliability optimization problems. To list a few of them, Coit and Smith (1996) were the first to employ a genetic algorithm to solve reliability optimization problems. Later, Ravi *et al.* (1997) developed an improved version of non-equilibrium simulated annealing called INESA and applied it to solve a variety of reliability optimization problems. Further, Ravi *et al.* (2000) first formulated various complex system reliability optimization problems with single and multi objectives as fuzzy global optimization problems. They also developed and applied the non-combinatorial version of another meta-heuristic viz. threshold accepting to solve these problems. Threshold accepting (Dueck and Sheurer, 1990) is a faster variation of the simulated annealing and often leads to superior optimal solutions than does the simulated annealing. Recently, Shelokar *et al.* (2002) applied the ant colony optimization algorithm (Dorigo *et al.*, 1997) to these problems and obtained superior results compared to those reported by Ravi *et al.* (1997). Most recently, Ravi (2004) developed an extended version of the great

V. Ravi: *Modified Great Deluge Algorithm versus Other Metaheuristics in Reliability Optimization*, Computational Intelligence in Reliability Engineering (SCI) **40**, 21–36 (2007)
www.springerlink.com   © Springer-Verlag Berlin Heidelberg 2007

deluge algorithm and demonstrated its effectiveness in solving the reliability optimization problems.

The objective of the chapter is primarily to discuss the relative performance of various metaheuristics such as the modified great deluge algorithm (MGDA), simulated annealing (SA), improved non-equilibrium simulated annealing (INESA) and ant colony optimization (ACO) on the reliability optimization problems in complex systems. The performance of other methods such as generalized Lagrange function approach, sequential unconstrained minimization technique, a random search technique and an integer programming approach would also be discussed.

The remainder of the chapter is arranged as follows: Section 2 describes the problems studied here. In section 3 a brief description of the algorithms SA, INESA and MGDA is presented. Section 4 compares the performance of these algorithms on three reliability optimization problems occurring in complex systems. Section 5 concludes the chapter. The numerical problems solved are described in the appendix.

Notation

R_S, C_S	[reliability , cost] of the system
r_i, C_i	[reliability , cost] of the ith component
R_i	Reliability of the ith stage
m	number of constraints
n	number of decision variables (number of components in a complex system or the number of stages in a multi stage mixed system)
$r_{i, min}$	lower bound on the reliability of the ith component
$R_{S,min}$	lower bound on the system reliability
K_i , α_i	constants associated with cost function of the ith component
x_i	number of the redundancies of the ith component
g_i	i^{th} constraint
$itr, limit$	number of [global, inner] iterations
xll_i, xul_i	lower and upper bounds on the ith decision variable
f^o, f^c	Objective function value of the old and candidate solutions respectively
$rnd(0,1) \ or \ u$	uniform random number generated between 0 and 1
x_i^c , x_i^o	i^{th} decision variable of the candidate and old solution vectors respectively
p	Pre-specified odd-integer
$LEVEL$	Water level , a parameter used in the solution acceptance criterion

UP	A parameter reduces or increases the *LEVEL* according as it is a minimization or maximization problem
old, new	Dummy variables represent the objective function of the old and current solutions respectively
itrmax	Maximum number of global iterations
SA	Simulated Annealing
INESA	Improved Non-Equilibrium Simulated Annealing
ACO	Ant Colony Optimization
MGDA	Modified Great Deluge Algorithm

2.2 Problem Description

A complex system in the field of reliability engineering consists of several components connected to one another neither purely in series nor purely in parallel. The block diagrams of two such complex systems and a multistage mixed system studied here are depicted in Figures 1, 2 and 3.

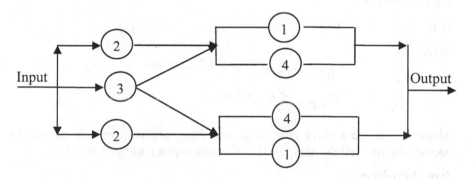

Fig. 1. Life support system in a space capsule

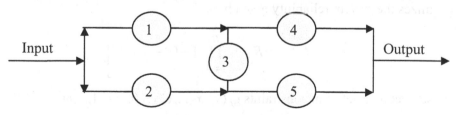

Fig. 2. Bridge network problem

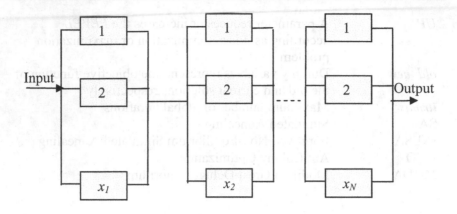

Fig. 3. Multi-stage mixed system

The number in circles (rectangles) in each of the figures represents the type of component in the system. In this chapter, two types of problems are studied.

Type 1 Problem:

Minimize C_s

subject to

$$r_{i,\min} \leq r_i \leq 1.0, i = 1,2,...,n \text{ and}$$

$$R_{S,\min} \leq R_S \leq 1.0$$

where C_s is the system cost, $r_{i,min}$ and $R_{s,min}$ are respectively the lower bounds on the reliabilities of the ith component and system.

Type 2 Problem:

Find the optimal number of components $x_i \geq 1$, ($i = 1,..,n$) which maximizes the system reliability given by

$$R_S = \prod_{i=1}^{n}\left[1-(1-r_i)^{x_i}\right]$$

subject to a set of m constraints $g_j (x_1, x_2,...,x_n) \leq 0$, $j = 1,...,m$.

2.3 Description of Various Metaheuristics

2.3.1 Simulated Annealing (SA)

SA is developed based on the principles of statistical mechanics. It found a number of applications in diverse disciplines such as science, engineering and economics in finding global solution to highly nonlinear constrained optimization problems and combinatorial optimization problems. SA is very much analogous to the physical process of annealing. Annealing refers to the physical thermal process of melting a solid by first heating it and then cooloing it slowly in order to allow the molecules in the material to attain the lowest energy level (stable or ground state). If the cooling rate is not carefully controlled or the initial temperature is not sufficiently high, then the cooling solid does not attain thermal equilibrium at each temperature. Therefore, under such circumstances, local optimal lattice structures may occur that translate into lattice imperfections, also known as, metastable state. Thermal equilibrium at a given temperature is characterized by Boltzmann distribution of the energy states. Under these conditions, even at a low temperature, a transition can occur from low energy level to a high energy level, although with a small probability. Presumably, such transitions are responsible for the system reaching a global minimum energy state instead of being trapped in a local metastable state (Cardoso *et al.*, 1993). It was Metropolis *et al.* (1953) who first to proposed an SA algorithm to simulate the process. While applying the SA to determine global optimum of a multivariable function, the following observation can be made:

- The energy state of the system is analogous to the objective function in the problem;
- the molecular positions are the analogues of decision variables;
- the ground state corresponds to the global minimum;
- attaining a metastable state implies reaching a local minimum.

Kirkpatrick *et al.* (1983) rejuvinated interest in SA by formally building the connection between statistical mechanics and combinatorial optimization problems. They applied SA to solve two combinatorially large optimization problems: 1) traveling salesman problem, and 2) designing the layout of very large scale integration (VLSI) computer chips. The SA also found applications in (i) Chemical sciences such as heat exchanger network, pressure relief header networks (Dolan *et al.*, 1989) and global optimization of molecular geometry (Dutta *et al.*, 1991) (ii) Biology such as multiple sequence alignment for studying molecular evolution and analyzing structure sequence relationships (Kim *et al.*, 1994) (iii) Economics such as determining optimal portfolio considering all possible utility functions of an investor (Dueck and Winker, 1992).

2.3.2 Improved Non-equilibrium Simulated Annealing (INESA)

Cardoso *et al.* (1993) presented an improved version of the SA that resulted in reduced computation time as well as improved convergence aspects. They introduced non-equilibrium simulated annealing (NESA) by modifying the original Metropolis *et al.* (1953) and Glauber (1963) algorithms. They argued that it is not necessary to achieve equilibrium at each temperature level in order to obtain near-global optimal solutions. Unlike the original SA algorithm, NESA operates at a non-equilibrium condition, i.e., the cooling schedule is enforced as soon as an improved solution is obtained, without waiting for the occurrence of near-equilibrium condition at each temperature. This feature overcomes the slowness of the SA algorithm, without actually comprising on the quality of the global optimal solution (Cardoso *et al.* 1993). Further, this aspect significantly lowers the computational time.

Later, Ravi *et al.* (1997) developed an extended version of the NESA, called INESA, by proposing a two-phase approach. In INESA, the phase-1 implements the NESA with relaxed temparature conditions and the phase-2 employs a simplex-like heuristic which works on the sampled solutions obtained from the progress of phase-1 along with the best solutions obtained before the termination of phase-1. They applied INESA to solve the relaibility optimization problems in complex systems and optimal redundancy allocation problems in a multistage mixed system. They reported that INESA using the Glauber algorithm and exponential cooling schedule outperformed the SA and NESA by yielding superior optimal solutions and improving the speed of convergence.

2.3.3 Modified Great Deluge Algorithm (MGDA)

Ravi (2004) developed the modified great deluge algorithm (MGDA) as an extended version of the great deluge algorithm (GDA). Here a brief description of the GDA is first presented. Then, the MGDA is described.

2.3.3.1 Great Deluge Algorithm

Dueck and Scheuer (1990) proposed a faster and superior variant of simulated annealing namely the threshold accepting (TA) algorithm by employing a deterministic criterion compared to the probabilistic one used in SA while accepting/rejecting a candidate solution. Later, Dueck (1993) further extended the ideas proposed in TA and suggested two new optimization meta-heuristics viz. the great deluge algorithm (GDA) and record-to-record travel (RRT). He observed that GDA outperformed the original TA in the case of some hard, benchmark instances of the traveling salesman problem. Then, Sinclair (1993) applied GDA and RRT to 37 real examples of hydraulic turbine runner balancing problem, which is a special case of quadratic assignment problem. In this study, he compared their performance with that of SA, genetic algorithms and tabu search. For more information on RRT, the reader is referred to Sinclair (1993).

The motivation for the GDA is as follows: Imagine the goal is to find the highest point in a country. Assume that it rains incessantly on this piece of land. The algorithm moves around on the uncovered land without getting its feet wet. As the *water level* increases, the algorithm will eventually end up on some high point of land. The similarity with simulated annealing is that the temperature parameter is analogous to the *water level* in GDA (Dueck, 1993; Sinclair 1993). An important feature of GDA is that it is governed by a single parameter unlike SA.

2.3.3.2 The MGDA

While developing the MGDA the following issues are carefully considered. As with any other metaheuristic, the choice of a powerful neighborhood search scheme and initialization and updating of controlling parameters is critical to the performance of the GDA. Accordingly, in MGDA, a new neighborhood search scheme is devised along with the addition of a new parameter *limit*. Further, the initialization scheme and updating formula of the parameter *LEVEL* is also redesigned.

(i)   Generate the initial feasible solution randomly in the range (xll_i, xul_i) using a uniform random number generator, where xll_i *and* xul_i are respectively the lower and upper limits on the ith decision variable given in the problem.

(ii)   Compute the objective function value for this initial feasible solution and store it in f^o. Let the initial values of some parameters be as follows: $itr = 0$; $old = 9999$ and $LEVEL = f^o$.

(iii) Increment the global iteration counter: $itr = itr + 1$.

(iv)   The inner iterations essentially perform a neighbourhood search. To accomplish this, the following stochastic procedure is deas compared withed and employed resulting in a neighbouring solution of the original (old) one.

$$x_i^c = x_i^o + (10 * u - 5)^p ; i = 1,2,...,n$$

where u is a random number drown from uniform distribution in the range $(0,1)$, p is a pre-specified odd integer and the superscripts c and o indicate the candidate solution and the old solution respectively.

This stochastic perturbation does not necessarily result in a feasible solution. Hence, this perturbation is perfomed several times until a feasible solution is obtained. If the number of such trials, say, *limit*, exceeds a large pre-specified number, then it is understood/assumed that the algorithm is unable to find a feasible solution and hence the algorithm is forced to stop and the old solution at this stage is reported as the optimal solution. This parameter value is, however, problem-dependent.

(v)   Compute the value of the objective function for the candidate solution and store it in f^C.

(vi)   If ($f^C < LEVEL$) then,

$$f^o = f^c$$

$$\mathbf{x}_i^o = \mathbf{x}_i^c , \ i = 1,2,...,n$$

$$LEVEL = LEVEL - UP * (LEVEL - f^C)$$

$$new = f^o \quad \text{and go to step (vii)}.$$

Else go to step (iv)

(vii) If ($itr < itrmax$)
           If ((new-old)/old<0.000001)
           Report x_i^c, $i = 1, 2, ..., n$ as the global optimal solution with f as the global optimum.
               Else go to step (iii)
       Else

Report x_i^c, $i = 1, 2, ..., n$ as the global optimal solution with f as the global optimum.

In the above algorithm, the convergence criterion in step (vii) normally takes care of convergence in the case of many problems. However, in the case of difficult problems, when sufficiently large number of global iterations have passed and generation of feasible solution itself becomes difficult in the neighborhood search, then the parameter *limit* will be invoked and convergence is achieved. Here, *LEVEL* is the *"water level"* value; *UP* >0, is the factor used in the reduction of the *LEVEL*; *old* is the arbitrarily specified initial large value; *itr* is the global iteration counter; *itrmax* is the maximum number of global iterations and *limit* is the number of trials allowed to find a feasible solution in the neighbourhood search. The most important parameters that govern the accuracy and speed of the MGDA are *UP* and *p* used in Step (iv) to perform the neighborhood search. The parameter *limit* is also important to increase the speed of convergence. Thus, MGDA has an additional parameter compared to the original GDA. In GDA, *LEVEL* is treated as a parameter, but in MGDA, the objective function value corresponding to the initial solution is assigned to *LEVEL*. This trick saved us one parameter.

The modifications suggested to the GDA by Ravi (2004) are as follows: (1) devising a new neighborhood search scheme different from the one used in Ravi *et al.* (2000); (2) assigning the objective function value corresponding to the initial solution to the parameter *LEVEL* (3) employing the method of Sinclair (1993) to update *LEVEL* in step (vi) and finally (4) introducing the parameter *limit*.

The second modification reduces the need to specify the initial value of *LEVEL*. In the fourth modification, the parameter *limit* restricts the number of trials to find a feasible solution in the neighbourhood of a candidate solution to a fixed number. This is necessitated owing to the inherent property of the original GDA and RRT algorithm to perform several "unproductive" trials in the neighbourhood search. This phenomenon was observed in the case of examples 1 and 2 where the algorithm took excessively long time even after reaching a better optimal solution than the ones reported in literature. First and third modifications improved the speed and accuracy of the algorithm.

2.4 Discussion of Results

The results of all the algorithms are presented in Tables 1, 2, 3 and 4. The results obtained by MGDA are compared to those obtained by the ant colony optimization (ACO) algorithm (Shelokar *et al.* 2002), simulated annealing (SA), improved non-equilibrium simulated annealing (INESA) (Ravi *et al.*, 1997) and generalized Lagrange function approach, sequential unconstrained minimization technique (Tillman *et al.*, 1980), a random search technique (Mohan and Shanker, 1988) and an integer programming approach (Luus, 1975). The function evaluations are presented for the MGDA and the ACO algorithm only. This information is missing for other algorithms.

In case (i) of Problem 1 (see Table 1), MGDA obtained 641.823608 as the optimal system cost with 65,603 function evaluations, while ACO obtained 641.823562 with 20,100 function evaluations. The system reliability of 0.9 is obtained in both cases. Thus, for this problem, ACO obtained a marginal improvement over MGDA in terms of accuracy and also consumed far less function evaluations. However, MGDA outperformed the INESA, SA (Ravi *et al.*, 1997) and generalized Lagrange function approach (Tillman *et al.*, 1980) in terms of both accuracy and speed.

Table 1. Results of Problem 1- Case (i)

Solution	MGDA Ravi (2004)	ACO Shelokar *et al.* (2002)	INESA Ravi *et al.* (1997)	SA Ravi *et al.* (1997)	Tillman *et al.* (1980)
R_1	0.50001	0.5	0.50006	0.50095	0.50001
R_2	0.838919	0.838920	0.83887	0.83775	0.84062
R_3	0.5	0.5	0.50001	0.50025	0.50000
R_4	0.5	0.5	0.50002	0.50015	0.50000
R_S	0.9	0.9	0.90001	0.90001	0.90005
C_S	641.823608	641.823562	641.8332	641.903	642.04
FE*	65,603	20,100	NA	NA	NA

*: Numer of Function Evaluations

In case (ii) of Problem 1 (see Table 2), MGDA yielded 390.570190 as the optimal system cost with 188,777 function evaluations, while ACO yielded 390.570892 with 54,140 function evaluations. Both algorithms obtained the same optimal system reliability of 0.99. Thus, in this case, MGDA outperformed the ACO in terms of accuracy but consumed more CPU time. Once again, MGDA outperformed the INESA, SA (Ravi *et al.*,

1997) and a sequential unconstrained minimization technique (Tillman *et al.*, 1980) in terms of both accuracy and speed.

Table 2. Results of Problem 1- Case (ii)

Solution	MGDA Ravi (2004)	ACO Shelokar *et al.* (2002)	INESA Ravi *et al.* (1997)	SA Ravi *et al.* (1997)	Tillman *et al.* (1980)
R_1	0.825808	0.825895	0.82516	0.82529	0.825895
R_2	0.890148	0.890089	0.89013	0.89169	0.890089
R_3	0.627478	0.627426	0.62825	0.62161	0.627426
R_4	0.728662	0.728794	0.72917	0.72791	0.728794
R_S	0.99	0.99	0.99	0.990003	0.99041
C_S	390.570190	390.570892	390.572	390.6327	397.88
FE*	188,777	54,140	NA	NA	NA

*: Numer of Function Evaluations

As regards Problem 2 (see Table 3), MGDA yielded 5.019919 as the optimal system cost and 0.99 as the optimal system reliability with 50,942 function evaluations thereby outperforming the ACO, which obtained an optimal system cost of 5.019923 and optimal system reliability of 0.990001 with 80,160 function evaluations. Yet again, the MGDA outperformed the INESA, SA (Ravi *et al.*, 1997) and a random search technique (Mohan and Shanker, 1988) in terms of both accuracy and speed. Thus, in this problem, the MGDA yielded the best performance over all the algorithms that are compared in this study.

Table 3. Results of Problem 2

Solution	MGDA Ravi (2004)	ACO Shelokar *et al.* (2002)	INESA Ravi *et al.* (1997)	SA Ravi *et al.* (1997)	Mohan and Shanker (1988)
R_1	0.935400	0.933869	0.93747	0.93566	0.93924
R_2	0.935403	0.935073	0.93291	0.93674	0.93454
R_3	0.788027	0.798365	0.78485	0.79299	0.77154
R_4	0.935060	0.935804	0.93641	0.93873	0.93938
R_5	0.934111	0.934223	0.93342	0.92816	0.92844
R_S	0.99	0.990001	0.99	0.99001	0.99004
C_S	5.019919	5.019923	5.01993	5.01997	5.02001
FE*	50,942	80,160	NA	NA	NA

*: Numer of Function Evaluations

In the case of Problem 3 (see Table 4), which is an optimal redundancy allocation problem, modeled as a non-linear integer-programming problem, both MGDA and ACO produced the identical optimal system reliability of 0.945613, but MGDA fared much better than ACO in terms of speed. However, as in the case of other problems, MGDA scored over the INESA, SA (Ravi et al., 1997) and an integer programming technique (Luus, 1975) in terms of both accuracy and speed.

Table 4. Results of Problem 3

Algrithm / Solution	MGDA	ACO	INESA	SA	LUUS (1975)
X_1	3	3	3	3	3
X_2	4	4	4	4	4
X_3	6	6	5	5	5
X_4	4	4	3	4	3
X_5	3	3	3	3	3
X_6	2	2	2	2	2
X_7	4	4	4	4	4
X_8	5	5	5	5	5
X_9	4	4	4	4	4
X_{10}	2	2	3	3	3
X_{11}	3	3	3	3	3
X_{12}	4	4	4	4	4
X_{13}	5	5	5	5	5
X_{14}	4	4	5	5	5
X_{15}	5	5	5	4	5
R_S	0.945613	0.945613	0.944749	0.943259	0.944749
C_S	392	392	389	380	389
Ws	414	414	414	414	414
FE*	217,157	244,000	NA	NA	NA

*: Numer of Function Evaluations

Thus, in summary, for all the problems, the MGDA comprehensively outperformed all the algorithms SA, INESA and generalized Lagrange function approach, sequential unconstrained minimization technique, a random search technique and an integer programming approach except the ACO algorithm in terms of both accuracy and speed. When accuracy and speed are considered simultaneously, MGDA and ACO are clear winners in one problem each. In the problem, where they produced the same solution, MGDA scored over ACO in terms of speed. However, in another problem, MGDA outperformed ACO in terms of accuracy, but lost in terms of speed. Thus, based on the numerical experiments conducted in this study it can be inferred that in choosing between MGDA and ACO, one ends up with mixed results, when the criteria accuracy and speed are considered simultaneously.

2.5 Conclusions

This chapter reviews the application of several metaheuristics to solve reliability optimization problems occurring in complex systems. Further, it presents a new global optimization meta-heuristics, the modified great deluge algorithm (MGDA) and compares its performance with that of other metaheuristics namely SA, INESA, ACO and some derivative-based optimization techniques on some reliability optimization problems of complex systems. Two different kinds of problems (i) Reliability optimization of a complex system with constraints on component and system reliabilities (ii) Optimal redundancy allocation in a multi-stage mixed system with constraints on cost and weight are solved to illustrate the effectiveness of the algorithm. Based on the results, it is concluded that MGDA succeeded in yielding better optimal solutions in two instances when compared to ACO. However, MGDA comprehensively outperformed the SA and INESA and other optimization algorithms reported in literature in terms of accuracy and speed. Hence, it can be inferred that the MGDA can be used as a sound alternative to ACO and other optimization techniques in reliability optimization.

References

Cardoso MF, Salcedo RL, de Azevedo SF (1994) Non-equilibrium simulated annealing: a faster approach to combinatorial minimization. Industrial Engineering Chemical Research 33: 1908-1918.

Coit DW, Smith AE (1996) Reliability optimization of series parallel systems using a genetic algorithm. IEEE Transactions on Reliability 45: 254-260

Dolan WB, Cummings PT, Levan MD (1989) Process optimization via simulated annealing: Applications to network design. American Institute of Chemical Engineering Journal 35: 725-736

Dorigo M, Di Caro G, Gamberdella LM (1997) Ant algorithms for discrete optimization. Artificial Life 5: 137-172

Dueck G, Scheuer T (1990) Threshold Accepting: a general purpose optimization algorithm appearing superior to simulated annealing. Journal of Computational Physics 90: 161-175

Dueck G (1993) New optimization heuristics. Journal of Computational Physics 104: 86-92.

Dutta P, Majumdar D, Bhattacharya SP (1991) Global optimization of molecular geometry. Chemical Physics Letters 181: 293-297

Glauber RJ (1963) Time dependent statistics of the Ising model. Journal of Mathematical Physics 4: 294

Kim J, Pramanik S, Chung MJ (1994) Multiple sequence alignment using simulated annealing. Computer Applications in Biological Sciences 10: 419-426

Kirkpatrick S, Gelatt Jr CD, Vecchi MP (1983) Optimization by simulated annealing. Science 220: 671-680

Luus R (1975) Optimization of system reliability by a new nonlinear integer programming procedure. IEEE Transactions on Reliability 24:14-16

Metropolis N, Rosenbluth AW, Rosenbluth MN (1953) Equation of state calculations by fast computing machines. Journal of Chemical Physics 21: 10-16

Mohan C, Shanker K (1988) Reliability optimization of complex systems using random search techniques. Microelectronics and Reliability 28: 513-518

Nelder JA, Mead R (1965) A simplex method for function optimization. Computer Journal 7: 308

Ravi V, Murty BSN, Reddy PJ (1997) Nonequilibrium simulated annealing-algorithm applied to reliability optimization of complex systems. IEEE Transactions on Reliability 46: 233-239

Ravi V, Reddy PJ, Zimmermann H.-J (2000) Fuzzy global optimization of complex system reliability. IEEE Transactions on Fuzzy Systems 8: 241-248

Ravi V (2004) Optimization of complex system reliability by modified great deluge algorithm. Asia-Pacific Journal of Operational Research 21: 487-497

Shelokar PS, Jayaraman VK, Kulkarni BD (2002) Ant algorithm for single and multi objective reliability optimization problems. Quality and Reliability Engineering International 18: 497-514

Sinclair M (1993) Comparison of the performance of modern heuristics for combinatorial optimization problems of real data. Computers and Operations Research 20: 687-695.

Tillman FA, Hwang C.-L, Kuo W (1980) Optimization of System Reliability. Marcel Dekker Inc, New York

Appendix

Problem 1: Life support system in a space capsule

The block-diagram is presented in Figure 1. The system reliability is given by (Tillman *et al.*, 1980):

$$R_S = 1 - r_3 \left[(1 - r_1)(1 - r_4) \right]^2 - (1 - r_3) \left[1 - r_2 \left\{ 1 - (1 - r_1)(1 - r_4) \right\} \right]^2$$

Minimize C_s

subject to $r_{i,min} \le r_i \le 1.0$, $i = 1,2,3,4$ and $R_{S,min} \le R_S \le 1.0$

where, C_S is the system cost, $r_{i,min}$ (= 0.5) and $R_{S,min}$ are respectively the lower bounds on the reliabilities of the ith component and the system.

Two different forms of system cost functions are considered as follows

Case (i):
$$C_S = 2K_1 r_1^{\alpha_1} + 2K_2 r_2^{\alpha_2} + K_3 r_3^{\alpha_3} + 2K_4 r_4^{\alpha_4}$$

where, $K_1 = 100$, $K_2 = 100$, $K_3 = 200$, $K_4 = 150$, $R_{S,min} = 0.9$ and $\alpha_i = 0.6$ for $i = 1,2,3,4$.

Case (ii):
$$C_S = \sum_{i=1}^{4} K_i \left[\tan(\frac{\pi}{2} r_i) \right]^{\alpha_i}$$

where, $K_1 = 25$, $K_2 = 25$, $K_3 = 50$, $K_4 = 37.5$, $r_{i,min} = 0.5$, $R_{S,min} = 0.99$ and $\alpha_i = 1.0$ for all i.

Problem 2: Bridge network.

A bridge network system as shown in figure 2 is considered, with a component reliability, r_i, $i = 1,2,...,5$. The system reliability, R_S, is given by (Mohan and Shanker, 1997)

$$R_S = r_1 r_4 + r_2 r_5 + r_2 r_3 r_4 + r_1 r_3 r_5 + 2\, r_1 r_2 r_3 r_4 r_5 - r_1 r_2 r_4 r_5 - r_1 r_2 r_3 r_4 - r_1 r_3 r_4 r_5$$
$$- r_2 r_3 r_4 r_5 - r_1 r_2 r_3 r_5$$

Minimize

$$C_S = \sum_{i=1}^{5} a_i \exp\left[\frac{b_i}{1-r_i}\right]$$

subject to $0 \le r_i < 1$, $i = 1,..,5$ and $0.99 \le R_s \le 1$,

where $a_i = 1$ and $b_i = 0.0003$, for $i = 1,..,5$.

Problem 3: Fifteen-stage mixed system (Luus, 1975)

The block-diagram for this problem is shown in Figure 3. Find the optimal number of components $x_j \geq 1$, $j = 1,..,15$ which maximizes system reliability

$$R_S = \prod_{j=1}^{15}\left[1-(1-r_j)^{x_j}\right]$$

subject to $g_1 = \sum_{j=1}^{15} C_j x_j \leq 400$ and $g_2 = \sum_{j=1}^{15} W_j x_j \leq 414$

The constants for the fifteen stage problem are as follows

i	r_i	C_i	W_i	i	r_i	C_i	W_i
1	0.9	5	8	9	0.78	4	7
2	0.75	4	9	10	0.91	5	8
3	0.65	9	6	11	0.79	6	9
4	0.8	7	7	12	0.77	7	7
5	0.85	7	8	13	0.67	9	6
6	0.93	5	9	14	0.79	8	5
7	0.78	6	9	15	0.67	6	7
8	0.66	9	6				

Applications of the Cross-Entropy Method in Reliability

Dirk P. Kroese

University of Queensland, Australia

Kin-Ping Hui

Defence Science and Technology Organisation, Australia

3.1 Introduction

Telecommunication networks, oil platforms, chemical plants and airplanes consist of a great number of subsystems and components that are all subject to failure. Reliability theory studies the failure behavior of such systems in relation to the failure behavior of their components, which is often easier to analyze. However, even for the most basic reliability models, the overall reliability of the system can be difficult to compute. In this chapter we give an introduction to modern Monte Carlo methods for fast and accurate reliability estimation. We focus in particular on Monte Carlo techniques for network reliability estimation, and network design.

3.1.1 Network Reliability Estimation

It is well known that for large networks the exact calculation of the network reliability is difficult (indeed, this problem can be shown to be #P-complete [6,24]), and hence simulation becomes an option. However, in highly reliable networks such as modern communication networks, network failure is very infrequent, and direct simulation – also called *crude Monte Carlo* (CMC) simulation – of such rare events is computationally expensive. Various techniques have been developed to speed up the estimation procedure. For example, Kumamoto proposed a very simple technique called *Dagger Sampling* to improve the CMC simulation [20]. Fishman proposed *Procedure Q,* which can provide reliability estimates as

D.P. Kroese and K.-P. Hui: *Applications of the Cross-Entropy Method in Reliability*, Computational Intelligence in Reliability Engineering (SCI) **40**, 37–82 (2007)
www.springerlink.com

well as bound [13]. Colbourn and Harms proposed a technique that will provide progressive bounds that will eventually converge to an exact reliability value [7]. Easton and Wong proposed a sequential construction method [10]. Elperin, Gertsbakh and Lomonosov proposed *Evolution Models* for estimating the reliability of highly reliable networks [11, 12, 22]. Hui *et al.* [18] proposed a hybrid scheme that provides bounds and can provide a speed-up by several orders of magnitude in certain classes of networks. They also proposed another scheme [19] which employs the Cross-Entropy technique to speed-up the estimation in general classes of networks. Other relevant references on network reliability include [15, 31, 32]. We note that the network reliability in this chapter is always considered in the *static*, that is, non-repairable, case. However, for repairable systems a similar approach can be employed if instead of the system reliability the system *availability* is used. We will briefly discuss this issue in Section 2.

3.1.2 Network Design

Accurate reliability estimation is essential for the proper design, planning and improvement of an unreliable network, such as a telecommunications network. A typical question in network design is, for example, which components (links, nodes) to purchase, subject to a fixed budget, in order to achieve the most reliable network. There are several reasons why network planning is difficult. Firstly, the problem in question is a complex constrained integer programming problem. Secondly, for large networks the value of the objective function – that is, the network reliability – is difficult or impractical to evaluate [6, 24]. Thirdly, when Monte Carlo simulation is used to estimate the network reliability, the objective function becomes noisy (random). Finally, for highly reliable networks, sophisticated variance reduction techniques are required to estimate the reliability accurately. The literature on network planning is not extensive, and virtually all studies pertain to networks for which the system reliability can either be evaluated exactly, or sharp reliability bounds can be established. Colbourn and Harms [7] proposed a technique that provides progressive bounds that eventually converge to an exact reliability value. Cancela and Urquhart [4] employed a Simulated Annealing scheme to obtain a more reliable alternative network, given a user-defined network topology. Dengiz *et al.* used a Genetic Algorithm to optimize the design of communication network topologies subject to the minimum reliability requirement [9]. Yeh *et al.* [33] proposed a method based on a Genetic Algorithm to optimize the k-node set reliability subject to a specified capacity constraint. Reichelt *et al.* used a Genetic Algorithm in combination with a repair heuristic to minimize the network cost with specified network reliability constraints [25].

We show that the *Cross-Entropy* (CE) method [28] provides an effective way to solve both the network reliability estimation and the network planning/improvement problems. The CE method was first introduced in [26] as an adaptive algorithm for estimating probabilities of rare events in complex stochastic networks. It was soon realized [27, 29] that it could be used not only for rare event simulations but for solving difficult combinatorial optimization problems as well. Moreover, the CE method is well-suited for solving noisy optimization problems. A tutorial on the CE method can be found in [8], which is also available on-line from the CE homepage: http://www.cemethod.org.

The rest of the chapter is organized as follows. Network reliability is introduced in Section 3.2. Various Monte Carlo simulation techniques are described in Section 3.3. In Section 3.4, we present how the CE method can be applied to improve the Monte Carlo simulations. The more challenging problem of reliability optimization is tackled in Section 3.5 using the CE method. In Section 3.6, we show how to adapt the CE method to a generalized optimization problem in the context of network recovery and extension.

3.2 Reliability

The most basic mathematical model to describe complex unreliable systems is the following (see for example [1]): Consider a system that consists of m components. Each component is either functioning or failed. Suppose that the state of the system is also only functioning or failed. We wish to express the state of the system in term of the states of the components. This can be established by defining binary variables x_i, $i = 1,\ldots,m$ representing the states of the components: $x_i = 1$ if the i-th component works, and 0 else. The state of the system, s say, is a binary variable as well (1 if the system works and 0 else). We assume that s is completely determined by the vector $\mathbf{x}=(x_1,\ldots,x_m)$ of component states. In other words, we assume that there exists a function $\varphi: \{1,0\}^m \rightarrow \{1,0\}$ such that

$$s = \varphi(\mathbf{x}).$$

This function is called the *structure function* of the system. To determine the structure function of the system it is useful to have a graphical representation of system.

Example 1

Suppose a four-engine airplane is able to fly on just one engine on each wing. Number the engines 1, 2 (left wing) and 3, 4 (right wing). The network in Fig.1 represents the system symbolically. The structure function can be written as

$$\varphi(\mathbf{x}) = \left(1 - (1 - x_1)(1 - x_2)\right)\left(1 - (1 - x_3)(1 - x_4)\right).$$

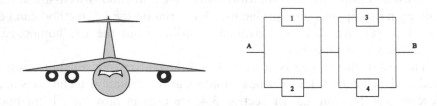

Fig. 1. An airplane with 4 engines, the system works if there is a "path" from A to B.

A system that only functions when *all* components are operational is called a *series* system. The structure function is given by

$$\varphi(\mathbf{x}) = \min\{x_1, \ldots, x_m\} = x_1 \cdots x_m = \prod_{i=1}^{m} x_i.$$

A system that functions as long as at least one component is operational is called a *parallel* system. Its structure function is

$$\varphi(\mathbf{x}) = \max\{x_1, \ldots, x_m\} = 1 - (1 - x_1) \cdots (1 - x_m) \equiv \coprod_{i=1}^{m} x_i.$$

A *k*-out-of-*m* system is a system, which works if and only if at least *k* of the *m* components are functioning.

We mention two well-know techniques for establishing the structure function. The first is the *modular decomposition* technique. Often a system consists of combinations of series and parallel structures. The determination of the structure function for such systems can be handled in stages. The following example explains the procedure. Consider the upper-most system in Fig. 2. We can view the system as consisting of component 1 and modules 1* and 2*. This gives the second system in Fig. 2. Similarly, we can view this last system as a series system consisting of component 1 and module 1** (last system in Fig. 2).

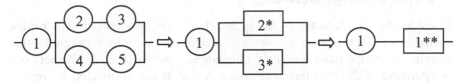

Fig. 2. Decomposition into modules

Now define s as the state of the system, z_1 as the state of module 1^{**}, y_i the state of module i^* and x_i the state of component i. Then, working "backwards", we have

$$s = x_1 z_1,$$

$$z_1 = 1 - (1 - y_1)(1 - y_2),$$

$$y_1 = x_2 x_3 \quad \text{and} \quad y_2 = x_4 x_5.$$

Successive substitution gives

$$\varphi(\mathbf{x}) = s = x_1 \left(1 - (1 - x_2 x_3)(1 - x_4 x_5) \right).$$

A second technique for determining structure functions is the *method of paths and cuts*. Here, the structure function is assumed to be monotone, that is, $\mathbf{x} < \mathbf{y} \Rightarrow \varphi(\mathbf{x}) \le \varphi(\mathbf{y})$, for all vectors $\mathbf{x}$ and $\mathbf{y}$, where $\mathbf{x} < \mathbf{y}$ means that $x_i \le y_i$ for all i and $x_i < y_i$ for at least one i. A *minimal path vector* (MPV) is a vector $\mathbf{x}$ such that $\varphi(\mathbf{x}) = 1$ and $\varphi(\mathbf{y}) = 0$ for all $\mathbf{y} < \mathbf{x}$. A *minimal cut vector* (MCV) is a vector $\mathbf{x}$ such that $\varphi(\mathbf{x}) = 0$ and $\varphi(\mathbf{y}) = 1$ for all $\mathbf{y} > \mathbf{x}$. The *minimal path set* corresponding to the MPV $\mathbf{x}$ is the set of indices i for which $x_i - 1$. The *minimal cut set* corresponding to the MCV $\mathbf{x}$ is the set of indices i for which $x_i = 0$.

The minimal path and cut sets determine the structure function. Namely, let $P_1,\ldots,P_p$ be the minimal path sets and $K_1,\ldots,K_k$ be the minimal cut sets of a system with structure function φ. Then,

$$\varphi(\mathbf{x}) = \coprod_{j=1}^{p} \prod_{i \in P_j} x_i,$$

and

$$\varphi(\mathbf{x}) = \prod_{j=1}^{k} \coprod_{i \in K_j} x_i.$$

The first equation is explained by observing that the system works if and only if there is at least one minimal path set with all components working. Similarly, the second equation means that the system working if and only if at least one component is working for each of the cut sets.

Example 2 (*Bridge Network*)

Consider the simple network in Fig.3, called a *bridge network*. The bridge network will serve as a convenient reference example throughout this chapter. Here we have five unreliable edges, labelled 1,...,5. The network is operating if the two terminal nodes A and B are connected by operational edges.

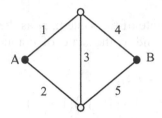

Fig. 3. Two-terminal bridge network

The minimal path sets are {1,4}, {2,5}, {1,3,5}, {2,3,4}, and the minimal cuts sets are {1,2}, {4,5}, {1,3,5}, {2,3,4}. The minimal path vector corresponding to the minimal path set {1,4} is the vector (1,0,0,1,0). The minimal cut vector corresponding to the minimal cut set {1,2} is the vector (0,0,1,1,1). It follows that the structure function φ is given by

$$\varphi(\mathbf{x}) = 1 - (1 - x_1 x_3 x_5)(1 - x_2 x_3 x_4)(1 - x_1 x_4)(1 - x_2 x_5). \qquad (1)$$

3.2.1 Reliability Function

We now turn to the case where the system's components are random. The *reliability* of a component is defined as the probability that the component will perform a required function under stated conditions for a stated period of time. Consider a system with m components and structure function φ, where the state of each component i is represented by a random variable X_i, with

$$X_i = \begin{cases} 1 & \text{with probability } p_i \\ 0 & \text{with probability } 1 - p_i \end{cases} \qquad i = 1,...,m$$

We gather the component reliabilities p_i into a vector $\mathbf{p} = (p_1,...,p_m)$. The reliability of the system, i.e., the probability that the system works, is

$$P[\text{System works}] = P[\varphi(X) = 1] = E[\varphi(X)],$$

where $\mathbf{X}$ is the random vector $(X_1, \ldots, X_m)$. Under the assumption that the component states are independent, $\Pi[\varphi(\mathbf{X}) = 1]$ can be expressed in terms of $p_1, \ldots, p_m$. The function $r(\mathbf{p}) = \Pi[\varphi(\mathbf{X}) = 1]$ is called the *reliability function* of the system. Note that for the series and parallel systems the reliability function is $r(\mathbf{p}) = \prod_{i=1}^{m} p_i$ and $r(\mathbf{p}) = \coprod_{i=1}^{m} p_i$, respectively.

Example 3

For the bridge system we have by (1)

$$\varphi(\mathbf{X}) = 1 - (1 - X_1 X_3 X_5)(1 - X_2 X_3 X_4)(1 - X_2 X_5)(1 - X_1 X_4).$$

Using the fact that $X_i = X_i^2$, the expansion of $\varphi(\mathbf{X})$ can be written as

$$\begin{aligned}
\varphi(\mathbf{X}) = {} & X_1 X_3 X_5 + X_2 X_3 X_4 + X_2 X_5 + X_1 X_4 - X_1 X_2 X_3 X_5 \\
& - X_1 X_2 X_4 X_5 - X_1 X_3 X_4 X_5 - X_1 X_2 X_3 X_4 - X_2 X_3 X_4 X_5 \\
& + 2 X_1 X_2 X_3 X_4 X_5.
\end{aligned}$$

Because all the terms are products of independent random variables the reliability $r = r(\mathbf{p})$ is given by

$$\begin{aligned}
r = {} & p_1 p_3 p_5 + p_2 p_3 p_4 + p_2 p_5 + p_1 p_4 - p_1 p_2 p_3 p_5 - p_1 p_2 p_4 p_5 \\
& - p_1 p_3 p_4 p_5 - p_1 p_2 p_3 p_4 - p_2 p_3 p_4 p_5 + 2 p_1 p_2 p_3 p_4 p_5.
\end{aligned}$$

For highly reliable networks it is sometimes more useful to analyze or estimate the system *unreliability*

$$\bar{r} = 1 - r.$$

In this case the system unreliability is equal to

$$\begin{aligned}
\bar{r} = {} & q_1 q_2 + q_2 q_3 q_4 - q_1 q_2 q_3 q_4 + q_1 q_3 q_5 - q_1 q_2 q_3 q_5 + q_4 q_5 \\
& - q_1 q_2 q_4 q_5 - q_1 q_3 q_4 q_5 - q_2 q_3 q_4 q_5 + 2 q_1 q_2 q_3 q_4 q_5,
\end{aligned}$$

where $q_i = 1 - p_i$ is the unreliability of component i, $i = 1, \ldots, 5$.

Remark 1 (Availability)

A concept closely related to reliability is the availability of a repairable system, defined as the long-run average fraction of the time that the system works.

Consider the simplest model of a repairable system where the system state is, as before, given by a structure function $\varphi(\mathbf{x})$. Suppose that each component i has a lifetime with cdf F_i and is being repaired according to a repair time cdf G_i. Assume that all the life and repair times are independent of each other. Note that the component state process $\{X_i(t),\ t{\geq}0\}$ alternates between "up" and "down" (1 and 0), and forms a so-called *alternating renewal process*. The *availability, $a_i(t)$*, of component i time t is defined as the probability that it works at time t, that is, $a_i(t) = \mathrm{P}[X_i(t) = 1]$. From renewal theory the long-run average fraction of the time that i works, the *limiting availability* of i, is given by

$$a_i = \lim_{r \to \infty} a_i(t) = \lim_{r \to \infty} \mathrm{P}[X_i(t) = 1] = \frac{u_i}{u_i + d_i},$$

where u_i is the expected lifetime or *Mean Time To Failure* (MTTF), and d_i is the expected repair time or *Mean Time To Repair* (MTTR) of component i. Thus, the limiting availability depends only on the *means* of the distributions.

The system availability at time t, $a(t)$ say, is given by

$$a(t) = \mathrm{P}[X(t) = 1] = r(a_1(t),...,a_m(t)).$$

For each i, $a_i(t)$ converges to a constant a_i, as $t{\to}\infty$. Since r is a continuous function, we therefore have

$$a(t) \to r(a_1(t),...,a_m(t)).$$

as $t{\to}\infty$. This means that all the theory in this chapter for non-repairable or *static* systems can be applied to repairable systems, provided the notion of reliability is replaced by that of (limiting) availability.

3.2.2 Network Reliability

The reliability modeling of systems such as the airplane engines in Fig. 1 and the bridge network in Fig. 3 can be generalized to *network reliability systems* in the following way. Consider an undirected graph (or network) $\Gamma(\varsigma,\mathrm{E},\mathrm{K})$, where ς is the set of n vertices (or nodes), E is the set of m edges (or links), and $\mathrm{K} \subseteq \varsigma$ is a set of *terminal* nodes, with $|\mathrm{K}| \geq 2$. Associated with each edge $e \in \mathrm{E}$ is a binary random variable X_e, denoting the *failure state* of the edge. In particular, $\{X_e = 1\}$ is the event that the edge is operational, and $\{X_e = 0\}$ is the event that it has failed. We label the edges from

1 to m, and call the vector $\mathbf{X} = (X_1,\ldots,X_m)$ the (failure) state of the network, or the state of the set E. Let Σ be the set of all 2^m possible states of E.

Notation A

For $A \subseteq E$, let $\mathbf{x} = (x_1,\ldots,x_m)$ be the vector in $\{0,1\}^m$ with

$$x_i = \begin{cases} 1, & i \in A \\ 0, & i \notin A. \end{cases}$$

We can identify $\mathbf{x}$ with the set $\mathbf{A}$. Henceforth we will use this identification whenever this is convenient.

Next, we assume that the random variables $\{X_e, e \in E\}$ are mutually independent. Let p_e and q_e denote the reliability and unreliability of $e \in E$ respectively. That is $p_e = P[X_e = 1]$, and $q_e = P[X_e = 0] = 1 - p_e$. Let $\mathbf{p} = (p_1,\ldots,p_m)$. The reliability $r = r(\mathbf{p})$ of the network is defined as the probability of K being connected by operational edges. Thus,

$$r = E\left[\varphi(\mathbf{X})\right] = \sum_{\mathbf{x} \in S} \varphi(\mathbf{x}) P[\mathbf{X} = \mathbf{x}]$$

$$= \sum_{\mathbf{x} \in S} \varphi(\mathbf{x}) \prod_{i=1}^{m} \left[p_i^{x_i} (1 - p_i)^{1-x_i} \right], \qquad (2)$$

where

$$\varphi(\mathbf{x}) = \begin{cases} 1 & \text{if } K \text{ is connected} \\ 0 & \text{otherwise} \end{cases}. \qquad (3)$$

3.3   Monte Carlo Simulation

The evaluation of network reliability in general is a #P-complete problem. When direct evaluation, e.g., via (2), is not feasible, estimation via Monte Carlo simulation becomes a viable option. The easiest way to estimate the reliability r (or unreliability $\bar{r}$) is to use CMC simulation, that is, let $\mathbf{X}_{(1)},\ldots,\mathbf{X}_{(N)}$ be independent identically distributed random vectors with the same distribution as $\mathbf{X}$. Then

$$\hat{r}_{CMC} = \frac{1}{N} \sum_{i=1}^{N} \varphi(\mathbf{X}_{(i)})$$

is an unbiased estimator of r. Its sample variance is given by

$$\mathrm{var}(\hat{r}_{CMC}) = r(1-r)/N.$$

An important measure for the efficiency of an estimator $\hat{\ell}$ is its *relative error*, defined as $\mathrm{Std}(\hat{\ell})/\mathrm{E}[\hat{\ell}]$. The relative error for $\hat{r}_{\mathrm{CMC}}$ is thus given by

$$\mathrm{re}(\hat{r}_{\mathrm{CMC}}) = \sqrt{\frac{\mathrm{var}(\hat{r}_{\mathrm{CMC}})}{(\mathrm{E}[\hat{r}_{\mathrm{CMC}}])^2}} = \sqrt{\frac{r(1-r)/N}{r^2}} = \sqrt{\frac{1-r}{Nr}}.$$

Similarly, the relative error for $\hat{\bar{r}}_{\mathrm{CMC}}$ is

$$\mathrm{re}(\hat{\bar{r}}_{\mathrm{CMC}}) = \sqrt{\frac{1-\bar{r}}{N\bar{r}}}.$$

This shows that for small $\bar{r}$ (which is typical in communication networks), a large sample size is needed to estimate $\bar{r}$ accurately. When $\bar{r}$ is small, the event that the terminal nodes are not connected is a *rare event*. Next, we discuss methods to increase the accuracy of simulation procedures that work well for rare events.

3.3.1 Permutation Monte Carlo and the Construction Process

A more efficient way than CMC for estimating the static network unreliability is *Permutation Monte Carlo* simulation [11]. This approach can be applied to estimating equilibrium renewal parameters (see [21]) such as availability described in Remark 1. The idea is as follows. Consider the network $\Gamma(\varsigma,\varepsilon)$ in which each edge e has an exponential repair time with repair rate $\lambda(e) = -\ln(q_e)$ where q_e is the failure probability of e. That is, the repair time of edge e is exponentially distributed with mean $1/\lambda(e)$. We assume that at time $t = 0$ all edges are failed and that all repair times are independent of each other. The state of e at time t is denoted by $X_e(t)$ and the state of the edge set ε at time t is given by the vector $\mathbf{X}(t)$, defined in a similar way as before. Thus, $(\mathbf{X}(t), t \geq 0)$ is a Markov process [2, 16, 23] with state space $\{0,1\}^m$ or, in view of Notation A, a Markov process on the subsets of ε. This process is called the *Construction Process* (CP) of the network.

Let Π denote the *order* in which the edges come up (become operational), and let $S_0, S_0+S_1,\ldots, S_0+\ldots+S_{m-1}$ be the times when those edges are constructed. Hence, the S_i are sojourn times of $(\mathbf{X}(t), t \geq 0)$. Π is a random variable which takes values in the space of permutations of ε denoted by Ξ. For any permutation $\pi = (e_1,\ldots,e_m)$ define

$$\varepsilon_0 = \varepsilon, \quad \varepsilon_i = \varepsilon_{i-1} \setminus \{e_i\}, \quad 1 \le i \le m-1,$$
$$\lambda(\varepsilon_i) = \sum_{e \in \varepsilon_i} \lambda(e)$$

and let

$$b(\pi) = \min_{i} \{\varphi(\varepsilon \setminus \varepsilon_i) = 1\}$$

be the *critical number* of π, that is, the number of repairs required to bring up the network. From the general theory of Markov processes it is not difficult to see that

$$\mathbf{P}[\Pi = \pi] = \prod_{j=1}^{m} [\lambda(e_j) / \lambda(\varepsilon_{j-1})].$$

Moreover, conditional on Π, the sojourn times $S_0,\ldots,S_{m-1}$ are independent and each S_i is exponentially distributed with parameter $\lambda(\varepsilon_i)$, $i = 0,\ldots,m-1$.

Note that the probability of each edge e being operational at time $t = 1$ is p_e. It follows that the network reliability at time $t = 1$ is the same as in equation (2). Hence, by conditioning on Π we have

$$r = \mathrm{E}\left[\varphi(\mathbf{X}(1))\right] = \sum_{\pi \in \Xi} \mathrm{P}[\Pi = \pi]\, \mathrm{P}\left[\varphi(\mathbf{X}(1)) = 1 \mid \Pi = \pi\right],$$

or

$$\bar{r} = 1 - r = \sum_{\pi \in \Xi} \mathrm{P}[\Pi = \pi]\, \mathrm{P}\left[\varphi(\mathbf{X}(1)) = 0) \mid \Pi = \pi\right]. \tag{4}$$

Using the definitions of S_i and $b(\pi)$, we can write the last probability in terms of convolutions of exponential distribution functions. Namely, for any $t \ge 0$ we have

$$\mathrm{P}\left[\varphi(\mathbf{X}(t)) = 0 \mid \Pi = \pi\right] = \mathrm{P}\left[S_0 + \cdots S_{b(\pi)-1} > t \mid \Pi = \pi\right]$$
$$= 1 - \operatorname*{Conv}_{0 \le i < b(\pi)} \left\{1 - \exp\left[-\lambda(E_i)t\right]\right\}. \tag{5}$$

Let

$$g_C(\pi) = \mathrm{P}\left[\varphi(\mathbf{X}(1)) = 0 \mid \Pi = \pi\right], \tag{6}$$

as given in equation (5). Equation (4) can be rewritten as

$$\bar{r} = \mathrm{E}\left[g_C(\Pi)\right], \tag{7}$$

and this shows how the CP simulation scheme works. Namely, for $\Pi_{(1)},\ldots,\Pi_{(N)}$ independent identically distributed random permutations, each distributed according to Π, one has

$$\hat{\bar{r}}_{CP} = \frac{1}{N}\sum_{i=1}^{N} g_c\left(\Pi_{(i)}\right) \tag{8}$$

as an unbiased estimator for $\bar{r}$. We note that the CP estimation scheme is a particular instance of *conditional Monte Carlo* estimation. It is well known that conditioning always reduces variance; see for example Section 4.4 of [30]. As a result, the CP estimator has a smaller variance than the corresponding CMC estimator. However this accuracy comes at the expense of more complex computations.

3.3.2 Merge Process

A closer look at the evolution of the CP process reveals that many of the above results remain valid when we merge various states into "super states" at various stages of the process. This is known as the *Merge Process* (MP). We briefly describe the ideas below (see [22] for a detailed description and [21] for its application to estimating equilibrium renewal parameters)

To begin with, any subset $\Phi \subseteq E$ of edges partitions the nodeset ς into connected components known as a *proper partition*. Let $\sigma = \{\varsigma_1,\ldots,\varsigma_k\}$ be the proper partition of the subgraph $\Gamma(\varsigma,\Phi)$ (including isolated nodes, if any). Let I_i denote the edge-set of the induced subgraph $\Gamma(\varsigma_i)$. The set $I_\sigma = I_1\cup\ldots\cup I_k$ of *inner* edges, that is, the edges within the components, is the *closure* of Φ (denoted by $\langle\Phi\rangle$). Denote its complement (the *inter-component* edges) by $E_\sigma = E\backslash I_\sigma$. Fig. 4 shows an example of a complete 6-node graph (K_6), a subgraph, and its corresponding closure.

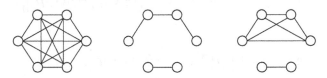

Fig. 4. K_6, a subgraph, and its corresponding closure.

Let $\Lambda(\Gamma)$ be the collection of all proper partitions of $\Gamma(\varsigma,E)$. The states in $\Lambda(\Gamma)$ are ordered by the relation $\pi\sigma \Leftrightarrow I_\tau\subset I_\sigma$ (that is, σ is obtained by merging components of τ). Any state σ in $\Lambda(\Gamma)$ has a transition path to the terminal state $\sigma_\omega = \Gamma$. Therefore $\Lambda(\Gamma)$ is a *lattice*.

We consider now the CP ($\mathbf{X}(t)$) of the network. By restricting the process ($\mathbf{X}(t)$) to $\Lambda(\Gamma)$ we obtain another Markov process ($\Xi(t)$), called the *Merge Process* (MP) of the network. This process starts from the initial state σ_0 of isolated nodes and ends at the terminal state σ_ω corresponding to $\Gamma(\varsigma,E)$. Fig. 5 shows $\Lambda(K_4)$, the lattice of all proper partitions of the complete 4-node graph K_4, grouped into 4 different levels according to the number of components. The arrows show the direct successions in $\Lambda(K_4)$, thus forming the transition graph of the Markov process ($\Xi(t)$).

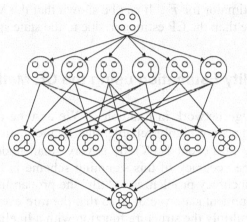

Fig. 5. State transition diagram for merge process of K_4

For each $\sigma \in \Lambda(\Gamma)$ the sojourn time in σ has an exponential distribution with parameter $\lambda(\sigma) = \sum_{e \in E_\sigma} \lambda(e)$, independent of everything else. Moreover, the transition probability from τ to σ (where σ is a direct successor of τ) is given by:

$$[\lambda(\tau) - \lambda(\sigma)] / \lambda(\tau).$$

Next, in analogy to the results for the CP, we define a *trajectory* of ($\Xi(t)$) as a sequence $\theta = (\sigma_0, \ldots, \sigma_b)$, where $b = b(\theta)$ is the first index i such that σ_i is "up", that is, the network is operational. Since $\varphi(\Xi(t)) = \varphi(\mathbf{X}(t))$, we have

$$\bar{r} = P\left[\varphi(\mathbf{X}(1)) = 0\right] = E\left[g_M(\Theta)\right],$$

where Θ is the random trajectory of ($\Xi(t)$). For each outcome $\theta = (\sigma_0, \ldots, \sigma_b)$ of Θ, $g_M(\theta)$ is given by

$$\mathbf{g_M}(\theta) = \mathbf{P}[\varphi(\mathbf{X}(1)) = 0 \mid \Theta = \theta] = \mathbf{P}[S_0 + \dots + S_{b(\theta)-1} \mid \Theta = \theta], \qquad (9)$$

where S_i is the sojourn time at σ_i. Therefore, $g_M(\theta)$ is given by the value

$$1 - \operatorname*{Conv}_{0 \le i < b(\theta)} \left\{ 1 - \exp\left[-\lambda(\sigma_i)t \right] \right\},$$

at $t = 1$. Let $\Theta_{(1)}, \dots, \Theta_{(N)}$ be independent identically distributed random trajectories distributed according to Θ. Then,

$$\widehat{\bar{r}}_{\mathrm{MP}} = \frac{1}{N} \sum_{i=1}^{N} g_M\left(\Theta_{(i)} \right) \qquad (10)$$

is an unbiased estimator for $\bar{r}$. It can be shown that the MP estimator has a smaller variance than the CP estimator, due to the state space reduction.

3.4   Reliability Estimation using the CE Method

Consider the bridge network in Example 2. We assume the typical situation where the edges are highly reliable, that is, the q_i are close to 0. The probability of the rare event $\{\varphi(\mathbf{X})=0\}$ is very small under CMC simulation and hence the accuracy of this sampling scheme is low. One way to combat the low accuracy problem is to "tilt" the probability mass function (pmf) of the component state vector $\mathbf{X}$ so that the rare event happens more often, and then multiply the structure function with a likelihood ratio to unbias the estimate. More precisely, let $f(\mathbf{x}) = \Pi[\mathbf{X} = \mathbf{x}]$ be the original pmf of $\mathbf{X}$ and $g(\mathbf{x})$ be a new pmf. Then,

$$\bar{r} = \sum_{x \in S} \mathbf{f}(x)(1 - \varphi(x)) = \sum_{x \in S} \mathbf{g}(x) \frac{\mathbf{f}(x)}{\mathbf{g}(x)} (1 - \varphi(x))$$

$$= \mathbf{E}_g \left[\frac{\mathbf{f}(X)}{\mathbf{g}(X)} (1 - \varphi(X)) \right] = \mathbf{E}_g \left[\mathbf{W}(X)(1 - \varphi(X)) \right],$$

where $W(\mathbf{x})$ is the likelihood ratio for an outcome $\mathbf{x}$, and E_g is the expectation under the pmf g. This indicates that we can estimate $\bar{r}$ also via

$$\frac{1}{N} \sum_{i=1}^{N} W\left(\mathbf{X}_{(i)} \right) \left(1 - \varphi\left(\mathbf{X}_{(i)} \right) \right),$$

where $\mathbf{X}_{(1)}, \dots, \mathbf{X}_{(N)}$ is a random sample from g. This is the well-know concept of *Importance Sampling* (IS).

Example 4 (Bridge Example (Continued))

Suppose that the edge failure probabilities are all the same, q_i=0.001. After choosing to "tilt" their probabilities to q_i'=0.5, the likelihood ratio becomes

$$W(X) = \prod_{i=1}^{5}(0/999X_i + 0.001(1-X_i))/0.5^5.$$

Table 1 shows a simulation result with 10^5 samples. With a network failure probability of about two in a million, the CMC with 10^5 samples cannot estimate $\bar{r}$ accurately, while the CMC with IS (CMC-IS) delivers a much better result. This simple example demonstrates how Importance Sampling can help improve estimate accuracy.

Table 1. Results for CMC and CMC-IS

Scheme	$\hat{\bar{r}}$	$\hat{re}$
CMC	0.000e-00	undefined
CMC-IS	2.022e-06	2.22e-00
True $\bar{r}$	2.002e-06	

However, the question of how we should tilt the parameters still remains open and that is where the CE technique can help. Before we discuss the CE technique, we first look at how to construct the ideal probability measure.

Consider the scenario where one wants to estimate the expectation ℓ of a positive function H(**X**). In the case of network reliability estimation, $\ell = \bar{r}$ and H(**X**) = 1−φ(**X**). It is possible to construct a probability measure such that one can accurately estimate ℓ with zero sample variance. Namely, since

$$\ell = \sum_{x \in S} f(x)H(x) = \sum_{x \in S} f(x)W(x)H(x),$$

if we take $g(x) = g^*(x) \equiv f(x)H(x)/\ell$, then $W(x) = \ell/H(x)$ and thus under g* we have $W(X)H(X) = \ell$ with probability 1. In other words, we only need one sample from g* to obtain an exact estimate. Obviously such construction is of little practical use, as we need to know ℓ in order to construct g*(x). Moreover, f often comes from a parametric family f(**x**;**u**) where **u** is a parameter vector. One would like to keep g in the same family, that is g(**x**)=f(**x**;**v**) such that the likelihood ratio W(**x**;**u**,**v**)=f(**x**;**u**)/f(**x**;**v**) is easier to compute.

3.4.1 CE Method

It is not our intention to give a detailed account of the CE method for estimation – for this we refer to [8] and [28] – but in order to keep this chapter
self-contained, we mention the main points. Consider the problem of estimating

$$\ell = E_u\left[H(Y)\right] = \int H(y)dF(y;u). \tag{11}$$

Here H(y) is some positive function of $y = (y_1,\dots,y_m)$, and $F(y;u)$ is a probability distribution function with pmf or probability density function (pdf)
f(y;u) that depends on some *reference parameter* u. We consider for simplicity only the pdf case. The expectation operator under which the random
vector $Y = (Y_1,\dots,Y_m)$ has pdf f(y;u) is denoted by E_u. We can estimate
ℓ using IS as

$$\hat{\ell} = \frac{1}{N}\sum_{i=1}^{N}H\left(Y_{(i)}\right)W\left(Y_{(i)};u,v\right), \tag{12}$$

where $Y_{(1)},\dots,Y_{(N)}$ is a random sample from f(·; v) – using a *different* reference parameter v – and

$$W(Y;u,v) = W(f;u)/f(Y;v), \tag{13}$$

is the likelihood ratio. We can choose *any* reference vector v in
equation (12) but we would like to use one that is in some sense "close" to
the ideal (zero variance) IS pdf

$$g*(y) = H(y)f(y;u)/\ell.$$

One could choose the parameter that minimizes the sample variance

$$var_v\left(H(Y)W(Y;u,v)\right)$$

through the optimization program

$$\min_v E\left[H^2(Y)W^2(Y;u,v)\right]. \tag{14}$$

The optimal solution of program (14) is typically hard to find and often not
available analytically since the variance of H(Y)W(Y;u,v) is not either.
One can view the variance as a "distance" measure, and the program (14)
is to find a parameter vector v that minimizes the "variance distance" between g*(y) and g(y). An alternate distance measure is the *Kullback-Leiber
CE distance* (or simply CE distance). The CE distance between two probability densities f(y) and g(y) can be written as

$$D(f,g) = \int f(y) \ln \frac{f(y)}{g(y)} dy = \int \ln \frac{f(y)}{g(y)} dF(y).$$

The optimal reference parameter in the CE sense (that is, for which $\Delta(g^*, f(\cdot; v))$ is minimal) is then

$$v^* = \arg\min_v \int \frac{H(y)}{\ell} \ln \frac{H(y) f(y; u)}{f(y; v) \ell} dF(y; u)$$

$$= \arg\min_v E_u \left[H(y) \ln \frac{H(y) f(y; u)}{f(y; v)} \right]$$

$$= \arg\min_v \left\{ E_u \left[H(y) \ln (H(y) f(y; u)) \right] - E_u \left[H(y) \ln f(y; v) \right] \right\}.$$

Since $H(y) \ln(H(y) f(y; u))$ does not depend on v, we have

$$v^* = \arg\max_v E_u \left[H(y) \ln f(y; v) \right]$$

Note that the expectation is under the original pdf $f(y; u)$. However, we can apply the IS technique and use any pdf with parameter w to get the same optimal solution

$$v^* = \arg\max_v E_w \left[H(y) W(y; u, w) \ln f(y; v) \right]. \tag{15}$$

Therefore, we can estimate the optimal CE reference vector as the solution of the iterative procedure

$$v_t = \arg\max_v \frac{1}{N} \sum_{i=1}^{N} H\left(Y_{(i)}\right) W\left(Y_{(i)}; u, v_{t-1}\right) \ln f\left(Y_{(i)}; v\right), \tag{16}$$

where at each iteration t a random sample from $f(\cdot; v_{t-1})$ is taken. The solution of program (16) can often be determined *analytically*. One example is when f is in an exponential family of distributions.

3.4.2 Tail Probability Estimation

In the *rare-event* setting, $H(Y)$ is of the form $H(Y) = I\{S(Y) \geq \gamma\}$ where I is the indicator function, γ is a constant, and

$$\ell = P\left[S(Y) \geq \gamma \right] \tag{17}$$

is a small tail probability. The function S is called the *performance function*. For rare-event estimation problems, program (16) is difficult to carry out because the rareness of the event causes most of the indicators $H(Y_{(i)})$ to be zero. For such problems, a two-phase CE procedure is employed: Apart from the reference parameter $\mathbf{v}$, we also choose to update the *level* γ, creating a sequence of pairs $\{(\mathbf{v}_t,\gamma_t)\}$ with the goal of estimating the optimal CE reference parameter $\mathbf{v}^*$. Starting with $\mathbf{v}_0 = \mathbf{u}$ (the original or nominal parameter vector), the updating formulas are as follows:

Given a random sample $Y_{(1)},...,Y_{(N)}$ from $f(\cdot;\, \mathbf{v}_{t-1})$, we use the best performing ρ–portion of the samples. Let γ_t be the sample $(1-\rho)$–quantile of the performances $S(Y_{(i)})$, $i = 1,...,N$, provided the sample quantile is less than γ; otherwise we set γ_t equal to γ. In other words, set

$$\gamma_t = \min\left\{\gamma, S_{(\lceil (1-\rho)N\rceil)}\right\}, \tag{18}$$

where $S_{(j)}$ is the j-th *order-statistic* of the performances. Using the *same sample*, we let

$$\mathbf{v}_t = \arg\max_{\mathbf{v}} \frac{1}{N}\sum_{i=1}^{N} I\left\{S\left(Y_{(i)}\right) \ge \gamma_t\right\} W\left(Y_{(i)};\mathbf{u},\mathbf{v}_{t-1}\right)\ln f\left(Y_{(i)};\mathbf{v}\right). \tag{19}$$

When γ_t reaches γ, we stop the iteration procedure and take $\mathbf{v}_t$ as the estimate of $\mathbf{v}^*$.

Again, it is important to understand that in many cases an explicit formula for $\mathbf{v}_t$ can be given, that is, we do not need to "solve" the optimization problem (19). Provided ρ is small and N is large enough, $\mathbf{v}_t$ in program (19) converges to the optimal $\mathbf{v}^*$ in program (15) (see [28]).

3.4.3 CMC and CE (CMC-CE)

The standard interpretation of the CMC scheme does not naturally allow for the application of the CE method. However, if we interpret CMC sampling using the CP framework, the CE method can be applied naturally. Instead of sampling the up/down state of individual edges, we can sample the up time of each edge. Then we check if the network is functioning at time $t=1$, and this probability is the network reliability estimate. In order to maintain the CMC features, we treat all edges independently and do not consider the concept of permutations.

In other words, we translate the original problem (estimating $\bar{r}$), which involves independent Bernoulli random variables $X_1,...,X_m$, into an estimation problem involving independent exponential random variables $Y_1,...,Y_m$. Specifically, imagine that we have a time-dependent system in which at time 0 all edges have failed and are under repair, and let

$Y_1, \ldots, Y_m$, with $Y_i \sim \mathsf{Exp}(u_i^{-1})$ and $u_i = 1/\lambda(i) = -1/\ln q_i$, be the independent *repair times* of the edges. Note that, by definition

$$P\left[Y_i \geq 1\right] = e^{-1/u_i} = q_i \quad i = 1, \ldots, m.$$

Now, for each $\mathbf{Y} = (Y_1, \ldots, Y_m)$, let $S(\mathbf{Y})$ be the (random) time at which the system "comes up" (the terminal nodes become connected). Then, we can write

$$\bar{r} = P\left[S(\mathbf{Y}) \geq 1\right].$$

Hence, we have written the estimation of $\bar{r}$ in the standard rare event formulation of equation (17) and we can thus apply the CE method from [8], as described above. Note that $\gamma = 1$ in this situation.

Instead of sampling independently for each i from $\mathsf{Exp}(-1/u_i)$, we sample from $\mathsf{Exp}(-1/v_i)$. The vector $\mathbf{v} = (v_1, \ldots, v_m)$ is thus our reference parameter. We now construct a sequence of pairs $\{(\mathbf{v}_t, \gamma_t)\}$ such that $\mathbf{v}_t$ converges to a reference vector close to the optimal CE reference parameter and γ_t eventually reaches one. Starting with $\mathbf{v}_0 = \mathbf{u} = (u_1, \ldots, u_m)$, at each iteration t we draw a random sample $\mathbf{Y}_{(1)}, \ldots, \mathbf{Y}_{(N)}$ from the pdf $f(\cdot; \mathbf{v}_t\})$ of $\mathbf{Y}$ and update the level parameter γ_t using equation (18) and the reference parameter $\mathbf{v}_t$ using equation (19), which in this case has the analytical solution

$$v_{t,j} = \frac{\sum_{i=1}^{N} I\left\{S\left(\mathbf{Y}_{(i)}\right) \geq \gamma_t\right\} W\left(\mathbf{Y}_{(i)}; \mathbf{u}, \mathbf{v}_{t-1}\right) Y_{(i)j}}{\sum_{i=1}^{N} I\left\{S\left(\mathbf{Y}_{(i)}\right) \geq \gamma_t\right\} W\left(\mathbf{Y}_{(i)}; \mathbf{u}, \mathbf{v}_{t-1}\right)}, \tag{20}$$

where W is the likelihood ratio

$$W(\mathbf{y}; \mathbf{u}, \mathbf{v}) = \frac{f(\mathbf{y}; \mathbf{u})}{f(\mathbf{y}; \mathbf{v})} = \exp\left(-\sum_{j=1}^{m} y_j\left(\frac{1}{u_j} - \frac{1}{v_j}\right)\right) \prod_{j=1}^{m} \frac{v_j}{u_j}.$$

After iteration T, when γ_T reaches one, we estimate $\bar{r}$ using IS as

$$\hat{\bar{r}} = \frac{1}{N} \sum_{i=1}^{N} I\left\{S\left(\mathbf{Y}_{(i)}\right) \geq 1\right\} W\left(\mathbf{Y}_{(i)}; \mathbf{u}, \mathbf{v}_T\right).$$

Example 5 (Bridge Network, CMC with CE)

Consider now the bridge network in Fig. 3. Suppose the "nominal" parameter vector is $\mathbf{u} = (0.3, 0.1, 0.8, 0.1, 0.2)$, that is $\mathbf{q} = (3.57\text{e-}2, 4.54\text{e-}5,$

2.87e-1, 4.54e-5, 6.74e-3). A result of simulations is given in Table 2. The following CE parameters were used: (initial) sample size N=2000 and rarity parameter ρ=0.01 in equation (18). In both CMC and CMC-CE, a final sample size of one million was used to estimate $\bar{r}$.

Table 2. Results for CMC and CMC-CE

Scheme	$\hat{\bar{r}}$	$\hat{re}$
CMC-CE	6.991e-05	1.67e-02
CMC	6.100e-05	1.28e-01
True $\bar{r}$	7.079e-05	

By using the CE method we have achieved, with minimal effort, a 98% reduction in variance compared to the CMC method. The CMC-CE algorithm required two iterations only to converge, as illustrated in Table 3. Notice that the algorithm tilted the parameters of the *mincut* elements {1,3,5} to higher values, which means they will fail much more often in the simulations. One can interpret this as the algorithm placing more importance on the mincut elements than on the remaining edges.

Table 3. Convergence of the parameters

t	γ_t	$\mathbf{v}_t$				
0	–	0.3	0.1	0.8	0.1	0.2
1	0.507	0.964833	0.216927	1.20908	0.0892952	0.567551
2	1.000	1.19792	0.120166	1.57409	0.0630103	1.15137

3.4.4 CP and CE (CP-CE)

We now apply the CE method to the CP simulation using reference parameters determined by the CE method rather than the nominal parameters. There are many ways to define a distribution on the space of permutations. However, note that the original distribution of Π is determined by the exponential distribution of $\mathbf{Y}$. In fact, Π can be viewed as a function of $\mathbf{Y}$. To see this, we generate $Y_1,...,Y_m$ independently according to $Y_i \sim \mathsf{Exp}(u_i^{-1})$ and order the Y_i's such that $Y_{\Pi_1} \leq Y_{\Pi_2} \leq \cdots \leq Y_{\Pi_m}$. Then we take $\Pi(\mathbf{Y}) = (\Pi_1,...,\Pi_m)$ as our random permutation.

We can express the network failure probability by

$$\bar{r} = E_u \left[g_C \left(\Pi(\mathbf{Y}) \right) \right] = E_u \left[S(\mathbf{Y}) \right], \tag{21}$$

where we *redefine* S(**Y**) as $g_C(\Pi(\mathbf{Y}))$, with g_C being the Markov process function for the permutation defined in equation (6). A natural way of defining a change of measure is to choose different parameters v_i (instead of the nominal u_i) for the exponential distributions of the edge lifetimes, in a similar way to Section 3.4.3. Thus $\mathbf{v} = (v_1,...,v_m)$ is still the vector of mean "repair" times. However, we have a slightly different situation from Section 3.4.3, because instead of having to estimate a rare event probability $\Pi[S(\mathbf{Y}) \geq 1]$ we now have to estimate the (small) expectation $E[S(\mathbf{Y})]$. We can no longer use a *two-phase* procedure (updating γ and **v**), but instead use the one-phase procedure in which we only update $\mathbf{v}_t$. The analytic solution to program (16) for the *i*-th component of $\mathbf{v}_t$ is

$$v_{t,i} = \frac{\sum_{k=1}^{N} S\left(\mathbf{Y}_{(k)}\right) W\left(\mathbf{Y}_{(k)}; \mathbf{u}, \mathbf{v}_{t-1}\right) Y_{(k)i}}{\sum_{k=1}^{N} S\left(\mathbf{Y}_{(k)}\right) W\left(\mathbf{Y}_{(k)}; \mathbf{u}, \mathbf{v}_{t-1}\right)}, \tag{22}$$

where $Y_{(k)i}$ is the *i*-th component of $\mathbf{Y}_{(k)}$. To improve convergence in random sampling situations, it is often beneficial to use a smoothing parameter α to blend the old with the new estimates. That is we take

$$\mathbf{v}'_t = \alpha \mathbf{v}_t + (1-\alpha)\mathbf{v}_{t-1}$$

as the new parameter vector for the next iteration.

3.4.5 MP and CE (MP-CE)

The situation can be further generalized by employing CE for MP simulation. Recall that for each permutation π in the CP, there is a corresponding trajectory θ in the MP. Let $\Theta: \pi \mapsto \theta$ be the mapping that assigns to each permutation π the corresponding unique trajectory θ. Then equation (21) can be rewritten as

$$\bar{r} = E_u \left[g_M \left(\Theta \left(\Pi(\mathbf{Y}) \right) \right) \right] = E_u \left[S(\mathbf{Y}) \right],$$

where S(**Y**) has been redefined as $g_M(\Theta(\Pi(\mathbf{Y})))$, with g_M being the Markov process function for the trajectory θ defined in equation (9). The same CE procedure (22) described in Section 3.4.4 can be applied to the MP as well.

Example 6 (Bridge Network, MP and CP with CE)

We return to the bridge network of Example 5. Table 4 lists the results for the standard MP and CP simulations, compared with their counterparts with IS in which the reference parameters are determined by the CE method. The nominal reference parameter remains unchanged. That is, $\mathbf{u} =$ (0.3, 0.1, 0.8, 0.1, 0.2), and we use the CE parameters $\alpha = 0.7$ and $N = 2000$. The final sample sizes are $N = 10^5$ in all the original and CE simulations.

Table 4. Results for CP-CE and MP-CE

Scheme	$\hat{\bar{r}}$	$\hat{re}$
MP-CE	7.082e-05	1.16e-03
MP	7.081e-05	1.32e-03
CP-CE	7.079e-05	1.21e-03
CP	7.079e-05	1.32e-03
CMC-CE	6.991e-05	1.67e-02
True $\bar{r}$	7.079e-05	

We have repeated this experiment numerous times and have consistently found that the Merge Process and the CP have very close performance in such a small example network. We also found that the CE technique provides an improvement (reduction) in variance of roughly 20% in both cases.

Note that the CMC simulation with CE still has over 100 times the variance of that in MP, MP-CE, CP or CP-CE simulations. This shows that no matter how much one modifies the CMC scheme with smart sampling techniques, the scheme still cannot compare to the simple CP sampling. In other words, it is the "structure" of sampling in the CP that makes it superior.

With the MP-CE or CP-CE sampling, there is no parameter γ to indicate when to stop the CE parameter tuning, therefore we need to use other strategies. Since we have imprecise knowledge of the performance function, we have to resort to simulation to evaluate that function at each point in order to optimize program (16). On the other hand, we do not want to spend too long on the CE parameter estimation effort, compared to the real simulation. As a result, we cannot use classic convergence criteria such as "stop when two consecutive vectors are ε close in some norm". Fortunately, permutation (and trajectory) sampling depends on the relative weight of each edge and hence the sampling is fairly insensitive to the precise values of the Importance Sampling parameter $\mathbf{v}_t$. Therefore we only require a vector that is in the "right" region. As a rule of thumb, we recommend 5% to 10% of the final estimation effort to be spent on the CE parameter estimation.

Table 5 displays the evolution of the reference parameters for the MP-CE, where we stopped the CE algorithm after only three iterations, when the estimates "stabilized" (the values stop fluctuating). Again the algorithm allocated more attention to the mincut elements {1,3,5} and treated the rest as less important.

Table 5. Evolution of the reference parameters

t	$\mathbf{v}_t$				
0	0.3	0.1	0.8	0.1	0.2
1	0.35768	0.07378	0.86343	0.06899	0.2548
2	0.37752	0.06507	0.86312	0.05950	0.2718
3	0.38688	0.05956	0.85218	0.05764	0.2785

3.4.6 Numerical Experiments

In this section we give a few larger examples that might be found in communication networks. Fig. 6 shows a 3×3 and a 6×6 grid network, each network has four terminals at the corners. All links have the same failure probability. All experiments use a final sample size of 10^6 and the CE tuning batch sample size of 5000. In the tables, T denotes the CE tuning iteration and α denotes the smoothing parameter. The CMC-CE had a rarity parameter $\rho=0.02$.

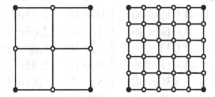

Fig. 6. A 3×3 and a 6×6 grid network

The estimated relative error ($\hat{re}$), sample variance ($\hat{var}$), simulation time (t) as well as Relative Time Variance (RTV) are also provided for comparisons. The RTV is defined as the product of the simulation time t (in seconds) and the estimated sample variance $\hat{var}$. It can be used as a metric to compare different algorithms. For a large number of samples N, the simulation time is proportional to N and the sample variance is inversely proportional to N. Therefore the RTV is a number that largely depends on the network and the performance of the algorithm under investigation rather than on N. The smaller the RTV value, the more efficient is the simulation algorithm.

For verification purposes, the exact network failure probabilities are evaluated and listed as well. Note that in these trivial examples, alternative approaches such as approximation [3, 14] can also be used to obtain fairly accurate results.

Example 7 (3×3 unreliable grid)

In this example, all links of the 3×3 grid network have the same failure probability $q = 10^{-3}$. A result of the simulation is given in Table 6.

The CMC method gives a poor variance and relative error as expected. The CMC-CE shows a 95% reduction in variance but the variance is still too high to make this scheme very useful. In fact, the CMC-CE scheme has not converged in this example (also indicated by a relatively high $\hat{re}$). The CP method gives a much smaller variance (0.1% of CMC-CE) while the CE method achieved a further reduction of 20-25% on average. The MP method has an even smaller variance (10% of CP) and the CE method provides roughly a 10% further reduction. Taking into account the computation overhead introduced by the CE method (approximately 10%), the MP-CE has a slight overall speed advantage over the MP algorithm, making the MP-CE the most efficient method to use.

Table 6. Simulation results for the 3×3 unreliable grid network

Scheme	T	α	$\hat{\bar{r}}$	$\hat{re}$	$\hat{var}$	t	RTV
MP-CE	10	0.1	4.012e-06	1.07e-03	1.85e-17	18	3.34e-16
MP	-	-	4.012e-06	1.12e-03	2.03e-17	17	3.39e-16
CP-CE	10	0.1	4.011e-06	3.42e-03	1.88e-16	10	1.96e-15
CP	-	-	4.016e-06	3.89e-03	2.45e-16	9	2.24e-15
CMC-CE	4	1	2.830e-06	1.51e-01	1.81e-13	9	1.66e-12
CMC	-	-	4.000e-06	5.00e-01	4.00e-12	8	3.14e-11
True value			4.012e-06				

Example 8 (6×6 reliable grid)

This is a larger network example consisting of 36 nodes and 60 edges with equal link failure probability $q=10^{-6}$. A result of the simulation is given in Table 7.

Table 7. Simulation results for the 6×6 reliable grid network

Scheme	T	α	$\hat{\bar{r}}$	$\hat{re}$	$\hat{var}$	t	RTV
MP-CE	10	0.1	3.999e-12	1.53e-03	3.76e-29	122	4.60e-27
MP	–	–	3.998e-12	1.75e-03	4.89e-29	113	5.55e-27
CP-CE	10	0.1	4.005e-12	1.28e-02	2.61e-27	66	1.72e-25
CP	–	–	4.006e-12	2.10e-02	7.07e-27	56	3.99e-25
CMC-CE	5	1	8.625e-14	9.08e-01	6.13e-27	41	2.51e-25
CMC	–	–	0	undefined	0	34	–
True value			4.000e-12				

The CMC and CMC-CE methods cannot handle such a low probability with a million samples. The CP provides good estimates and yet the CE method reduces the sample variances further by about 65% in the CP-CE. The MP starts with a much lower (1%) sample variance than that of CP and the CE further reduces it by 25% in the MP-CE. The RTV of the MP-CE and the CP-CE show a 20% and 130% speed up over the MP and the CP, respectively.

Example 9 (20-node 30-link unreliable network)

The next example is a 20-node 30-link network shown Fig. 7 with equal link failure probability of 3%. Two terminal reliability (the two terminal nodes are marked by thick circles in the figure) is to be estimated using different simulation schemes. A result of the simulation is given in Table 8. Note that in this example, the simple Minimum Cut Approximation method described in [6] will estimate a network failure probability of 5.4×10^{-5}, a 12% deviation from the true value of 6.138×10^{-5}.

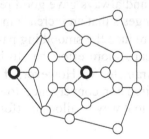

Fig. 7. A 20-node 30-link network

The CMC method performed poorly as reflected in high variance and relative error. The CMC-CE method has significant improvement over CMC but is still far from ideal. Both CP and MP provide good estimates and the

CE method improves them much further. In terms of RTV, the MP-CE and CP-CE have around 50% and 270% speed up over the MP and CP respectively.

Table 8. Simulation results for the 20-node 30-link network

Scheme	T	α	$\hat{\bar{r}}$	$\hat{re}$	$\hat{var}$	t	RTV
MP-CE	10	0.1	6.128e-5	3.15e-3	3.72e-14	54	2.00e-12
MP	–	–	6.111e-5	3.99e-3	5.94e-14	50	2.94e-12
CP-CE	10	0.1	6.124e-5	1.13e-2	4.80e-13	26	1.24e-11
CP	–	–	6.192e-5	2.32e-2	2.06e-12	22	4.55e-11
CMC-CE	7	0.5	6.091e-5	6.46e-2	1.55e-11	18	2.80e-10
CMC	–	–	6.500e-5	1.24e-1	6.50e-11	16	1.04e-09
True value			6.138e-5				

3.4.7 Summary of Results

With a better "sampling structure" and smart conditioning, the MP and CP schemes are superior to the CMC scheme. The CE technique further improves the performance of the MP and the CP schemes; the degree of improvement becomes more prominent as the network size grows. Close inspection of the IS parameter $\mathbf{v}_T$ reveals that the *bottleneck-cut* edges have been allocated higher importance than the rest.

Another point to note is the smoothing parameter α. If we keep $\alpha = 0.7$ as in the bridge example, the IS parameters $\mathbf{v}$ might oscillate instead of converge to the optimal $\mathbf{v}^*$ and as a consequence give poor estimates. We found that in larger networks, a smaller smoothing parameter such as $\alpha = 0.1$ is much more robust and always gave good results in our experiments. Numerical experience suggests that an increase in the tuning sample size N can alleviate the need to reduce the smoothing parameter α in larger problems. Of course, this means more effort has to be spent estimating each Importance Sampling parameter $\mathbf{v}_t$. However, if we leave α very small, more iterations are required for convergence towards $\mathbf{v}^*$. This raises the question of the most efficient way to allocate effort in estimating $\mathbf{v}^*$.

3.5   Network Design and Planning

This section is concerned with a network planning problem where the objective is to maximize the network's reliability subject to a fixed budget. More precisely, given a fixed amount of money and starting with a nonexistent network, the question is which network links should be purchased, in order to maximize the reliability of the finished network. Each link carries a pre-specified price and reliability. This Network Planning Problem (NPP) is difficult to solve, not only because it is a constrained integer programming problem, which complexity grows exponentially in the number of links, but also because for large networks the value of the objective function – that is, the network reliability – becomes difficult or impractical to evaluate.

3.5.1 Problem Description

As before, consider a network represented as an undirected graph $\Gamma(\varsigma,E)$, with set ς of nodes (vertices), and set E of links (edges). The number of links is $|E| = m$. Without loss of generality we may label the links $1,\ldots,m$. Let $K \subseteq \varsigma$ be the set of terminal nodes. With each of the links is associated a *cost* c_e and reliability p_e. The objective is to buy those links that optimize the reliability of the network – defined as the probability that the terminal nodes are connected by functioning links – subject to a total budget C_{max}. Let $\mathbf{c} = (c_1,\ldots,c_m)$ denote vector of link costs, and $\mathbf{p} = (p_1,\ldots,p_m)$ the vector of link reliabilities.

We introduce the following notation. For each link e let y_e be such that $y_e = 1$ if link e is purchased, and 0 otherwise. We call the vector $\mathbf{y} = (y_1,\ldots,y_m)$ the *purchase vector* and $\mathbf{y}^*$ the *optimal purchase vector*. Similarly, to identify the operational links, we define for each link e the link *state* by $x_e = 1$ if link e is bought and is functioning, and 0 otherwise. The vector $\mathbf{x} = (x_1,\ldots,x_m)$ is thus the state vector. For each purchase vector $\mathbf{y}$ let φ_y be the structure function of the purchased system. Thus, φ_y assigns to each state vector $\mathbf{x}$ the state of the system (working = terminal nodes are connected = 1, or failed = 0). Let X_e be random state of link e, and let $\mathbf{X}$ be the corresponding random state vector. Note that for each link e that is *not* bought, the state X_e is per definition equal to 0. The reliability of the network determined by $\mathbf{y}$ is (see Equation (2)) given by

$$r(\mathbf{y}) = E\left[\varphi_y(\mathbf{X})\right] = \sum_{\mathbf{x}} \varphi_y(\mathbf{x}) \, P[\mathbf{X} = \mathbf{x}]. \qquad (23)$$

We assume from now on that the links fail independently, that is, $\mathbf{X}$ is a vector of independent Bernoulli random variables, with success probability p_e for each purchased link e and 0 otherwise. Defining $\mathbf{p}_y = (y_1 p_1,\ldots,y_m p_m)$, we write $\mathbf{X} \sim \mathsf{Ber}(\mathbf{p}_y)$. Our main purpose is to determine

$$\max_{\mathbf{y}} r(\mathbf{y}), \quad \text{subject to} \quad \sum_{e \in E} y_e c_e \leq C_{\max}. \tag{24}$$

Let $r^* \equiv r(\mathbf{y}^*)$ denote the optimal reliability of the network.

3.5.2 The CE Method for Combinatorial Optimization

The CE method is not only useful for rare-event estimation problems, but can also be applied to solve difficult discrete and continuous optimization problems. In this context, the method involves the following main steps, which are iterated:

1. Generate random states in the search space according to some specified random mechanism.
2. Update the parameters of this mechanism in order to obtain better scoring states in the next iteration. This last step involves minimizing the CE distance between two distributions.

We now specify these two steps for the NPP.

3.5.2.1 Random Network Generation

A simple method to generate the random purchase vectors is describe below: Let $\mathbf{a} = (a_1, \ldots, a_m)$ be the probability vector where a_e is the probability of purchasing edge e. Further let $\mathbf{Y}_{(k)}$ be the k-th random purchase vector where $Y_{(k),e} = 1$ denotes edge e is purchased or else 0. Following is a simple algorithm to generate K random purchase vectors by rejecting the invalid (cost exceed maximum) ones.

Algorithm 1 (Generation Algorithm)

1. Generate a uniform random permutation $\pi = (e_1, \ldots, e_m)$. Set $k = 1$.
2. Calculate $C = c_{e_k} + \sum_{i=1}^{k-1} Y_{e_i} c_{e_i}$.
3. If $C \leq C_{\max}$, draw $Y_{e_k} \sim \text{Ber}(a_{e_k})$. Otherwise set $Y_{e_k} = 0$.
4. If $k = m$, then stop; otherwise set $k = k + 1$ and reiterate from step 2.

Remark 2

We note that, when drawing via Algorithm 1, the purchase vectors have some correlation bias. Theoretically, in order to generate random networks without such bias, one should sample from the conditional Bernoulli distribution (see [5]). However this is significantly more involved than the present algorithm/heuristic, and from our experience does not yield much gain.

3.5.2.2 Updating Generation Parameters

The usual CE procedure [28] proceeds by constructing a sequence of reference vectors $\{\mathbf{a}_t, t \geq 0\}$ (i.e., purchase probability vectors), such that $\{\mathbf{a}_t, t \geq 0\}$ converges to the degenerate (i.e., binary) probability vector $\mathbf{a}^* = \mathbf{y}^*$. The sequence of reference vectors is obtained via a two-step procedure, involving an auxiliary sequence of reliability levels $\{\gamma_t, t \geq 0\}$ that tend to the optimal reliability $\gamma^* = r^*$ at the same time as the $\mathbf{a}_t$ tend to $\mathbf{a}^*$. At each iteration t, for a given $\mathbf{a}_{t-1}$, γ_t is the sample $(1-\rho)$-quantile of performances (reliabilities). Typically ρ is chosen between 0.01 and 0.1. That is, generate a random sample $\mathbf{Y}_{(1)}, \ldots, \mathbf{Y}_{(K)}$ using the generation algorithm above; compute the performances $r(\mathbf{Y}_{(i)})$, $I = 1, \ldots, K$ and let $\gamma_t = r_{(\lceil (1-\rho)K \rceil)}$, where $r_{(1)} \leq \ldots \leq r_{(K)}$ are the order statistics of the performances. The reference vector is updated via CE minimization, which (see [28]) reduces to the following: For a given fixed $\mathbf{a}_{t-1}$ and γ_t, let the j-th component of $\mathbf{a}_t$ be

$$a_{t,j} = \frac{\sum_{i=1}^{K} I\{r(\mathbf{Y}_{(i)}) \geq \gamma_t\} Y_{(i)j}}{\sum_{i=1}^{K} I\{r(\mathbf{Y}_{(i)}) \geq \gamma_t\}}, \quad j = 1, \ldots, m, \qquad (25)$$

where we use the *same* random sample $\mathbf{Y}_{(1)}, \ldots, \mathbf{Y}_{(k)}$ and where $Y_{(i)j}$ is the j-th coordinate of $\mathbf{Y}_{(i)}$.

The main CE algorithm for optimizing Equation (24) using the above generation algorithm is thus summarized as follows.

Algorithm 2 (Main CE Algorithm for Optimization)

1. Initialize a_0. Set $t = 1$ (iteration counter).

2. Generate a random sample $\mathbf{Y}_{(1)}, \ldots, \mathbf{Y}_{(K)}$ using Algorithm with $\mathbf{a} = \mathbf{a}_{t-1}$. Compute the sample $(1-\rho)$-sample of performances γ_t.

3. Use the *same* sample to update $\mathbf{a}_t$, using Equation (25).

4. If $\max\left(\min\left(\mathbf{a}_t, 1 - \mathbf{a}_t\right)\right) \leq \beta$ for some small fixed β then stop (let T be the final iteration); otherwise set $t = t + 1$ and reiterate from step 2.

3.5.2.3 Noisy Optimization

As mentioned earlier, for networks involving a large number of links the exact evaluation of the network reliability is in general not feasible, and simulation becomes a viable option.

In order to adapt Algorithm 2 to *noisy* NPPs, we again, at iteration t, generate a random sample $\mathbf{Y}_{(1)},\dots,\mathbf{Y}_{(N)}$ according the $\mathsf{Ber}(\mathbf{a}_{t-1})$ distribution. However, the corresponding performances (network reliabilities) are now not computed exactly, but estimated by means of Monte Carlo simulations such as Equation (10). During the optimization process, one might need to estimate the reliability of a large number of networks using a limited number of samples. An efficient estimation algorithm is the MP described in previous section. It works well with relatively small sample size even for highly reliable networks.

3.5.3 Numerical Experiment

To illustrate the effectiveness of the proposed CE approach, consider the 6-node fully-connected graph with 3 terminal nodes given in Figure 8. The links costs and reliabilities are given in Table 9. Note that the direct links between the terminal nodes have infinite costs. We have deliberately excluded such links to make the problem more difficult to solve. The total budget is set to $C_{\max} = 3000$.

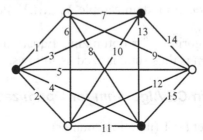

Fig. 8. Network with 3 terminal nodes, denoted by black vertices.

Table 9. Link costs and reliabilities

i	c_i	p_i	i	c_i	p_i	i	c_i	p_i
1	382	0.990	6	380	0.998	11	397	0.990
2	392	0.991	7	390	0.997	12	380	0.991
3	∞	0.992	8	395	0.996	13	∞	0.993
4	∞	0.993	9	396	0.995	14	399	0.992
5	320	0.994	10	381	0.999	15	392	0.994

Note that for a typical purchase vector **y** the network reliability $r(\mathbf{y})$ will be high, since all links are quite reliable. Consequently, to obtain an accurate estimate of the network reliability, or better, the network unreliability $\overline{r}(\mathbf{y}) = 1 - r(\mathbf{y})$, via conventional Monte Carlo methods, would require a large simulation effort. The optimal purchase vector for this problem – computed by brute force – is **y*** = (1,1,0,0,1,0,1,1,0,1,0,0,0,0,1), which yields a minimum network unreliability of $\overline{r}\,* = 7.9762 \times 10^{-5}$.

We used the following parameters for our algorithm: the sample size in Step 2 of the CE algorithm $K = 300$; the sample size in Equation (10) $N = 100$; the initial purchase probability $\mathbf{a}_0 = (0.5,\ldots,0.5)$. The algorithm stops when all elements of $\mathbf{a}_t$ are less than $\beta = 0.01$ away from either 0 or 1. Let T denote the final iteration counter. We round $\hat{\mathbf{a}}_T$ to the nearest binary vector and take this as our solution **a*** to the problem.

Table 10 displays a typical evolution of the CE method. Here, t denotes the iteration counter, γ_t the sample $(1-\rho)$-quantile of the estimated unreliabilities, and $\mathbf{a}_t$ the purchase probability vector, at iteration t. The important thing to notice is that $\mathbf{a}_t$ quickly converges to the optimal degenerate vector **a*** = **y***. The simulation time was 154 seconds on a 3.0GHz computer using a Matlab implementation.

In repeated experiments, the proposed CE algorithm performed effectively and reliably in solving the noisy NPP, which constantly obtained the optimal purchase vector. Moreover, the algorithm only required on average 9 iterations with a CPU time of 180 seconds.

Table 10. A typical evolution of the purchase vector

t	γ_t	$\mathbf{a}_t$
0		0.50 0.50 0.50 0.50 0.50 0.50 0.50 0.50 0.50 0.50 0.50 0.50 0.50 0.50 0.50
1	4.0e-03	0.66 0.69 0.15 0.15 0.62 0.48 0.59 0.64 0.38 0.62 0.52 0.38 0.15 0.41 0.62
2	2.6e-04	0.69 0.63 0.05 0.05 0.72 0.21 0.88 0.71 0.33 0.75 0.58 0.26 0.05 0.38 0.77
3	1.4e-04	0.67 0.75 0.01 0.01 0.78 0.11 0.89 0.89 0.12 0.76 0.57 0.22 0.01 0.44 0.77
4	1.0e-04	0.76 0.76 0.00 0.00 0.89 0.03 0.97 0.90 0.06 0.83 0.43 0.11 0.00 0.41 0.84
5	8.1e-05	0.79 0.88 0.00 0.00 0.97 0.01 0.99 0.97 0.02 0.90 0.15 0.03 0.00 0.33 0.95
6	6.7e-05	0.94 0.96 0.00 0.00 0.97 0.00 1.00 0.99 0.01 0.97 0.07 0.01 0.00 0.10 0.99
7	6.3e-05	0.98 0.99 0.00 0.00 0.99 0.00 1.00 1.00 0.00 0.99 0.02 0.00 0.00 0.03 1.00
8	5.8e-05	0.99 1.00 0.00 0.00 1.00 0.00 1.00 1.00 0.00 1.00 0.01 0.00 0.00 0.01 1.00

3.6   Network Recovery and Expansion

This section looks at the optimal network design problem in an incremental sense, that is, one starts with a *baseline* network and has to decide which additional links should be bought, in order to optimally improve the reliability of the network. This situation occurs for example in military networks, where as a result of components failures or attacks, part of the network has become isolated and must be reconnected. Another example is when new nodes are being deployed, and one has to choose between many available options to connect them to the existing network.

3.6.1 Problem Description

Consider an existing network (may be non-functional) as a base network. Additional links are bought to improve the network reliability. In situations where multiple possible configurations are available, the goal is to find the configuration those results in highest network reliability. Let E_B denotes the edge set of the base network and E_i denotes the additional links in the i-th network $\Gamma_i = \Gamma(\varsigma, E_B \cup E_i, K)$. Here we assume the node set and terminals are the same among all the networks. If r_i denotes the reliability of the network Γ_i, then the goal is to find the network Γ_o among all possible configurations such that $r_o = \max_i \{ r_i \}$.

In some situations where the connection points are limited – for example satellite terminals may be available to only a few nodes – the choices of alternate bearer (link) locations may not be extensive. In that case, the simplest approach is to use exhaustive search. Thus, the process can be divided into two steps:

1. generate all the valid configurations (or candidate networks Γ_i); and
2. compare their reliabilities and find the optimal network Γ_o.

Fig. 9 shows an example network being separated into two groups. In order to rejoin the two groups, at least one link is needed and there are 12 possible ways to connect the node sets {A1, A2, A3} and {B1, B2, B3, B4}. It is easy enough to generate all the 12 candidate networks and compare their reliability to find the optimal location of the alternate bearer.

If the existing network is large and/or multiple links are being added, the number of possible candidate networks grows very quickly. Often the computational cost of enumerating all networks and comparing their reliability becomes prohibitive. Another approach is to use simulation-based combinatorial optimization techniques under multiple constraints. The constraints of the optimization program can be as simple as fixing the maximum number of additional links, or specifying some maximum limit on the total cost of adding the links.

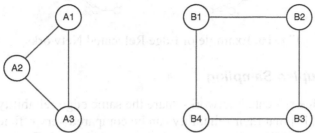

Fig. 9. Example with small number of candidates

3.6.2 Reliability Ranking

As for large networks the exact calculation of network reliability is difficult, estimating it via Monte Carlo simulation becomes favourable. Note that in applications like network reliability optimization, one needs to compare the reliability of multiple *similar* networks. When simulation is used to estimate network reliability, sampling error (noise) is introduced. If the two networks are similar, the noise can become so significant that it may impair the accuracy of the comparisons. Hui *et al.* proposed a coupled approach to estimate reliability difference of two similar networks with high confidence [17]. In this section we demonstrate how a similar concept can be used and specialized to compare many similar networks very effectively. The concept is somewhat similar to using common random numbers to reduce variance in Monte Carlo simulation.

3.6.2.1 Edge Relocated Networks

The concept of *Edge Relocated Networks* [17] refers to networks having the same number of edges with matched link reliabilities[1]. The differences between the networks are thus restricted to a few links being reconnected to different nodes. In other words, if Γ_i and Γ_j are edge relocated networks, they share the same edge reliability vector **p**. Fig. 10 shows two edge relocated networks derived from the same base network.

---

[1] Note that if there are unmatched links between the networks, redundant self-loops can be introduced to bring the networks to edge relocated versions of each other (see [17]).

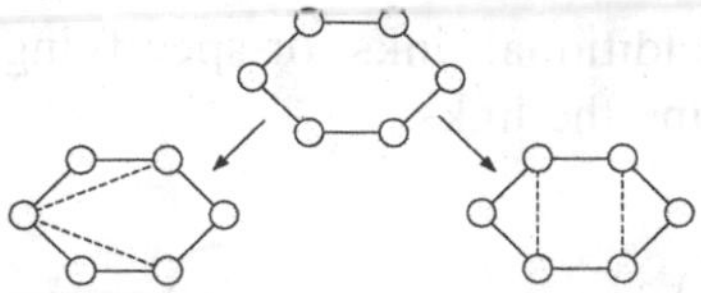

Fig. 10. Example of Edge Relocated Networks

3.6.2.2 Coupled Sampling

Since the edge relocated networks share the same edge reliability vector **p**, it is proven [17] that their reliability can be compared very efficiently. The idea is to sample the edge states and observe them in different edge relocated networks. When dealing with many edge-relocated networks simultaneously, one sampling scheme is of particular interest, namely *edge permutation sampling*.

Edge permutation sampling starts from the Construction Process in Section 3.3.1. In particular, we imagine the network is constructed dynamically by repairing links independently and with an exponential repair time. At time $t = 0$ each edge e is failed and is being repaired at a repair rate $\lambda(e)$ $= -\ln(q_e)$. The network unreliability is equal to the probability of the dynamic network not being operational at time $t = 1$, and is of the form (see Equation (7)) $\bar{r} = \mathrm{E}\left[\mathrm{g}(\Pi)\right]$, where g is a know function involving convolutions and Π is a permutation describing the random order in which the links come up. By sampling from Π, $\bar{r}$ can be efficiently estimated via Equation (8).

This edge permutation sampling (or simply permutation sampling) scheme is superior to other combinatorial sampling schemes because it elegantly avoids the rare event problem. In highly reliable networks such as communication networks, the networks are functioning most of the time, and hence it is hard for the combinatorial schemes to sample the failure state and estimate its probability. As a consequence, it is more difficult to compare the reliability of similar networks. In permutation sampling, however, the networks always start at the same failure state and will eventually come up. The only question is when will they come up or what is their operational probability at time $t=1$.

When comparing reliability of networks, the *Coupled CP* proposed in [17] is very efficient in finding the reliability difference of two networks. It can achieve 10^{11} times speedup over the best known independent sampling scheme. The scheme uses the simple observation of the following equation

$$\overline{r_{i,j}} = \overline{r_i} - \overline{r_j} = \sum_{\pi} P[\Pi = \pi]\left(g_i(\pi) - g_j(\pi)\right) = E\left[g_i(\Pi) - g_j(\Pi)\right].$$

The method takes samples of permutations and observes how the two networks behave under the same edge permutation. Figure 11 shows an example edge construction sequence on two edge-relocated networks. Assuming the black nodes are the terminals, network Γ_A will come up (i.e. all terminal nodes become connected) on the 5-th edge while network Γ_B will be on the 6-th edge. Hence their probability functions $g_A(\pi)$ and $g_B(\pi)$ will be different.

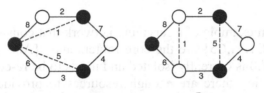

A   B
Fig. 11. Example of Coupled CP

If we take N random permutations $\{\Pi_{(1)},\ldots,\Pi_{(N)}\}$ and observe them on both networks, their reliability difference can be estimated by

$$\hat{r}_{A,B} = \sum_{i=1}^{N} \frac{g_A(\Pi_{(i)}) - g_B(\Pi_{(i)})}{N}.$$

3.6.2.3 Synchronous Construction Ranking (SCR)

In our reliability optimization problem, the prime interest is searching for the most reliable network among the candidates. Therefore the actual difference in reliability is not our main concern. All we need to find is which network is more reliable than others.

In the edge permutation sampling scheme, the most reliable network is expected to come up earlier than others. Therefore one can sample edge permutations and observe how all the candidate networks evolve simultaneously. The most reliable network should come up first most often.

Let $b_i(\pi)$ denotes the critical number of graph Γ_i on permutation π, that is the ordinal number when the network comes up. For example in Fig. 11, $b_A(\pi) = 5$ and $b_B(\pi) = 6$. Let $b^*(\pi)$ be the smallest critical number among the candidate networks on a given permutation π, that is $b^*(\pi) = \min_i b_i(\pi)$.

Then Γ_o is the most reliable network if and only if it has the highest chance of coming up before any other candidates. Mathematically, it is the network that corresponds to the solution of the program

$$\max_i \mathbf{P}[b_i(\Pi) = b*(\Pi)].$$

This is how the SCR scheme works: First randomly sample N permutations and then estimate the probability of network Γ_i being the best by

$$P_0(\mathbf{G}_i) = \sum_{j=1}^N I\{b_i(\Pi_{(j)}) = b*((\Pi_{(j)})\}/N.$$

Finally, find the network with the maximal P_o and take it as the optimal Γ_o.

Example 10

Fig. 12 shows an example of a surviving network that consists of 18 nodes and 28 links. $\{C1,...,C5\}$ are the core routers and $\{A0,...,A12\}$ are the access routers. A0 is currently isolated and needs to be re-connected to the network. Assuming there are enough resources to provide two wireless links connecting any nodes, the question is where the extra links should be attached in order to achieve maximal all-terminal reliability. Table 11 lists the reliabilities of different types of link in the example.

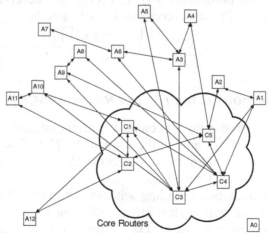

Fig. 12. Example isolated network

Since there are 18 nodes, there are $_{18}C_2 = 153$ ways to form a link. Hence there are $_{153}C_2 = 11628$ ways to add 2 different links. In order to reconnect A0 to the network, at least one of the links must attach to A0. Therefore, those configurations for which both links are not attached to A0 can be ruled out, and this leaves $_{153}C_2 - _{136}C_2 = 2448$ valid configurations to choose from.

We applied the SCR scheme to the 2448 candidate networks using 100,000 samples, it took 341 seconds on a 2.8GHz Pentium 4 machine to find the optimal network as shown in Fig. 13. The optimal network has a failure probability of 1.2485×10^{-4}. Intuitively, one might place the two links connecting A0 to its nearest core routers C3 and C4 as shown in Fig. 14. However, the intuitive network has a failure probability of 1.1069×10^{-3}, almost nine times that of the optimal network. It shows that spending a little time on searching for the optimal configuration can have significant benefits.

Table 11. Link reliabilities

Link	Core–core	Access–Core	Access–Access	Wireless links
Reliability	0.9999	0.999	0.999	0.99

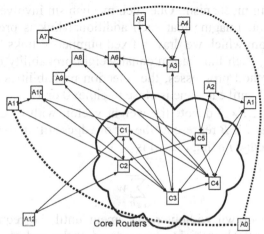

Fig. 13. Optimal network with two added links

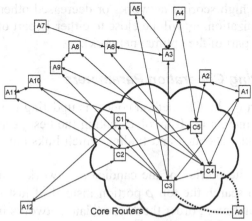

Fig. 14. Intuitive way to add two links

3.6.3 CE Method

Direct reliability comparison using the SCR scheme is effective when the number of candidates is small, however, it not practical to compare large number of networks even with the efficient SCR scheme. For example, there are 585,276 candidate networks if we want to add three links to the network in Figure 12. It quickly grows to nearly 22 million candidates if four links are to be added. Obviously we need a different approach to the problem and the CE-method is a good one.

3.6.3.1 Random Network Generation

The first step in the CE method is to generate random network according to some random mechanism. One such mechanism involves drawing without replacement. Imagine that each additional link is present in a lucky draw barrel from which we draw a fixed number of links to build our network. Initially, each link has an equal weight/probability of being picked. As the CE method progresses, the selection probabilities of the links are being modified until each one is close to either 0 or 1.

Let $\mathbf{w} = (w_1,\ldots,w_m)$ denote the weight vector, with $w_e \in [0,1]$ being the weight of edge e. If B represents the set of edges still in the barrel, then the probability of edge i being picked is

$$\frac{w_i}{\sum_{e \in B} w_e}.$$

The edges are drawn without replacement until the required number of edges is reached. The weights are updated at the end of each iteration of the CE method. Each weight w_e is increased if edge e is more likely to be involved in the high scoring networks or decreased otherwise. At the end of the CE optimization, w_e will be close to either 1 (part of the optimal network) or 0 (not part of the optimal network).

3.6.3.2 Updating Generation Parameters

The second part of the CE-method concerns updating the random network generation parameters w_e. It involves taking the best performing (e.g. 5%) random networks generated and finding which links are more likely being involved.

To find the elite portion of the candidate networks, one can extend the SCR scheme to search the top ρ portion instead of just the best network. For each permutation π, order the K candidate networks in ascending critical number. Then find the $\lceil \rho \times K \rceil$-th number and call it the elite critical number $b^\rho(\pi)$. With N random permutations, we can estimate the probability of network Γ_i in the elite ρ portion by

$$P_\rho(G_i) = \sum_{j=1}^{N} \frac{I\{b_i(\Pi_{(j)}) \le b^\rho(\Pi_{(j)})\}}{N}.$$

The elite network set Γ^ρ consists of the $\lceil \rho \times K \rceil$ networks that have the highest $P_\rho(\Gamma_i)$. Once we have the elite network set, we can update each edge weight by finding the probability of the edge being used in the elite networks, that is,

$$w'_e = \sum_{G_i \in G^\rho} \frac{I\{e \in G_i\}}{|G^\rho|}.$$

It is often beneficial to "smooth" the parameter update by incorporating part of the past history, especially when dealing with a noisy optimization problem. Lct $\mathbf{w}_t$ be the weight vector used in the t-th iteration of the CE-method. A smoothing parameter $\alpha \in [0,1]$ is used to update the weight for the next iteration:

$$\mathbf{w}_{t+1} = \alpha\mathbf{w}_t + (1-\alpha)\mathbf{w}'_t.$$

Putting the sample generation and updating together, the CE-method algorithm can be summarized as follows:

Algorithm 3 (Simple CE Algorithm)

1. **Initialization.** Set *all* edge weight to equal value $w_{0,e} = 0.5$.
2. **Generation.** Generate K (e.g. 1000) random networks by drawing m_a additional edges from the candidate edges without replacement.
3. **Elite Networks.** Rank the random networks using the SCR scheme to find the best ρ portion (e.g. 5%) for edge weight update.
4. **Updating.** Update the edge weight by

$$w_{t+1,e} = \alpha w_{t,e} + (1-\alpha) \sum_{G_i \in G^\rho} \frac{I\{e \in G_i\}}{|G^\rho|}. \tag{26}$$

5. **Termination.** Repeat from Step 2 until every element in $\mathbf{w}_t$ lies in the ranges $[0,0.01]$ or $[0.99,1]$, say. The m_a edges with $w_e \in [0.99,1]$ are the optimal edges to be added.

Example 11

In this example, three links are to be added to the base network in Fig. 12 and the result is shown in Fig. 15.

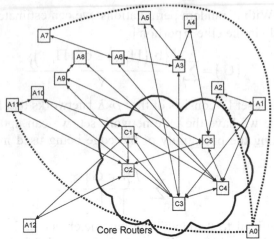

Fig. 15. Optimal network with three added links

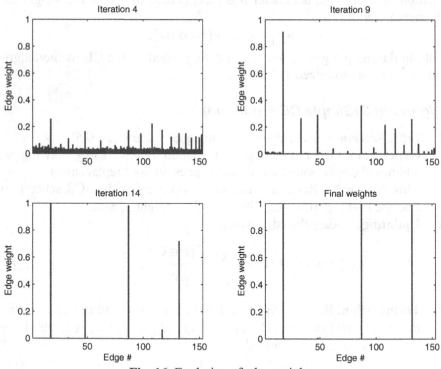

Fig. 16. Evolution of edge weights

In each iteration, $K = 1000$ random networks are generated and $\rho = 5\%$ or 50 elite networks are used to update the edge weights. The smoothing parameter of $\alpha = 0.5$ is used and 5000 random permutations are used to find the elite network set. It took 19 iterations and 142 seconds on a 2.8GHz Pentium 4 machine to find the optimal configuration. Fig. 16 shows how the edge weights evolve over the iterations. It shows how the weights of the prospective links grow while the others decay toward zero. Eventually, the weights become "degenerate", so that the edge weights of the optimal edge set stay at 1 while the remaining ones stay at 0.

This program was also used to find the optimal configuration for the two added links case. The same configuration as in Fig. 13 was found in 15 iterations using 105 seconds on the same machine using the same CE parameters. It demonstrates that the CE-method is an efficient and effective approach to finding the optimal configuration.

3.6.4 Hybrid Optimization Method

Algorithm 3 described above is tailored to finding a single optimal solution. If there is more than one solution, the algorithm has trouble deciding which one is better and as a consequence the weights "oscillate" and this keeps the algorithm from converging. Another situation where this occurs is when there are networks with performances that are very close to the optimal one.

Consider for instance in the previous examples the case where only one link is to be added to reconnect the network. There are 17 ways to reconnect node A0 to the rest of the network and each resultant network has exactly the same reliability. In this multiple optima situation, the Simple CE algorithm above will not converge.

3.6.4.1 Multi-optima Termination

One way to avoid the non-converging situation in the Simple CE algorithm is to terminate the algorithm once oscillating behavior is detected. In the case where the candidate networks have the same reliability and are perfectly matched, the SCP ranking scheme can detect the non-converging situation effectively, unlike independent sampling schemes, which will have real trouble telling whether two networks have the same reliability. If each of the N (e.g. 5000) samples indicates that every network comes up at exactly the same point, it is quite certain that all the K networks indeed have the same reliability. In that case the CE-method can stop and take the distinct networks as multiple optimal solutions.

Take the single additional link example described above using $K = 1000$ networks in each iteration and take $N = 5000$ samples to estimate the ranking. After 18 iterations that took 97 seconds, all the networks appeared to have the same reliability and the program finished. Among the 1000 generated networks, only 17 distinct networks exist and they are the 17 possible optimal solutions.

3.6.4.2 Mode Switching

In the situation where only one single optimal solution exists but there are other networks with reliabilities very close to the optimal, the Multi-optima Termination method may not work. One may try to terminate a prolonged simulation by detecting edge weight fluctuations, but unfortunately sometimes edge weight fluctuation before converging is part of the normal process. Therefore it is not easy to decide how long one should wait before the simulation is deemed to be non-converging. Even if it can be detected, it still does not help in searching for the optimal solution.

A more practical approach is to stop the CE iterations once the number of prospective links drops below a certain threshold, and then generate all the candidate networks using the prospective links and search for the optimal network using the SCR scheme. Since the edge weights will be polarized (close to either 0 or 1) in the CE-method, prospective links are simply those that have higher than the mean weight. Effectively, the CE-method is used as a filter removing those edges that are unlikely to be part of the optimal network.

To demonstrate how the scheme works, we repeat the two additional link examples with a much smaller sample size. Instead of 5000 samples, 100 were used to rank the 1000 generated networks. With such small number of samples, the confidence of the ranks is much reduced or, in other words, there is much more noise in the elite estimates. There is a high chance that the confidence intervals of top few candidate networks overlap each other. Therefore the algorithm is more likely to oscillate or even pick a sub-optimal solution if let to run indefinitely. However, we set the program to switch to SCR scheme when the number of candidate links drop below 15. Then 100,000 samples are used to search for the optimal configuration. The algorithm took eight iterations to filter out 140 of the 153 possible links and overall took 17 seconds to decide the optimal configuration is to add (A0, A11) and (A0, A7) links as shown in Fig. 13. Compare to 341 seconds required by the SCR algorithm alone, this hybrid scheme is much more robust and efficient. We applied the same scheme to the three additional link example with the switch-over point set to 10 links or below. The simulation took 21 seconds to find the same optimal solution depicted in Fig. 15.

To summarize, the Hybrid procedure is as follows:

Algorithm 4 (Hybrid CE Algorithm)

1. **Initialization.** Set all edge weight to an equal value $w_{0,e} = 0.5$.
2. **Generation.** Generate K (e.g. 1000) random networks by drawing m_a additional edges from the candidate edges without replacement.
3. **Elite Networks.** Rank the random networks using the SCR scheme to find the best ρ portion (e.g. 5%) for edge weight update.
4. **Multi-optima Condition.** Check if all generated networks have identical reliability. If yes, output distinct networks and terminate procedure.
5. **Mode Switching.** If the number of prospective links drops to or below a threshold m_p, generate all candidate network using the prospective links and use SCR to search for the optimal network. Output the optimal network and terminate the procedure.
6. **Updating.** Update the edge weight using Equation (26).
7. **Termination.** Repeat from Step 2 until every element in $\mathbf{w}_t$ lies in the ranges [0,0.01] or [0.99,1]. The m_a edges with w_e in the [0.99,1] range are the optimal edge set to be added.

3.6.5 Comparison Between the Methods

To compare the different schemes, the examples of 1, 2 and 3 additional links are re-run with more comparable parameters. For the SCR scheme, 500,000 samples are used to compare different networks. In the CE-methods, each iteration uses 10,000 samples to compare 1000 generated networks. The elite portion is set to 5% and a smoothing factor of 0.5 is used as well. With the Hybrid CE algorithm, the CE parameters are the same as for the Simple CE except a sample size of 1000 is used. The maximum prospective link is set so that the prospective network count is no more than 100. Once the threshold is reached, the prospective networks are generated for a final comparison using SCR with 500,000 samples. The average run-time in seconds are tabulated in Table 12.

Table 12. Comparison of different schemes

	1-Link	2-Link	3-Link
SCR scheme	16s	1572s	n/a
Simple CE	non-converging	234s	288s
Hybrid CE	18s	56s	53s

In the 1-Link case, which has 17 multiple optimal solutions, the Simple CE algorithm does not converge and will run indefinitely if allowed. The SCR scheme quickly determined that the 17 candidate networks have the same reliability. Note that in this case the SCR scheme is a semi-automatic process. It requires manual filtering of the candidate networks that will not reconnect the node A0. If all possible candidates were used in the comparison, it would take 102 seconds. With the Hybrid CE algorithm, it took only 2 seconds to filter out the 136 candidates and another 16 seconds to find out that the remaining 17 networks have the same reliability. It is a fully automatic process without the need of manual filtering.

In the 2-Link case, there are 2448 networks to compare for the SCR scheme and it takes a fairly long time (over 25 minutes) to finish. On the other hand, the Simple CE algorithm converges to the optimal solution in less than 4 minutes. The Hybrid CE algorithm is able to cut the computation time by 75% to less than 1 minute. This is achieved by switching about half-way during the iterations and uses the SCR scheme to compare a small number of prospective candidates.

In the 3-Link case, the SCR method was not performed because it would have taken too long. Since the number of valid network grows exponentially with the added links, the 3-Link example is projected to take over 30 hours to compute using the SCR scheme. The Simple CE and the Hybrid CE algorithms behave similar to the 2-Link case. It is interesting to note that the Hybrid CE algorithm is faster in the 3-Link case than the 2-Link case. This is due to quantization of combinations: the three highest 2-Link combinations under 100 are {91, 78, 55} while that of the 3-Link case are {84, 56, 35}. In fact at the point of switch over, the 2-Link case has an average of 67 networks to compare while the 3-Link case has 48.

Acknowledgements

We thank Sho Nariai for providing numerical results for the network planning problem. The first author acknowledges the financial support of the Australian Research Council, under grant number DP0558957.

References

1. R.E. Barlow and F. Proschan. *Statistical Theory of Reliability and Life Testing*. Holt, Rinehart and Wilson., 1975.
2. A.T. Bharucha-Reid. *Elements of the Theory of Markov Processes and Their Applications*. McGraw-Hill, 1960.
3. Yu Burtin and B. Pittel. Asymptotic estimates of the reliability of a complex system. *Engrg. Cybern.*, 10(3):445 – 451, 1972.

4. H. Cancela and M.E. Urquhart. Simulated annealing for communication network reliability improvement. In *Proceedings of the XXI Latin American Conference on Informatics*. CLEI-SBC, July 1995.

5. S.X. Chen and J.S. Liu. Statistical Applications of the Poisson-Binomial and Conditional Bernoulli Distributions. *Statistical Sinica*, 7:875–892, 1997

6. C.J. Colbourn. *The Combinatorics of Network Reliability*. Oxford University Press, 1987.

7. C.J. Colbourn and D.D. Harms. Evaluating performability: Most probable states and bounds. *Telecommunication Systems*, 2:275–300, 1994.

8. P.-T. de Boer, D.P. Kroese, S. Mannor, and R. Y. Rubinstein. A tutorial on the cross-entropy method. *Annals of Operations Research*, 134(1):19–67, 2005.

9. B. Dengiz, F. Altiparmak, and A. E. Smith. Local search genetic algorithm for optimal design of reliable networks. *IEEE Transactions on Evolutionary Computation*, 1(3):179–188, September 1997.

10. M.C. Easton and C.K. Wong. Sequential destruction method for Monte Carlo evaluation of system reliability. *IEEE Transactions on Reliability*, R-29:27–32, 1980.

11. T. Elperin, I.B. Gertsbakh, and M. Lomonosov. Estimation of network reliability using graph evolution models. *IEEE Transactions on Reliability*, 40(5):572–581, 1991.

12. T. Elperin, I.B. Gertsbakh, and M. Lomonosov. An evolution model for Monte Carlo estimation of equilibrium network renewal parameters. *Probability in the Engineering and Informational Sciences*, 6:457–469, 1992.

13. G.S. Fishman. A Monte Carlo sampling plan for estimating network reliability. *Operation Research*, 34(4):122–125, 1986.

14. I.B. Gertsbakh. *Reliability Theory with Application to Preventive Maintenance*. Springer, 2000.

15. I.B. Gertsbakh. *Statistical Reliability Theory*. Marcel Dekker, 1989.

16. I.I. Gikhman and A. V. Skorokhod. *Introduction to the Theory of Random Processes*. Dover Publications, 1996.

17. K-P. Hui, N. Bean, and M. Kraetzl. Network reliability difference estimation. *submitted for publication*, 2005.

18. K-P. Hui, N. Bean, M. Kraetzl, and D.P. Kroese. The tree cut and merge algorithm for estimation of network reliability. *Probability in the Engineering and Informational Sciences*, 17(1):24–45, 2003.

19. K-P. Hui, N. Bean, M. Kraetzl, and D.P. Kroese. The Cross-Entropy Method for Network Reliability Estimation. *The Annals of Operations Research*, 134:101–118, 2005.

20. H. Kumamoto, K. Tanaka, K. Inoue, and E. J. Henley. Dagger sampling Monte Carlo for system unavailability evaluation. *IEEE Transactions on Reliability*, R-29(2):376–380, 1980.

21. M. Lomonosov. An Evolution Model for Monte Carlo Estimation of Equilibrium Network Renewal Parameters. *Probability in the Engineering and Informational Sciences*, 6:457–469, 1992.

22. M. Lomonosov. On Monte Carlo estimates in network reliability. *Probability in the Engineering and Informational Sciences*, 8:245 –264, 1994.
23. J.R. Norris. *Markov Chains*. Cambridge University Press, 1997.
24. J.S. Provan and M.O. Ball. The complexity of counting cuts and of computing the probability that   a graph is connected. *SIAM Journal of Computing*, 12:777 –787, 1982.
25. D. Reichelt, F. Rothlauf, and P. Bmilkowsky. Designing reliable communication networks with a genetic algorithm using a repair heuristic. In *Evolutionary Computation in Combinatorial Optimization*.   Springer-Verlag Heidelberg, 2004.
26. R.Y. Rubinstein. Optimization of computer simulation models with rare events. *European Journal of Operations Research*, 99:89 –112, 1997.
27. R.Y. Rubinstein. The cross-entropy method for combinatorial and continuous optimization. *Methodology and Computing in Applied Probability*, 2:127 – 190, 1999.
28. R.Y. Rubinstein and D. P. Kroese. *The Cross-Entropy Method: A unified approach to Combinatorial   Optimization, Monte Carlo Simulation and Machine Learning*. Springer-Verlag, New York, 2004.
29. R.Y. Rubinstein. Combinatorial optimization, cross entropy, ants and rare events. *Stochastic Optimization: Algorithms and Applications*, pages   303 – 363, 2001.
30. R.Y. Rubinstein and B. Melamed. *Modern simulation and modeling*. Wiley series in probability and Statistics, 1998.
31. C. Srivaree-Ratana and A. E. Smith. Estimating all-terminal network reliability using a neural network. *Proceedings of the 1998 IEEE International Conference on   Systems, Man, and Cybernetics*, 5:4734 –4740, October 1998.
32. W.-C. Yeh. A new Monte Carlo method for the network reliability. *Proceedings of First International Conference on Information   Technologies and Applications(ICITA2002)*, November 2002.
33. Y-S. Yeh, C. C. Chiu, and R-S. Chen. A genetic algorithm for *k*-node set reliability optimization with   capacity constraint of a distributed system. *Proc. Natl. Sci. Counc. ROC(A)*, 25(1):27 –34, 2001.

Particle Swarm Optimization in Reliability Engineering

Gregory Levitin

The Israel Electric Corporation Ltd., Israel

Xiaohui Hu

Purdue School of Engineering and Technology, USA

Yuan-Shun Dai

Department of Computer & Information Science, Purdue University School of Science, USA

4.1 Introduction

Plenty of optimization meta-heuristics have been designed for various purposes in optimization. They have also been extensively implemented in reliability engineering. For example, Genetic Algorithm (Coit and Smith, 1996), Ant Colony Optimization (Liang and Smith, 2004), Tabu Search (Kulturel-Konak, et al., 2003), Variable Neighbourhood Descent (Liang and Wu, 2005), Great Deluge Algorithm (Ravi, 2004), Immune Algorithm (Chen and You, 2005) and their combinations (hybrid optimization techniques) exhibited effectiveness in solving various reliability optimization problems.

As proved by Wolpert and Macready (1997), no meta-heuristic is versatile, which could always outperform other meta-heuristics in solving all kinds of problems. Therefore, inventing or introducing new, good optimi-

G. Levitin et al.: *Particle Swarm Optimization in Reliability Engineering*, Computational Intelligence in Reliability Engineering (SCI) **40**, 83–112 (2007)
www.springerlink.com

zation approaches can be very helpful in some specific areas and benefit practitioners with more options.

Since the hybrid optimization technique becomes another promising direction, combining existing tools with new ones may produce robust and effective solvers. This consideration also encourages researchers to seek novel optimization meta-heuristics.

This chapter presents applications of a new Particle Swarm Optimization (PSO) meta-heuristic for single- and multi-objective reliability optimization problems.

Originally developed for the optimization of continuous unconstrained functions, PSO did not attract much attention from the reliability community because most reliability optimization problems are of discrete nature and have constraints. However, in this chapter we show that properly adapted PSO can be an effective tool for solving some discrete constrained reliability optimization problems.

4.2 Description of PSO and MO-PSO

PSO is a population-based stochastic optimization technique invented by Kennedy and Eberhart (Eberhart and Kennedy, 1995, Kennedy and Eberhart, 1995). PSO was originally developed to simulate the behavior of a group of birds searching for food in a cornfield. The early versions of the particle swarm model were developed for simulation purposes only. Later it was discovered that the algorithms were extremely efficient when optimizing continuous non-linear unconstrained functions. Due to its easy implementation and excellent performance, PSO has been gradually applied to many engineering fields in the last several years. Various improvements and modifications have been proposed and adapted to solve a wide range of optimization problems (Hu, et al., 2004).

4.2.1 Basic Algorithm

PSO is similar to Genetic Algorithm (GA) in that the system is initialized with a group of random particles (solutions) and each particle X_i $(1 \leq i \leq I)$ is represented by a string (vector of coordinates in the space of solutions): $X_i = \{x_{id}, 1 \leq d \leq D\}$. However, it is unlike GA in that a randomized velocity $V_i = \{v_{id}, 1 \leq d \leq D\}$ is assigned to each particle i and new solutions in every PSO iteration are not generated by crossover or mutation operators but by the following formula:

$$v_{id} = w \times v_{id} + c_1 \times rand_1() \times (p_{id} - x_{id}) + c_2 \times rand_2() \times (p_{nd} - x_{id}) \quad (1)$$

$$x_{id} = x_{id} + v_{id} \quad (2)$$

Eq. (1) calculates a new velocity for each particle based on its previous velocity v_{id}, the location at which it achieved the best fitness p_{id} so far, and the neighbor's location p_{nd} at which the best fitness in a neighborhood has been achieved so far. Eq. (2) updates the position of the particle in the problem space. In this equation, $rand_1()$ and $rand_2()$ are two random numbers independently generated, c_1 and c_2 are two learning factors that control the influence of p_{id} and p_{nd} on the search process. The weight w the particle inertia that prevents it from making undesired jumps in the solution space.

It can be learned from Eq. (1) that each particle is updated by the following two "best" values. The first one is the best solution *pBest* a particle has achieved so far. The second one is the best solution *nBest* that any neighbor of a particle has achieved so far. The neighborhood of a particle is defined as a fixed subset of particles in the population. When a particle takes the entire population as its neighbors, the best neighborhood solution becomes the global best (*gBest*).

The process of implementing the PSO is as follows:
1. Initialize the particle population (position and velocity) randomly.
2. Calculate fitness values of each particle.

3. Update *pBest* for each particle: if the current fitness value is better than *pBest*, set *pBest* to current fitness value.
4. Update *nBest* for each particle: set *nBest* to the particle with the best fitness value of all neighbors.
5. Update particle velocity/position according to equation (1) and (2).
6. If stop criteria is not attained, go back to step 2.
7. Stop and return the best solution found.

4.2.2 Parameter Selection in PSO

It can be learned from the particle update formula that particles search for better solutions by learning from their own and their neighbors' experiences. The two equations, Eq. (1) and (2), are the core part of the PSO algorithm. The parameters used in the formula will determine the performance of the algorithm.

4.2.2.1 Learning Factors

The learning factors c_1 and c_2 in Eq. (1) represent the weights of the stochastic acceleration terms that pull each particle toward *pBest* and *nBest* positions. From a psychological standpoint, the second term in Eq. (1) represents cognition, or the private thinking of the particle (tendency of individuals to duplicate past behavior that have proven successful) whereas the third term in Eq. (1) represents the social collaboration among the particles (tendency to follow the successes of others).

Both c_1 and c_2 were set to 2.0 in initial PSO works (Eberhart and Kennedy, 1995, Eberhart and Shi, 1998). The obvious reason is it will make the search, which is centered at the *pBest* and *nBest*, cover all surrounding regions. Clerc (Clerc, 1999) introduced the constriction coefficient, which might be necessary to ensure convergence of PSO. $c_1 = c_2 = 1.49445$ is also used according to the work by Clerc (Eberhart and Shi, 2001b)

In most cases, the learning factors are identical, which puts the same weights on cognitive search and social search. Kennedy (Kennedy, 1997) investigated two extreme cases: a cognitive-only model and a social-only model, and found out that both parts are essential to the success of particle swarm search.

4.2.2.2 Inertia Weight

In the original version of particle swarm, there was no inertia weight. Inertia weight w was first introduced by Shi and Eberhart (Shi and Eberhart, 1998). The function of inertia weight is to balance global exploration and local exploitation. Linearly decreasing inertia weights were recommended. Clerc (Clerc, 1999) introduced the constriction coefficient and suggested it to ensure convergence of PSO. Randomized inertia weight is also used in several reports (Eberhart and Shi, 2001b, Hu and Eberhart, 2002c, Hu and Eberhart, 2002b, Hu and Eberhart, 2002a). The inertia weight can be set to $[0.5 + (rand_1/2.0)]$, which is selected in the spirit of Clerc's constriction factor (Eberhart and Shi, 2001a).

4.2.2.3 Maximum Velocity

Particles' velocities are clamped to a maximum velocity $Vmax$, which serves as a constraint to control the global explosion speed of particles. It limits the maximum step change of the particle, thus adjusting the moving speed of the whole population in the hyperspace. Generally, $Vmax$ is set to the value of the dynamic range of each variable, which does not add any limit. If $Vmax$ is set to a lower value, it might slow the convergence speed of the algorithm. However, it would help to prevent PSO from local convergence.

4.2.2.4 Neighborhood Size

As mentioned before, $nBest$ is selected from a neighborhood. The neighborhood of a particle is usually pre-defined and does not change during iterations. The neighborhood size could vary from 1 to the maximum number of solutions in the population. This size affects the propagation of information about the best particle in the group. The bigger the neighborhood size, the faster the particles can learn from the global best solutions. In an extreme case, the global version of PSO, every particle knows every other particles' movements and can learn that within one step, making PSO converge very fast. However, it also causes premature convergence that can be avoided by the injection of new solutions. Small neighborhood size may prevent premature convergence at the price of slowing the convergence speed.

4.2.2.5 Termination Criteria

The PSO terminates when the pre-specified number of iterations has been performed or when no improvement of *gBest* has been achieved during a specified number of iterations.

4.2.3 Handling Constraints in PSO

Some studies that have reported the extension of PSO to constrained optimization problems (El-Gallad, et al., 2001, Hu, et al., 2003a, Hu and Eberhart, 2002a, Parsopoulos and Vrahatis, 2002, Ray and Liew, 2001). The goals of constrained optimization problems are to find the solution that optimizes the fitness function while satisfying a set of linear and non-linear constraints. The original PSO method needs to be modified in order to handle those constraints.

Hu and Eberhart (Hu and Eberhart, 2002a) introduced an effective method to deal with constraints based on a preserving feasibility strategy. Two modifications were made to the PSO algorithm: First, when updating the *pBest* values, all the particles consider only feasible solutions; Second, during the initialization process, only feasible solutions form the initial population. Various tests show that such modification of the PSO outperforms other evolutionary optimization techniques when dealing with optimization problems with linear or nonlinear inequity constraints (Hu, et al., 2003a, Hu and Eberhart, 2002a). The disadvantage of the method is that the initial feasible solution set is sometimes hard to find.

El-Gallad (El-Gallad, et al., 2001) introduced a similar method. The only difference is that when a particle gets outside of feasible region, it is reset to the last best feasible solution found for this particle. He, et al. (He, et al., 2004) reset the particle to a previous position instead of the last best feasible solution. However, if there are several isolated feasible regions, particles may be confined in their local regions with above approaches.

Parsopoulos, *et. al* (Parsopoulos and Vrahatis, 2002) converted the constrained optimization problem into a non-constrained problem by using a non-stationary multi-stage penalty function and then applied PSO to the converted problems. It was reported that the obtained PSO outperformed Evolution Strategy and GA on several benchmark problems (Parsopoulos and Vrahatis, 2002).

Ray, *et al.* (Ray and Liew, 2001) proposed a swarm metaphor with a multilevel information sharing strategy to deal with optimization problems. It is assumed that there are some better performers (leaders) in a swarm that set the direction of the search for the rest of the particles. A particle that does not belong to the better performer list (BPL) improves its performance by deriving information from its closest neighbor in BPL. The constraints are handled by a constraint matrix. A multilevel Pareto ranking scheme is implemented to generate the BPL based on the constraint matrix. In this case, the particle should be updated using a simple generational operator instead of the regular PSO formula. Tests of such PSO modifications have showed much faster convergence and much lower number of function evaluations compared to the GA approach (Ray and Liew, 2001)

The above mentioned works have showed that modified PSO can successfully handle linear or non-linear constraints.

4.2.4 Handling Multi-objective Problems with PSO

Multi-objective optimization addresses problems with several design objectives. In multi-objective optimization (MO) problems, objective functions may be optimized separately from one another and the best solution may be found for each objective. However, the objective functions are often in conflict among themselves and a Pareto front represents the set of optimal solutions. The family of solutions of a multi-objective optimization problem is composed of all those potential solutions such that the components of the corresponding objective vectors cannot be all simultaneously improved (concept of Pareto optimality). The Pareto optimum usually gives a group of solutions called non-inferior or non-dominated solutions instead of a single solution.

The traditional way of handling MO problems is to convert them to single objective problems by using weights. Multiple optimal solutions could be obtained through multiple runs with different weights. However, methods that find groups of Pareto optimal solutions simultaneously can save time and cost.

In PSO, a particle searches the problem space based on its own (*pBest*) and its peers' (*nBest*) experience. Both cognitive and social terms in Eq. (1) play crucial roles in guiding the search process. Thus, the selection of the cognitive and social leader (*pBest* and *nBest*) are key points of MO-PSO algorithms. The selections should satisfy two rules: first, it should provide effective guidance to the particle to reach the most promising Pareto front region; second, it should provide a balanced search along the Pareto front to maintain the population diversity.

The selection of cognitive leader (*pBest*) is almost the same as in the original PSO (Hu, et al., 2003b, Hu and Eberhart, 2002b). The only difference is that the comparison is based on Pareto optimality (*pBest* is updated only if the new solution dominates all solutions visited by the particle so far).

The selection of the social leader (*nBest*) consists of two steps. The first step is to define a candidate pool from which the leader is chosen, and the second step is to define the process of choosing the leader from the candidate pool. Usually the candidate pool is the collection of all particles' *pBest* positions or an external repository that includes all the Pareto optimal solutions found by the algorithm. For the selection procedure, two typical approaches have been suggested in the literature:

1. In the roulette wheel selection scheme approach (Coello Coello, et al., 2004, Coello Coello and Lechuga, 2002, Li, 2003, Ray and Liew, 2002), all candidates are assigned weights based on some criteria (such as crowding radius, crowding factor, niche count or other measures). The general rule is to distribute the particles evenly. If there are too many particles in a small region, the region is considered to be crowded, and the particles belonging to the crowd region have less chance to be selected.

Thus they do not attract particles to this region anymore. Then, random selection is used to choose the social leader. In this scheme, selection for a candidate is stochastic and proportional to the weights. This technique aims the process at maintaining the population diversity.

2. In the quantitative standard approach, the social leader is determined by some procedure without any random selection involved, such as dynamic neighborhood (Hu, et al., 2003b, Hu and Eberhart, 2002b), sigma distance (Mostaghim and Teich, 2003), dominated tree (Fieldsend and Singh, 2002), and *etc.*

4.3 Single-Objective Reliability Allocation

4.3.1 Background

Nowadays, considerable effort is concentrated on optimal system design that balances system reliability, cost and performance. Many systems perform their intended functions at multiple levels, ranging from perfectly working to completely failed. These kinds of systems are called multi-state systems.

In the case of a multi-state system, the concept corresponding to that of reliability in a binary system is state distribution. Having the system state distribution, one can determine its reliability as a probability of being in acceptable states and its expected performance named system utility (Aven, 1993).

There are two ways to improve the system reliability or utility: First, to provide redundancies of components at each subsystem; Second, to improve the component's performance/reliability, such as allowing a component to have more chances to stay at better states or allocating more test resources on the component for reliability growth (Dai, et al., 2003). Finding an optimal balance between these two factors is a classical reliability allocation problem that has been studied in many works (Hikita, et al., 1992, Prasad and Kuo, 2000, Tillman, et al., 1977) from different aspects and by various methods.

In this section, PSO has been tested on a single objective reliability allocation problem and then compared with GA that has been carefully tuned by Tian *et al.* (2005).

4.3.2 Problem Formulation

4.3.2.1 Assumptions

A multi-state series-parallel system consists of N subsystems connected in series. Each subsystem i has n_i identical components connected in parallel, as depicted in Fig. 1.

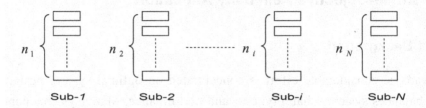

Fig. 1. Series-Parallel System

The components and the system have $M+1$ possible states: 0, 1, …, M. The states of the components in a subsystem are independent. The probability that the component belonging to subsystem i is in state j is p_{ij}. Since the component's states compose the complete group of mutually exclusive events $\sum_{j=0}^{M} p_{ij} = 1$. Therefore the state distribution of any element i is determined by M probabilities p_{ij} for $1 \leq j \leq M$ and $p_{j0} = 1 - \sum_{j=1}^{M} p_{ij}$.

The system behavior can be improved by changing the number of parallel components in some subsystems or by changing the component state distribution.

4.3.2.2 Decision Variables

Two types of decision variables are of concern in the reliability allocation problem: real numbers p_{ij} ($1 \leq i \leq N$, $1 \leq j \leq M$) representing state distribution of the components in each subsystem i, and integer numbers n_i $(1 \leq i \leq N)$ representing the number of components in each subsystem i. The total number of decision variables is $NM+N$.

4.3.2.3 Objective Function

In the optimization problem, the system utility should be maximized whereas its total cost should be limited within the given budget C_0. The system utility represents the expected performance of multi-state systems. It is assumed that certain utility (performance) value u_s corresponds to any system state s. Having the probability that a multi-state series-parallel system is in state s or above in the form

$$\Pr[\phi(x) \geq s] = \prod_{i=1}^{N} \left[1 - \left(1 - \sum_{k=s}^{M} p_{ik} \right)^{n_i} \right],$$

(3)

one can obtain the multi-state system utility U as

$$U = \sum_{s=0}^{M} u_s \cdot \Pr[\phi(x) = s], \quad \text{and}$$

(4)

$\Pr[\phi(x) = s]$ can be derived from Eq. (3) by

$$\Pr[\phi(x) = s] = \begin{cases} \Pr[\phi(x) \geq s] - \Pr[\phi(x) \geq s+1] & (0 \leq s < M) \\ \Pr[\phi(x) \geq s] & (s = M) \end{cases}$$

(5)

The cost model used in Tian, et $al.$ (2005) adopts the cost-reliability relationship function suggested by Tillman, et $al.$ (1977) which for subsystem i takes the form

$$C_i = c_i(r_i)[n_i + \exp(n_i / 4)],$$

(6)

$$c_i(r_i) = \alpha_i (-t / \ln r_i)^{\beta_i}$$

(7)

where r_i and $c_i(r_i)$ are reliability and cost of a single component, α_i and β_i are constants representing the inherent characteristics of components in subsystem i, and t is the operating time during which the component should not fail.

Eqs. (6) and (7) were adapted to fit the multi-state system model as follows: the cost of component i as a function of its state distribution is

$$c_i(p_{i1}, p_{i2}, ..., p_{iM}) = \sum_{j=1}^{M} (-t/\ln r_{ij})^{\beta_{ij}} \tag{8}$$

where

$$r_{ij} = p_{ij} / \sum_{k=0}^{M} p_{ik}, \quad \text{for } 1 \leq j \leq M, \tag{9}$$

α_{ij} and β_{ij} are characteristic constants with respect to state j, and t is the operating time. The total system cost is

$$C = \sum_{i=1}^{N} c_i(p_{i1}, p_{i2}, ..., p_{iM})[n_i + \exp(n_i/4)] \tag{10}$$

4.3.2.4 The Problem

The single objective optimization problem is formulated as follows:

$$\textbf{Maximize} \quad U = \sum_{s=0}^{M} u_s \cdot \Pr[\phi(x) = s] \tag{11}$$

$$\textbf{Subject to:} \quad C = \sum_{i=1}^{N} c_i(p_{i1}, p_{i2}, ..., p_{iM})\left[n_i + \exp\left(\frac{n_i}{4}\right)\right] \leq C_0$$

$$p_{ij} \in [0,1], \qquad 1 \leq i \leq N, \ 1 \leq j \leq M,$$

$$\sum_{j=1}^{M} p_{ij} \leq 1, \qquad 1 \leq i \leq N,$$

$$0 < n_i \qquad 1 \leq i \leq N,$$

where C_0 is the maximum allowed system cost (budget).

4.3.3 Numerical Comparison

The multi-state series-parallel system considered by Tian *et al.* (2005) contains three subsystems connected in series. Any individual component and the entire system can have one of three states. The values of system utility u_s corresponding to its states are $u_0=0$, $u_1=0.5$, $u_2=1.0$. The cost function parameters are presented in Table 1. The system operating time is

$t = 1000$.

Table 1. Cost function characteristic constants

Subsystem i	a_{i1}	a_{i2}	β_{i1}	β_{i2}
1	1.5E-5	4E-5	1.2	1.5
2	0.9E-5	3.2E-5	1.2	1.5
3	5.2E-5	9E-5	1.2	1.5

This problem was solved by GA in Tian, *et al.* (2005) using the physical programming framework. The optimal solution obtained by GA is shown in Table 2.

In order to compare the PSO results with results presented in Tian *et al.* (2005), C_0 was set to 89.476. The following PSO parameters were chosen: the population size of 40, the neighborhood size of 3. Maximum velocity was set to 20% of the dynamical range of the variables, the reason to choose a smaller maximum velocity is to control the convergence speed of the swarm. Learning factors c_1 and c_2 are set to 1.49445. The inertia weight w was set to $[0.5 + (rand_1/2.0)]$ as mentioned in previous section. The number of iterations was 10000.

The best solution achieved by the PSO is shown in Table 2. This solution provides greater utility than one obtained by the GA with the same budget. The distribution of solutions obtained in 200 runs of the PSO with population size of 20 is shown in Table 2. The best result over 200 runs is $U=0.9738$, the worst one is $U=0.9712$, the average is $U=0.9734$ and the standard deviation is 0.000515. The mean value obtained by the PSO runs is the same as the best solution obtained by the GA.

Table 2. Comparison of the best solutions achieved by PSO and GA

	Genetic Algorithm			Particle Swarm Optimization		
Subsystem i	1	2	3	1	2	3
p_{i1}	0.2030	0.2109	0.2100	0.2124	0.2208	0.2042
p_{i2}	0.4200	0.4300	0.4000	0.4579	0.4712	0.4066
n_i	8	8	7	7	7	7
System Utility		0.9734			0.9738	
System Cost		89.476			89.476	

4.4 Single-Objective Redundancy Allocation

The classical redundancy allocation problem belongs to the type of integer optimization problems. Many algorithms have been developed to solve the problem, including the GA (Coit and Smith, 1996), Ant Colony Optimization (Liang and Smith, 2004), Tabu Search (Kulturel-Konak, et al., 2003), Immune Algorithm (Chen and You, 2005), and Specialized Heuristic (You and Chen, 2005).

4.4.1 Problem Formulation

4.4.1.1 Assumptions

A system contains N subsystems connected in series. Each subsystem can contain multiple binary components connected in parallel (Fig. 1). Components composing each subsystem i can be different. They can be chosen from a list of M_i options. Different types of component are characterized by reliability, cost and weight. A subsystem fails if all its components fail. The entire system fails if any subsystem fails.

4.4.1.2 Decision Variable

The system structure in this problem is defined by integer numbers of components selected from the corresponding lists. The element σ_{ij} of the set of decision variables $\sigma = \{\sigma_{ij}, 1 \leq i \leq N, 1 \leq j \leq M_i\}$ determines the number of components of type j included in subsystem i. The total number of decision variables in the set σ is $\sum\limits_{i=1}^{N} M_i$.

4.4.1.3 Objective Function

The general objective in this problem is to maximize the system reliability R subject to constraints on the total system cost and weight. Suppose the component of type j in subsystem has reliability $r(i, j)$, cost $c(i, j)$, and

weight $w(i, j)$. For the given set of chosen components σ, the system re-

liability, cost and weight can be obtained by

$$R(\sigma) = \prod_{i=1}^{N}\left(1 - \prod_{j=1}^{M_i}[1 - r(i, j)]^{\sigma_{ij}}\right) \tag{12}$$

$$C(\sigma) = \sum_{i=1}^{N}\sum_{j=1}^{M_i} c(i, j)\sigma_{ij} \tag{13}$$

$$W(\sigma) = \sum_{i=1}^{N}\sum_{j=1}^{M_i} w(i, j)\sigma_{ij} \tag{14}$$

The optimization problem can be formulated as follows:

Maximize $\quad R(\sigma)$ $\hfill (15)$

Subject to: $\quad C(\sigma) \le C_0, \quad W(\sigma) \le W_0,$

$$0 \le \sigma_{ij} \le M_i, \quad \text{for } 1 \le i \le N,$$

$$\sum_{j=1}^{M_i} \sigma_{ij} \le K_i \quad , \quad 1 \le i \le N,$$

where C_0 and W_0 are maximal allowed system cost and weight, and K_i is a maximal allowed number of components in subsystem i.

4.4.2 Numerical Comparison

The Fyffe, *et al.* problems as devised by Nakagawa & Miyazaki (1981) are used for comparison among different algorithms. The results of this comparison can be found in chapter 1 of this book. PSO has been tested on the first 12 problems (W=191 to 180). For each problem the results of 100 runs were obtained. The worst, best and average results over 100 runs are shown in Table 3. It can be seen that PSO performs very poor compared to the algorithm by You & Chen (2005). PSO just slightly outperforms the random search algorithm running for the same time (the best PSO results are slightly better than the results of random search whereas the worst PSO results are even worse than the worst random search results).

Table 3. Results from PSO and Random Search

W	PSO Worst	Rand Worst	PSO Mean	Rand Mean	PSO Best	Rand Best	Y&C-05
191	0.96918	0.97413	0.97792	0.97711	0.98209	0.97916	0.98681
190	0.96900	0.97342	0.97772	0.97605	0.98238	0.97859	0.98642
189	0.97017	0.97137	0.97673	0.97494	0.98214	0.97783	0.98592
188	0.96668	0.97153	0.97570	0.97467	0.98121	0.97773	0.98538
187	0.96812	0.96923	0.97480	0.97340	0.98047	0.97574	0.98469
186	0.96554	0.96963	0.97344	0.97356	0.97974	0.97654	0.98418
185	0.96594	0.96879	0.97201	0.97149	0.97984	0.97627	0.98350
184	0.96562	0.96803	0.97163	0.97168	0.97846	0.97554	0.98299
183	0.95826	0.96706	0.97032	0.96951	0.97802	0.97163	0.98226
182	0.95713	0.96556	0.96960	0.96872	0.97538	0.97072	0.98152
181	0.95800	0.96347	0.96793	0.96745	0.97416	0.97063	0.98103
180	0.96030	0.96334	0.96696	0.96684	0.97374	0.96854	0.98029

The major reason why PSO has such a poor performance is due to the regular coding scheme of particles. There are poor correlations among neighbors in the solutions space. The main assumption of PSO is that the neighbors of a good solution are also good. However, it is not true in the considered redundancy allocation problem.

4.5 Single Objective Weighted Voting System Optimization

4.5.1 Problem Formulation

Voting systems are widely used in human organization systems as well as in technical decision making systems. The weighted voting systems (WVS) are generalizations of the voting systems. The applications of WVS can be found in imprecise data handling, safety monitoring and self-testing, multi-channel signal processing, pattern recognition and target detection, etc. (Levitin, 2005a).

A WVS makes a decision about propositions based on the decisions of n statistically independent individual units of which it consists (for example, in target detecting system speed detectors and heat radiation detectors provide the system with their individual decisions without communicating among themselves). Each proposition is *a priori* right or wrong, but this information is available for the units in implicit form. Therefore the units are subject to the following three errors:

1. Acceptance of a proposition that should be rejected (fault of being too optimistic),
2. Rejection of a proposition that should be accepted (fault of being too pessimistic),
3. Abstaining from voting (fault of being unavailable or indecisive).

This can be modeled by considering system input I being either 1 (proposition to be accepted) or 0 (proposition to be rejected), which is supplied to each unit. Each unit j produces its decision (unit output) $d_j(I)$ which can be 1, 0 or x (in the case of abstention). Inequality $d_j(I) \neq I$ means that the decision made by the unit is wrong. The listed above errors can be expressed as

1. $d_j(0)=1$ (unit fails stuck-at-1),
2. $d_j(1)=0$ (unit fails stuck-at-0),
3. $d_j(I)=x$ (unit fails stuck-at-x).

Accordingly, reliability of each unit j can be characterized by probabilities of these errors: $q_{01}^{(j)}$ for the first one, $q_{10}^{(j)}$ for the second one, $q_{1x}^{(j)}$ and $q_{0x}^{(j)}$

for the third one (stuck-at-x probabilities can be different for inputs $I=0$ and $I=1$).

Each voting unit j has two weights that express its relative importance in the WVS: "negative" weight w^0_j, which is assigned to the unit when it votes for the proposition rejection, and "positive" weight w^1_j, which is assigned to the unit when it votes for the proposition acceptance. To make a decision about proposition acceptance, the system incorporates all the unit decisions into a unanimous system output D. The proposition is rejected by the WVS ($D(I)=0$) if the total weight of units voting for its acceptance is less than a pre-specified fraction τ of total weight of not abstaining units (τ is usually referred to as the threshold factor). The WVS abstains ($D(I)=x$) if all of its voting units abstain.

The system fails if $D(I)\neq I$. The entire WVS reliability can be defined as $R=\Pr\{D(I)=I\}$. One can see that the system reliability is a function of reliabilities of units it consists of. It also depends on the unit weights and the threshold. While the units' reliabilities usually can not be changed when the WVS is built, the weights and the threshold can be chosen in such a way that maximizes the entire WVS reliability $R(w^0_1,w^1_1,\ldots, w^0_n,w^1_n,\tau)$.

In many technical systems the time when the output (decision) of each voting unit is available is predetermined. For example, the decision time of a chemical analyzer is determined by the time of a chemical reaction. The decision time of a target detection radar system is determined by the time of the radio signal return and by the time of signal processing by the electronic subsystem. In both these cases, the variation of the decision times is usually negligible.

On the contrary, the decision time of the entire WVS composed from voting units with different constant decision times can vary. Indeed, the system does not need to wait for decisions of slow voting units, as long, as the system can make a correct decision with reliability higher than a pre-specified level. Moreover, in some cases the decisions of the slow voting units do not affect the decision of the entire system since this decision becomes evident after the fast units have voted. This happens when the total weight of units voting for the proposition acceptance or rejection is enough to guarantee the system decision independently of the decisions of the units that have not voted yet. In such situations, the voting process can be terminated without waiting for slow units' decisions, and the WVS decision can be made in a shorter time.

The number of combinations of unit decisions that allow the entire system decision to be obtained before the outputs of all of the units become available depends on the unit weight distribution and on the threshold value. By increasing the weights of the fastest units one makes the WVS more decisive in the initial stage of voting and therefore reduces the mean system decision time at the price of making it less reliable.

Since the units' weights and the threshold affect both the WVS's reliability and its expected decision time, the problem of the optimal system tuning can be formulated as follows: find the voting units' weights and the threshold that maximize the system reliability while providing the expected decision time T not greater than a pre-specified value T^*:

Maximize $R(w^0_1, w^1_1, \ldots, w^0_n, w^1_n, \tau)$
Subject to: $T(w^0_1, w^1_1, \ldots, w^0_n, w^1_n, \tau) \leq T^*$

The method for calculating the WVS reliability and the expected decision time T, GA-based procedure for solving the optimization problem (16), was suggested in (Levitin 2005b). Here we compare PSO and GA optimization techniques on the numerical example presented in (Levitin 2005b).

4.5.2 Numerical Comparison

Experimentation was performed on a WVS consisting of six voting units with voting times and fault probabilities presented in Table 4.

Both GA and PSO require the solution to be coded as a finite length string. The natural representation of a WVS weight distribution is by an $2n+1$-length integer string $(s_1, \ldots, s_{2n+1})$ in which the values in $2j-1$ and $2j$ position corresponds to the weights w^0_j and w^1_j of j-th unit of the WVS and the value in position $2n+1$ corresponds to the threshold. The unit weights are further normalized in such a way that their total weight is always equal to a constant. As in (Levitin 2005b), the s_i elements take values in the range [0, 150], and the normalization takes the form:

$$w_j^0 = 10s_{2j-1} / \sum_{j=1}^{n} s_{2j-1}, \quad w_j^1 = 10s_{2j} / \sum_{j=1}^{n} s_{2j}, \quad \tau = s_{n+1}/150. \quad (17)$$

Table 4. Parameters of voting units

No of unit j	t_j	$q_{01}(j)$	$q_{0x}(j)$	$q_{10}(j)$	$q_{1x}(j)$
1	10	0.22	0.31	0.29	0.12
2	12	0.35	0.07	0.103	0.30
3	38	0.24	0.08	0.22	0.15
4	48	0.10	0.05	0.2	0.01
5	55	0.08	0.10	0.15	0.07
6	70	0.08	0.01	0.10	0.05

Each new solution is decoded, and its objective function (fitness) value is estimated. In order to find the solution of Eq. (16), the fitness function is defined as:

$$F = R - \alpha \cdot \min(T-T^*,0), \quad (18)$$

where α is a penalty coefficient. For solutions with $T<T^*$ the fitness of the solution depends only on WVS reliability.

The population size for both PSO and GA was chosen 50. Initial experimentation with the PSO showed that the best composition of parameters is: number of solution update cycles (PSO iterations) $N=4500$; $V_{max}=40$; $c_1=2$; $c_2=1.5$, w linearly decreases as PSO proceeds: $w = 0.8+0.4(1-i/N)$.

In order to improve the PSO performance and avoid its convergence to local optima, in each M-th solution update cycle, the solutions (besides the best one in the population), instead of updating, were replaced by new randomly generated solutions with probability p. These new solutions had velocity 0. It is similar to the "mutation" operator in Genetic Algorithms, the experiments show that "injection of new solutions" improves the performance and the composition $M=100$, and $p=1/3$ gives the best improvement.

In order to compare PSO with and without injection of random solutions the optimal voting unit weights and thresholds were obtained for WVS with parameters given in Table 4 and with different values of T^*. For each T^*, 100 solutions were obtained (for the same problem) by both modifications of the PSO starting with different initial randomly generated sets of solutions. The average obtained reliability A, the coefficient of variation V of the obtained reliability over 100 solutions and the average running time Tr were calculated. These indices are presented in Table 5, as well as relative increase of average obtained reliability $\delta A = 100 \cdot (A_i A)/A$ (%), increase in coefficient of variation $\Delta V = V_i V$, and relative increase of average running time $\delta Tr = 100\ (Tr_i Tr)/Tr$ (%).

Table 5. Comparison of PSO with and without injection of random solutions

T^*	No injection			Injection			Comparison		
	A	V	Tr	A_i	V_i	Tr_i	δA	ΔV	δTr
50	0.9498	0.0999	18.20	0.9501	0.0906	18.49	0.023	-0.009	1.593
48	0.9447	0.1400	18.29	0.9450	0.1313	19.01	0.033	-0.009	3.937
46	0.9436	0.1761	18.31	0.9440	0.1680	19.05	0.045	-0.008	4.042
44	0.9377	0.2431	16.28	0.9388	0.0669	17.37	0.110	-0.176	6.695
42	0.9321	0.7384	14.98	0.9337	0.2244	14.78	0.172	-0.514	-1.335
40	0.9222	1.2532	10.93	0.9254	0.3576	11.10	0.344	-0.896	1.555
38	0.9123	1.4015	13.14	0.9155	0.2526	13.20	0.350	-1.149	0.457
36	0.9053	2.6553	12.96	0.9147	0.2954	14.53	1.033	-2.360	12.114
34	0.9019	2.7998	12.31	0.9128	0.2686	14.09	1.208	-2.531	14.460
32	0.8846	4.3692	9.82	0.9106	0.0604	12.02	2.934	-4.309	22.403
30	0.8734	4.6231	7.05	0.8954	2.8278	8.25	2.515	-1.795	17.021
28	0.8711	3.2375	5.69	0.8834	0.8038	7.04	1.410	-2.434	23.726
26	0.8612	3.1374	4.20	0.8740	1.6292	4.08	1.489	-1.508	-2.857

It can be seen that the PSO with injection of random solutions always outperforms the regular PSO. It produces better solutions with less variation at the price of 8% increase of running time (on average).

In order to compare PSO with injection of random solutions with GA, 100 solutions were obtained by GA starting with different initial populations (100 seeds) for each one of the optimization problems. The parameters of the GA were the same as in (Levitin, 2005b). The results of this comparison are presented in Table 6.

Table 6. Comparison of PSO and GA results

T*	GA					PSO					Comparison			
	A_g	max_g	min_g	V_g	Tr_g	A_p	max_p	min_p	V_p	Tr_p	δA	δmax	ΔV	δTr
50	0.9502	0.9509	0.9478	0.0834	21.34	0.9501	0.9509	0.9466	0.0906	18.49	-0.018	0.0000	0.007	-13.355
48	0.9450	0.9466	0.9414	0.1661	21.96	0.9450	0.9466	0.9416	0.1313	19.01	0.003	0.0004	-0.035	-13.434
46	0.9420	0.9444	0.9265	0.2950	14.43	0.9440	0.9444	0.9324	0.1680	19.05	0.221	0.0000	-0.127	32.017
44	0.9379	0.9392	0.9336	0.1438	15.81	0.9388	0.9392	0.9358	0.0669	17.37	0.095	0.0004	-0.077	9.867
42	0.9317	0.9349	0.9225	0.3612	11.67	0.9337	0.9349	0.9232	0.2244	14.78	0.217	0.0005	-0.137	26.650
40	0.9252	0.9276	0.9192	0.3640	12.62	0.9254	0.9276	0.9181	0.3576	11.10	0.019	0.0000	-0.006	-12.044
38	0.9151	0.9180	0.9103	0.2154	16.15	0.9155	0.9180	0.9091	0.2526	13.20	0.045	0.0003	0.037	-18.266
36	0.9151	0.9159	0.9040	0.2867	16.74	0.9147	0.9160	0.9029	0.2954	14.53	-0.047	0.0008	0.009	-13.202
34	0.9130	0.9131	0.9123	0.0128	9.49	0.9128	0.9131	0.8887	0.2686	14.09	-0.020	0.0003	0.256	48.472
32	0.9107	0.9108	0.9086	0.0250	16.70	0.9106	0.9108	0.9074	0.0604	12.02	-0.013	0.0003	0.035	-28.024
30	0.9021	0.9036	0.8037	1.1327	12.13	0.8954	0.9037	0.8037	2.8278	8.25	-0.746	0.0049	1.695	-31.987
28	0.8736	0.8850	0.7950	3.0113	5.86	0.8834	0.8850	0.8214	0.8038	7.04	1.115	0.0000	-2.208	20.137
26	0.8695	0.8779	0.7460	2.8087	6.27	0.8740	0.8779	0.7893	1.6292	4.08	0.510	0.0010	-1.179	-34.928

For each problem, maximal, minimal and average reliability obtained over 100 seeds (max, min and A respectively) are presented in Table 6, as well as the coefficient of variation V and average running time Tr (seconds). Relative indices $\delta A = 100 \cdot (A_p - A_g)/A_g$ (%), δmax$=100 \cdot (max_p - max_g)/max_g$ (%), $\Delta V = V_p - V_g$ and $\delta Tr = 100 (Tr_p - Tr_g)/Tr_g$ (%) are calculated. The comparison shows that GA and PSO produce very close results. However, PSO usually produces better solutions; the best PSO solutions over 100 seeds are always better or the same as the best GA solutions (δmax≥ 0). The average value of δA over all of the problems tested is 0.106%. PSO produces solutions with less variation (average value of ΔV over all of the problems tested is -0.133) in less running time (average value of δTr over all of the problems tested is -2.16%).

The difference in variability over all of the test problems between the PSO and the GA is shown in Fig. 2.

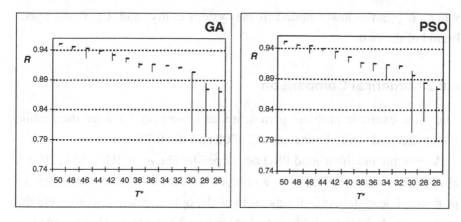

Fig. 2. Range of obtained WVS reliability over 100 seeds with mean shown as horizontal dash

4.6 Multi-Objective Reliability Allocation

4.6.1 Problem Formulation

In order to demonstrate PSO's ability to solve multi-objective optimization problems, it was applied to the multi-objective formulation of the reliability allocation problem presented in section 4.3. The problem formulation is as follows:

Maximize

$$U = \sum_{s=0}^{M} u_s \cdot \Pr[\phi(x) = s] \tag{19}$$

and Minimize

$$C = \sum_{i=1}^{N} c_i (p_{i1}, p_{i2}, ..., p_{iM})[n_i + \exp(n_i / 4)]$$

Subject to:

$$U \ge U_0, \quad C \le C_0,$$

$$p_{ij} \in [0,1], \qquad 1 \le i \le N, \ 1 \le j \le M,$$

$$\sum_{j=1}^{M} p_{ij} \le 1, \qquad 1 \le i \le N,$$

$$0 < n_i \qquad 1 \le i \le N,$$

where U_0 is the lower bound of the system utility, and C_0 is the upper bound of the cost.

4.6.2 Numerical Comparison

The same example problem parameters as in section 4.3 were used while the budget and utility limits were $C_0=200$ and $U_0=0.75$. and.

A dynamic neighborhood PSO developed by Hu, *et al.* (Hu, et al., 2003b, Hu and Eberhart, 2002b) was employed to deal with this two-objective problem. The system utility was set to be the optimization objective and the cost was set to be the neighborhood objective. As mentioned before, the key point is to find the cognitive leader and the social leader. The cognitive leader (*pBest*) is updated when the particle is better than the old *pBest* in both system utility and cost values. All the Pareto optimal solutions found by PSO form the candidate pool for the selection of social leader *nBest*. First, the differences of cost between the particle and all candidates in the pool are calculated, then the particles with closest cost are chosen to be the neighbors of the particle, finally the candidate with the greatest system utility value became the social leader. Fig. 3 illustrates the selection process. The black curve is the final Pareto front. The dots are the current Pareto optimal solutions. The circle is a particle. The particle finds several nearest Pareto optimal solutions in term of cost as neighbors. Then the neighbor with highest utility fitness value is selected to be *nBest*.
The general procedure is as follows:
1. Initialize the population.
2. Calculate values of each objective function for the particles in the population.
3. For each solution (particle): if the current fitness values of the particle are better than any current solution in the Pareto optimal solution archive, then put the particle into the archive.
4. Find the *nBest* and *pBest* according to the dynamic neighborhood method.
5. Update particle velocity and position.

6. If maximum iteration number is not reached, go back to step 2.

7. Find all Pareto optimal solutions in the archive and generate solution set.

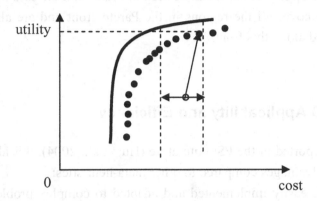

Fig. 3. Illustration of dynamic neighborhood strategy

The population size of 20 and neighborhood size of 3 were used in the optimization. Maximum velocity was set to 20% of the dynamical range of the variables. Learning factors c_1 and c_2 are set to 1.49445.

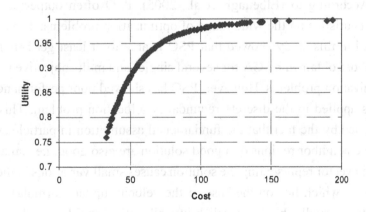

Fig. 4. Pareto optimal solutions obtained by MO-PSO

The inertia weight w was set to $[0.5 + (rand_1/2.0)]$ as mentioned in the previous section. Number of iterations was set to 10000. The average time for each run of the PSO on a HP Pentium IV 2.8GHz personal computer was less than 30 seconds.

The Pareto optimal solutions obtained from a single PSO run are presented in Fig. 4. A total of 200 Pareto optimal solutions have been found during the optimization. It can be seen in Fig. 4 that the solutions found by MO-PSO cover all the regions in the Pareto front and are almost evenly distributed along this front.

4.7 PSO Applicability and Efficiency

As it is reported in the PSO literature (Hu, et al., 2004), this algorithm has several advantages compared to other meta-heuristics:
-It can be easily implemented and adapted to complex problems (for the continuous optimization problems, the encoding is straightforward and does not need extra conversion);
- It allows simple constraint handling;
- Its convergence properties are almost insensitive to the design of fitness functions.

According to (Elbeltagi, et al., 2005), PSO often outperforms other meta-heuristics on the wide range of optimization problems. The tests performed in this study showed that PSO is able to at least get better results than those obtained by GA for several single- and multi-objective reliability optimization problems. However, PSO has showed poor performance when it was applied to the discrete redundancy allocation problem. This can be explained by the fact that the fundamental assumption in particle swarm is that the neighbor regions of a good solution are also good, i.e., small variations in vector representing the solution causes small variations in the fitness function (which lies on the base of the velocity update formula). This assumption usually holds in reliability allocation problems where fitness functions are relatively smooth. However, it is not always true in some discrete problems such as redundancy optimization.

It seems that the main cause preventing good performance of the PSO in solving discrete optimization problems with unsmooth fitness functions lies in its studying behavior based on gradual learning from the best solution. The basic PSO principle of emulating social behavior of learning from better examples becomes meaningless when approaching the best solution causes degradation.

The possible directions of PSO performance improvement are design of solution encoding schemes providing fitness function smoothness and combining the PSO with other heuristics and local search methods.

As an emerging stochastic optimization method, PSO exhibits great potential in solving reliability engineering problems. However, many issues remain unsolved, which requires further investigation. The adaptation of PSO for solving sequencing, partition and scheduling problems that arise in reliability engineering is the main challenge for further research.

References

Aven, T., "On performance-measures for multi-state monotone systems," *Reliability Engineering & System Safety*, vol. 41, no. 3, pp. 259-266, 1993.

Chen, T. C. and You, P. S., "Immune algorithms-based approach for redundant reliability problems with multiple component choices," *Computers In Industry*, vol. 56 pp. 195-205, 2005.

Clerc, M. The swarm and the queen: towards a deterministic and adaptive particle swarm optimization. Proceedings of IEEE Congress on Evolutionary Computation, Vol.3pp. -1957, 1999

Coello Coello, C. A. and Lechuga, M. S. MOPSO: a proposal for multiple objective particle swarm optimization. Proceedings of IEEE Congress on Evolutionary Computation, Vol.2pp. 1051-1056, 2002

Coello Coello, C. A., Pulido, G. T., and Lechuga, M. S., "Handling multiple objectives with particle swarm optimization," *IEEE Transactions on Evolutionary Computation*, vol. 8, no. 3, pp. 256-279, 2004.

Coello Coello, C. A., "A comprehensive survey of evolutionary-based multiobjective optimization techniques," *Knowledge and Information Systems*, vol. 1, no. 3, pp. 269-308, Aug.1999.

Coit, D. W. and Smith, A. E., "Reliability optimization of series-parallel systems using a genetic algorithm," *IEEE Transactions on Reliability*, vol. 5, no. 2, pp. 254-260, 1996.

Dai, Y., Xie, M., Poh, K. L., and Yang, B., "Optimal testing-resource allocation with genetic algorithm for modular software systems," *Journal of Systems and Software*, vol. 66, no. 1, pp. 47-55, Jan.2003.

Eberhart, R. C. and Kennedy, J. A new optimizer using particle swarm theory. Proceedings of International Symposium on Micro Machine and Human Science, pp. 39-43, 1995

Eberhart, R. C. and Shi, Y. Tracking and optimizing dynamic systems with particle swarms. Proceedings of IEEE Congress on Evolutionary Computation, Vol.1pp. 94-100, 2001b

Eberhart, R. C. and Shi, Y. Particle swarm optimization: developments, applications and resources. Proceedings of IEEE Congress on Evolutionary Computation, Vol.1pp. 81-86, 2001a

Eberhart, R. C. and Shi, Y. Evolving artificial neural networks. Proceedings of International Conference on Neural Networks and Brain, Beijing, P. R. China. pp. PL5-PL13, 1998

El-Gallad, A. I., El-Hawary, M. E., and Sallam, A. A., "Swarming of intelligent particles for solving the nonlinear constrained optimization problem," *Engineering Intelligent Systems for Electrical Engineering and Communications*, vol. 9, no. 3, pp. 155-163, Sept.2001.

Elbeltagi, E., Hegazy, T., and Grierson, D., "Comparison among five evolutionary-based optimization algorithms," *Advanced Engineering Informatics*, vol. 19 pp. 43-53, 2005.

Fieldsend, J. E. and Singh, S. A multi-objective algorithm based upon particle swarm optimisation, an efficient data structure and turbulence. Proceedings of U.K.Workshop on Computational Intelligence, Birmingham, UK. pp. 37-44, 2002

He, S., Prempain, E., and Wu, Q. H., "An improved particle swarm optimizer for mechanical design optimization problems," *Engineering Optimization*, vol. 36, no. 5, pp. 585-605, Oct.2004.

Hikita, M., Nakagawa, Y., Nakashima, K., and Narihisa, H., "Reliability optimization of systems by a surrogatecon-straints algorithm," *IEEE Transactions on Reliability*, vol. 41, no. 3, pp. 473-480, 1992.

Hu, X. and Eberhart, R. C. Multiobjective optimization using dynamic neighborhood particle swarm optimization. Proceedings of IEEE Congress on Evolutionary Computation, Vol.2pp. 1677-1681, 2002b

Hu, X. and Eberhart, R. C. Solving constrained nonlinear optimization problems with particle swarm optimization. Proceedings of World Multiconference on Systemics, Cybernetics and Informatics, Orlando, USA. 2002a

Hu, X. and Eberhart, R. C. Adaptive particle swarm optimization: detection and response to dynamic systems. Proceedings of IEEE Congress on Evolutionary Computation, Vol.2pp. 1666-1670, 2002c

Hu, X., Eberhart, R. C., and Shi, Y. Engineering optimization with particle swarm. Proceedings of IEEE Swarm Intelligence Symposium, pp. 53-57, 2003a

Hu, X., Eberhart, R. C., and Shi, Y. Particle swarm with extended memory for multiobjective optimization. Proceedings of IEEE Swarm Intelligence Symposium, pp. 193-197, 2003b

Hu, X., Shi, Y., and Eberhart, R. C. Recent advances in particle swarm. Proceedings of IEEE Congress on Evolutionary Computation, Vol.1pp. 90-97, 2004

Kennedy, J. Minds and cultures: particle swarm implications. Proceedings of AAAI Fall Symposium: Socially Intelligent Agents, Menlo Park, CA. pp. 67-72, 1997

Kennedy, J. and Eberhart, R. C. Particle swarm optimization. Proceedings of IEEE International Conference on Neural Networks, Vol.4pp. 1942-1948, 1995

Kulturel-Konak, S., Coit, D. W., and Smith, A. E., "Efficiently solving the redundancy allocation problem using tabu search," *IIE Transactions*, vol. 35, no. 6, pp. 515-526, 2003.

Li, X. A non-dominated sorting particle swarm optimizer for multiobjective optimization. Lecture Notes in Computer Science, Chicago, IL, USA. Vol.2723pp. 37-48, 2003

Liang, Y.-C. and Smith, A. E., "An ant colony optimization algorithm for the redundancy allocation problem (RAP)," *IEEE Transactions on Reliability*, vol. 53, no. 3, pp. 527-423, 2004.

Liang, Y.-C. and Wu, C.-C., "A variable neighborhood descent algorithm for the redundancy allocation problem," *Industrial Engineering and Management Systems*, vol. 4, no. 1, pp. 109-116, 2005.

Mostaghim, S. and Teich, J. Strategies for finding good local guides in multi-objective particle swarm optimization (MOPSO). Proceedings of IEEE Swarm Intelligence Symposium, pp. 26-33, 2003

Nakagawa, Y. and Miyazaki, S., "Surrogate constraints algo-rithm for reliability optimization problems with two constraints," *IEEE Transactions on Reliability*, vol. R-30 pp. 175-180, 1981.

Parsopoulos, K. E. and Vrahatis, M. N. Particle swarm optimization method for constrained optimization problems. Proceedings of Euro-International Symposium on Computational Intelligence, 2002

Prasad, V. R. and Kuo, W., "Reliability optimization of coherent systems," *IEEE Transactions on Reliability*, vol. 49, no. 3, pp. 323-330, 2000.

Ravi, V., "Optimization of complex system reliability by a modified great deluge algorithm.," *Asia-Pacific Journal of Operational Research*, vol. 21, no. 4, pp. 487-497, 2004.

Ray, T. and Liew, K. M., "A swarm metaphor for multiobjective design optimiza-tion," *Engineering Optimization*, vol. 34, no. 2, pp. 141-153, 2002.

Ray, T. and Liew, K. M. A swarm with an effective information sharing mechanism for unconstrained and constrained single objective optimization problem. Pro-ceedings of IEEE Congress on Evolutionary Computation, Seoul, Korea. pp. 75-80, 2001.

Shi, Y. and Eberhart, R. C. A modified particle swarm optimizer. Proceedings of IEEE Congress on Evolutionary Computation, pp. 69-73, 1998

Tian, Z., Zuo, M. J., and Huang, H. Reliability-redundancy allocation for multi-state series-parallel systems. Proceedings of European Safety and Reliability Conference, Tri City, Poland. pp. 1925-1930, 2005

Tillman, F. A., Hwang, C. L., and Kuo, W., "Determining component reliability and redundancy for optimum system reliability," *IEEE Transactions on Reliability*, vol. 26, no. 3, pp. 162-165, 1977.

Wolpert, D. H. and Macready, W. G., "No Free Lunch Theorems for Optimization," *IEEE Transactions on Evolutionary Computation*, vol. 1, no. 1, pp. 67-82, 1997.

You, P. S. and Chen, T. C., "An efficient heuristic for series-parallel redundant re-liability problems," *Computers and Operations Research*, vol. 32 pp. 2117-2127, 2005.

Cellular Automata and Monte Carlo Simulation for Network Reliability and Availability Assessment

Claudio M. Rocco S.

Facultad de Ingeniería, Universidad Central Venezuela, Caracas

Enrico Zio

Dipartimento di Ingegneria Nucleare Politecnico di Milano, Milano, Italy

5.1 Introduction

The aim of this chapter is to illustrate the computational benefits in network reliability assessment which results from combining the modeling power of Cellular Automata [23] and Monte Carlo sampling and simulation [10, 19].

In recent years, network reliability analysis has received considerable attention for the verification of the design and the evaluation of the performance of many real world distributed systems, such as computer and communication systems [2, 15, 24], power transmission and distribution systems [14, 28], rail and road transportation systems [4], oil/gas production systems [4, 5], among others.

The assessment of the reliability of a network system entails ascertaining the connectivity of a set of sources to a set of targets in the network. This could be done knowing the system cut or path sets or by a depth-first procedure [7, 12, 22]. However, these approaches lead to NP-hard problems, which require cumbersome and mathematically intensive methods of solution.

Furthermore, in practice the modification of an existing network may be required for expansion or reinforcement planning or occur inadvertently due to link failures. In such cases, the standard algorithms entail recomputing the network connection and reliability from scratch.

M.R.S. Claudio and E. Zio: *Cellular Automata and Monte Carlo Simulation for Network Reliability and Availability Assessment*, Computational Intelligence in Reliability Engineering (SCI) **40**, 113–144 (2007)
www.springerlink.com   © Springer-Verlag Berlin Heidelberg 2007

The above limitations can be overcome by resorting to the powerful modeling and computational framework offered by Cellular Automata (CA) and Monte Carlo (MC) sampling and simulation.

Cellular Automata form a general class of mathematical models, which are appealingly simple and yet capture a rich complexity of behavior of dynamical systems [26]. As such, CA have been used to study and model the dynamics of many real complex systems, including fluids, neural networks, molecular systems, ecological systems, economical systems, network systems. CA also offer a significant computational potential due to their spatially and temporally discrete nature characterized by local interaction and an inherently parallel form of evolution.

When applied for the reliability assessment of a network system, CA operate in a way to basically mimic traditional graph methods, such as Depth First Search or Breadth First Search [7, 12, 22]. However, CA algorithms allow for a straightforward parallel implementation and therefore enhance the performance of classical algorithms used in reliability evaluation.

The Monte Carlo method allows simulating the stochastic behavior of a complex system during its life by sampling the state of its components from given probability distributions. This entails modeling the components' failure/repair stochastic dynamics. Many realizations of system evolution are sampled and statistical estimates of its reliability and availability are computed.

Within the network reliability assessment problem, the combination of CA and MC techniques is used for different types of problems to:

1. Verify the existence of the connection between a single source and a single terminal target in a network of interconnected nodes;
2. Solve problem type 1 after a network connectivity reconfiguration, without the need of recomputing the whole network from scratch;
3. Evaluating the network reliability.
4. Evaluate the All-Terminal reliability of a network. This is defined as the probability that every node of the network can communicate with every other node through some path. This implies that the network forms at least a spanning tree [9], i.e. a set of arcs that connect all nodes. To address the problem by means of CA, a generalized class of Boolean networks is introduced. An algorithm to solve the k-terminal reliability problem is also provided as a simplified case of the All-Terminal one.
5. Solve the maximum unsplittable flow problem, a simplified case of the maximum flow distribution problem.
6. Determine the maximum reliability path in a network, by modeling the well know Dijkstra algorithm [11] used for short path determination.

7. Compute the availability of renewable network systems. This entails modeling the elements' failure/repair stochastic dynamics.

The solution to problems of types 1-6 is obtained by effectively combining CA for the connectivity evaluation with MC sampling of the network configurations. Problem 7 instead, requires that CA be coupled with a MC simulation model of the stochastic dynamics of the network states [10, 19]. In few qualitative words, several system life-histories are simulated by the MC approach and after each system stochastic transition, CA are used to check the connectivity between the source and the target nodes. Further, the approach allows extracting from the simulation and with no significant additional computational burden, the information regarding the importance of the elements with respect to the network availability.

An advantage of modeling the connectivity of a network by CA is that the computational complexity does not suffer from changing environments where the connectivity or the capacity of the links is modified. This computing efficiency is of utmost importance in Monte Carlo reliability/availability evaluations, where a large number of connectivity assessments are typically needed.

The chapter is organized as follows: in Sec. 2 we introduce the basics of CA computing. Sec. 3 presents the fundamentals of Monte Carlo sampling and simulation. Sec. 4 describes the application of CA and MC sampling for the reliability assessment of network systems whereas Sec. 5 presents the use of CA and MC simulation for network availability assessment. Finally, Sec. 6 contains the conclusions.

5.2 Basics of CA Computing

The state space upon which the dynamics of CA unfolds is a discrete lattice of cells $\mathcal{L}$, assumed homogeneous (all cells bear the same properties). For example, in a three-dimensional cellular state space, the state at the discrete time t of the generic cell ijl, of co-ordinates x_i, y_j, z_l with $i,j,l \in Z$, is described by the state variable $s_{ijl}(t)$. Each cell of $\mathcal{L}$ is a *finite automaton* which can assume one of a finite number of discrete values in a *local value space* $S = \{0,1,2,\ldots,k-1\}$.

The generic cell ijl interacts only with a fixed number of cells that belong to its predefined local neighborhood N_{ijl}. At the next discrete time $t+1$, the cell ijl updates its state $s_{ijl}(t+1)$ according to a transition rule $\phi : S^n \rightarrow S$, which is a function of the state variables at time t of the n cells in N_{ijl}, viz.

$$s_{ijl}(t+1) = \phi[s_{rsp}(t), \quad rsp \in N_{ijl}]$$ (2.1)

Notice that the functional form of the rule is assumed to be the same everywhere in the cellular state space, i.e. there is no space index attached to ϕ. Differences between what is happening at different locations are due only to differences in the values of the state variables of the local neighborhood, not to the update rule. The rule is also homogeneous in time. One "iteration step" of the dynamical evolution of the CA is achieved after the simultaneous application of the rule ϕ to each cell in the lattice L.

In the following, we introduce some of the basic concepts of CA computing with reference to one and two-dimensional lattices.

5.2.1 One-dimensional CA

Consider a generic cell i of a one-dimensional lattice, such as the one depicted in Figure 1. The size of the neighborhood N_i is defined by the *radius, r*, viz.

$$N_i = \{i-r, i-r+1, ..., i-1, i, i+1, ..., i+r-1, i+r\}$$ (2.2)

The dynamics of the system is governed by an arbitrary transition rule $\phi: S^{2r+1} \rightarrow S$

$$s_i(t+1) = \phi[s_{i-r}(t), ..., s_i(t), ..., s_{i+r}(t)]$$ (2.3)

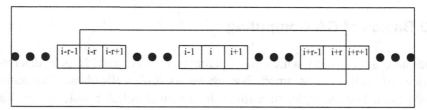

Fig. 1. Sketch of the neighborhood N_i of cell i with radius r

Since cells can take any one of k values in the local value space S={0,1,2,...,k-1}, to completely define ϕ one must assign a value in S to $s_i(t+1)$ for each of the k^{2r+1} possible $(2r+1)$-tuple configurations which can occur to the radius-r neighborhood N_i of the generic cell i.

Since in correspondence of each of the k^{2r+1} possible configurations of the *radius-r* neighborhood N_i any one of the k values in S can be assigned to $s_i(t+1)$, there are $k^{k^{2r+1}}$ possible rules.

For example, let $k=2$, so that $S \equiv \{0,1\}$, and $r=1$. To define a rule one must specify the values of the generic cell i corresponding to the 8 possible triplets of the neighborhood $N_i \equiv \{i-1, i, i+1\}$. For example, the *addition modulo 2* rule, symbolically denoted as $\oplus_2$, can be represented as follows:

$$s_i(t+1) = \mathrm{mod}_2\left[s_{i-1}(t) + s_i(t) + s_{i+1}(t)\right] \equiv \oplus_2\left\{s_{i-1}(t), s_i(t), s_{i+1}(t)\right\} \qquad (2.4)$$

and sets the value of the i-th cell to 1 if the number of cells with value 0 in the neighborhood is even and to 0 if this number is odd:

$i-1(t), i(t), i+1(t)$	1,1,1	1,1,0	1,0,1	1,0,0	0,1,1	0,1,0	0,0,1	0,0,0
$i(t+1)$	1	0	0	1	0	1	1	0

The temporal evolution of this CA is obtained by:
1. specifying the finite size M of the lattice $\mathcal{L}$
2. specifying the boundary conditions
3. specifying the initial condition $\bar{s}(0) = [s_1(0), s_2(0), ..., s_M(0)]$
4. simultaneously applying the rule ϕ to each of the M lattice cells, in an iterative manner.

Figure 2 shows the evolution obtained for a lattice of M=16 cells, with periodic boundary conditions (e.g. $s_{17} = s_1$) and initial condition $\bar{s}(0) = [1011000101001100]$. Following a sequence of states very different from each other, the CA recovers the initial configuration after eight updates. This cycle of length 8 results from a space periodicity of 4 [6].

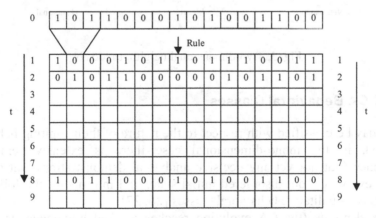

Fig. 2. Example of evolution of a one-dimensional CA under rule (2.4)

5.2.2 Two-dimensional CA

Two-dimensional CA provide a great opportunity for describing physical systems. A variety of different neighborhood structures can be defined in two-dimensional CA (Fig. 3). The *von-Neumann* neighborhood consists of the four cells, which are horizontally and vertically adjacent to the center cell of interest; the *Moore* neighborhood consists of all eight cells, which are immediately adjacent to the center cell. Both these neighborhood have unit radius. A rule defined on a von–Neumann neighborhood would take the following form:

$$s_{ij}(t+1) = \phi[s_{ij}(t), s_{i-1j}(t), s_{i+1j}(t), s_{ij-1}(t), s_{ij+1}(t)] \tag{2.5}$$

whereas on a Moore neighborhood it would read

$$s_{ij}(t+1) = \phi[s_{i-1j-1}(t), s_{ij-1}(t), s_{i+1j-1}(t), s_{i-1j}(t), s_{ij}(t), s_{i+1j}(t), s_{i-1j+1}(t), s_{ij+1}(t), s_{i+1j+1}(t)] \tag{2.6}$$

Triangular and hexagonal lattices, which are special cases of the Moore neighborhood, can also be used.

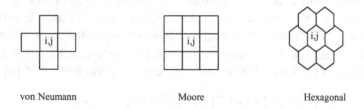

von Neumann   Moore   Hexagonal

Fig. 3. Two-dimensional neighborhoods N_{ij}.

5.2.3 CA Behavioral Classes

CA may be classified with respect to the nature of their limiting behavior. Indeed, in the mono-dimensional case there is extensive empirical evidence (but not yet any decisive analytical demonstration) that all CA rules evolving from disordered initial states fall into one of the following four basic qualitative behavioral classes [26, 27]:

1. Fixed points (the CA evolution reaches a fixed homogeneous lattice configuration in which each cell attains the same state value)
2. Inhomogeneous configuration or cycles (the CA evolution leads to simple stable configurations or to the emergence of periodic and separated structures)

3. Chaotic, aperiodic patterns
4. Complex, localized, propagating structures

All CA within a given class yield a qualitatively similar behavior, regardless of the specific underlying transition rule. The behaviors of the first three classes bear a strong resemblance to those observed in continuous dynamical systems. The homogeneous final configurations occurring for the CA in class 1, for example, are essentially the same as *fixed points* attractors. Class 2 automata usually create patterns that repeat periodically, similarly to continuous *limit cycles*. The aperiodic, chaotic patterns emerging from class 3 automata are analogous to the *strange attractors* appearing in continuous dynamical systems. The statistical properties of the limit patterns and of the starting one are almost identical, giving rise to a kind of *self-similar fractal curves*. The more complicated localized structures emerging from class 4 CA do not appear to have any obvious continuous analogue. This last class of CA is capable of performing *universal computation* and shows a high invariability in their time development.

5.3 Fundamentals of Monte Carlo Sampling and Simulation

The development of computer power has led to a strong increase in the use of MC methods for system engineering calculations.

In the past, restrictive assumptions had to be introduced to fit the system models to the numerical methods available for their solution, at the cost of drifting away from the actual system operation and at the risk of obtaining sometimes dangerously misleading results. Thanks to the inherent modeling flexibility of MC simulation, these assumptions can be relaxed, so that realistic operating rules can be accounted for in the system models for reliability, maintainability and safety applications.

This Section synthesizes the principles underlying the MC simulation method for application to the evaluation of the reliability and availability of complex systems [19].

5.3.1 The System Transport Model

Let us consider a system whose state is defined by the values of a set of variables, i.e. by a point P in the phase space Ω. The evolution of the system, i.e. the succession of the states occupied in time, is a stochastic process, represented by a trajectory of P in Ω. The system dynamics can be studied by calculating the ensemble values of quantities of interest, e.g. probability distributions of state occupancy and expected values.

The MC method allows generating the sample function of the ensemble by simulating a large number of system stochastic evolutions. Every MC trial, or history, simulates one system evolution, i.e. a trajectory of P in Ω. In the course of the simulation, the occurrences of events of interest (e.g. system failure) are accumulated in appropriate counters. At the end of the simulation, after the large number of trials has been generated, the sample averages of the cumulated quantities give the MC (ensemble) estimates of the quantities of interest (e.g. system unreliability) [10, 19].

In the case here under consideration, the network is made up of physical components (arcs) subject to failures and repairs that occur stochastically in time. Each arc can be in two states, e.g. working or failed, and for each arc we assign the probabilities for the transitions between different states. Each MC trial represents one possible realization of the ensemble of the stochastic process. It describes what happens to the system during the time interval of interest, called *mission time*, in terms of state transitions, i.e. the sequence of states randomly visited by the arcs, starting from a given initial configuration.

The next Section provides details on the actual MC procedure for generating the system life histories and thereby estimating the system reliability and availability.

5.3.2 Monte Carlo Simulation for Reliability Modeling

The problem of estimating the reliability and availability of a system can be framed in general terms as the problem of evaluating functionals of the kind [19]:

$$G(t) = \sum_{k \in \Gamma} \int_0^t \psi(\tau, k) R_k(\tau, t) d\tau \qquad (3.1)$$

where $\psi(\tau, k)$ is the probability density of entering a state k of the system at time τ and Γ is the set of possible system states which contribute to the function of interest $R_k(\tau, t)$.

In particular, the functionals we are interested in for reliability applications are the system unreliability and unavailability at time t, so that Γ is the subset of all failed states and $R_k(\tau, t)$ is unity, in the former case of unreliability, or the probability of the system not exiting before t from the failed state k entered at $\tau < t$, in the latter case of unavailability. Note that the above expression (3.1) is quite general, independent of any particular system model for which the $\psi(\tau, k)$'s are computed. In what follows we show the details on how Eq. (3.1) tanisolved by MC simulation, with reference to the system unavailability estimation problem which provides a more interesting case than unreliability because of the presence of repairs.

Let us consider a single trial and suppose that the system enters a failed state $k \in \Gamma$ at time τ_{in}, exiting from it at the next transition at time τ_{out}. The time is suitably discretized in intervals of length Δt and counters are introduced which accumulate the contributions to $G(t)$ in the time channels: in this case, we accumulate a unitary weight in the counters for all the time channels within $[\tau_{in}, \tau_{out}]$, to count that in this realization the system is unavailable. After performing a large number, M, of MC histories, the content of each counter divided by the time interval Δt and by the number of histories gives an estimate of the unavailability at that counter time. This procedure corresponds to performing an ensemble average of the realizations of the stochastic process governing the system life.

The system transport formulation of Eq. (3.1) suggests another analog MC procedure, consistent with the solution of definite integrals by MC sampling [10, 19]. In a single MC trial, the contributions to the unavailability at a generic time t are obtained by considering all the preceding entrances, during this trial, into failed states $k \in \Gamma$. Each such entrance at a time τ gives rise to a contribution in the counters of unavailability for all successive times t up to the mission time, represented by the probability $R_k(\tau, t)$ of remaining in that failed state at least up to t. In case of a system made up of components with exponentially distributed failure and repair times, we have

$$R_k(\tau, t) = e^{-\lambda^k (t - \tau)} \qquad (3.2)$$

where λ^k is the sum of the transition rates (repairs or further failures) out of state k.

Again, after performing all the MC histories, the contents of each unavailability counter are divided by the time channel length and by the total number of histories M to provide an estimate of the time-dependent unavailability.

The two analog MC procedures presented above are equivalent and both lead to satisfactory estimates. Indeed, consider an entrance in state $k \in \Gamma$ at time τ, which occurs with probability density $\psi(\tau, k)$, and a subsequent time t: in the first procedure a one is scored in the counter pertaining to t

only if the system has not left the state k before t and this occurs with probability $R_k(\tau,t)$. In this case, the collection of ones in t obeys a Bernoulli process with parameter $\psi(\tau,k) \cdot R_k(\tau,t)$ and after M trials the mean contribution to the unavailability counter at t is given by $M \cdot \psi(\tau,k) \cdot R_k(\tau,t)$. Thus, the process of collecting ones in correspondence of a given t over M MC trials and then dividing by M and Δt leads to estimating the quantity of interest $G(t)$. On the other hand, the second procedure leads, in correspondence of each entrance in state $k \in \Gamma$ at time τ, which again occurs with probability density $\psi(\tau,k)$, to scoring a contribution $R_k(\tau,t)$ in all the counters corresponding to $t > \tau$ so that the total accumulated contribution in all the M histories is again $M \cdot \psi(\tau,k) \cdot R_k(\tau,t)$. Dividing the accumulated score by M and Δt yields the estimate of $G(t)$. In synthesis, given $\psi(\tau,k)$, with the first procedure for all t's from τ up to the next transition time we collect a one with a Bernoulli probability $R_k(\tau,t)$, while with the second procedure we collect $R_k(\tau,t)$ for all t's from τ up to the mission time: the two procedures lead to equivalent ensemble averages, even if with different variances. We shall not discuss further the subject of the variance, for space limitation.

The MC procedures just described, which rely on sampling realizations of the random transitions from the true probability distributions of the system stochastic process, are called "analog" or "unbiased". Different is the case of a non-analog or biased MC computation in which the probability distributions from which the transitions are sampled are properly varied so as to render the simulation more efficient. The interested reader can consult Refs. [19, 20] for further details on this.

Finally, in the case that the quantity of interest $G(t)$, $t \in [0,T_M]$, is the system unreliability, $R_k(\tau,t)$ is set equal to one so that the above MC estimation procedure still applies with the only difference being that in each MC trial a one is scored only once in all time channels following the first system entrance at τ in a failed state $k \in \Gamma$.

5.4 Application of CA for the Reliability Assessment of Network Systems

In this Section, we first recall the algorithm introduced in [23] for the verification of the existence of the connection from a source node (S) to a target node (T). This will set the stage for the successive extensions introduced to treat the All-Terminal, k-Terminal, maximum unsplittable flow and maximum reliability path problems.

The results of the CA algorithms presented are obtained by simulating their parallel behavior on a serial machine.

5.4.1 S-T Connectivity Evaluation Problem

Consider a network of m binary nodes whose function is to deliver a given throughput from a source S to a target node$T$. Ascertaining the connectivity of the network from source to target would require knowledge of the system cut or path sets or a depth-first procedure [7, 12, 22]. These approaches may be cumbersome and mathematically intensive for realistic networks.

Let us map each node i into a spatial cell whose neighborhood N_i is the set of cells (i.e. network elements) which provide their input to it. The state variable s_i of cell i is binary, assuming the value of 1 when node is operating (active, i.e. receiving input) and of 0 when not operating (passive, i.e. not receiving input).

The transition rule governing the evolution of cell i consists of the application of the logic operator OR ($\vee$) to the states of the nodes in its neighborhood:

$$s_i(t+1) = s_p(t) \vee s_q(t) \vee ... \vee s_r(t), \qquad p,q,...,r \in N_i \qquad (4.1)$$

where t is the iteration step.

According to this rule, a cell is activated if there is at least one active cell in its neighborhood, i.e. if it receives input from at least one of its connected nodes.

At the beginning of the evolution, the source S is the only active cell. The successive application of the transition rule to each node of the network generates the paths of transmission through the network. The computation ends either when the target node is activated or the process stagnates. Since the longest possible path from active to T activates all the remaining m - 1 nodes, the S-T connection can be computed in $O(m)$ iterations, i.e. the computation ends after m-1 steps or the process stagnates.

The basic algorithm proceeds as follows:
1. $t = 0$
2. Set all the cells state values to 0 (passive)
3. Set $s_s(0) = 1$ (source activated)
4. While t < m-1
     $\{ t = t + 1$
     Update all cells states by means of rule (4.1)

If $s_T(t) = 1$, stop (target activated) }

5. $s_T(m-1) = 0$ (target passive): there is no connection path from S to T

For an example of application of this algorithm, the interested reader is referred to the original work of [23].

5.4.2 S-T Network Steady-state Reliability Assessment

Assume now that the generic connecting element (arc) ji from node j to i can be in two states, success ($w_{ji} = 1$) or failure ($w_{ji} = 0$), with probabilities p_{ji} and $q_{ji} = 1 - p_{ji}$ respectively. The ji arc state variable w_{ji} defines the operational state of the arc.

The transition rule governing the evolution of the generic cell i consists of the application of the logic operator OR ($\vee$) on the results of the logic operator AND ($\wedge$) applied to the states of the nodes in its neighborhood and to the states of their connection with i:

$$s_i(t+1) = [s_p(t) \wedge w_{pi}] \vee [s_q(t) \wedge w_{qi}] \vee ... \vee [s_r(t) \wedge w_{ri}]$$

$$p, q, ..., r \in N_i \tag{4.2}$$

The network reliability p_{ST}, i.e. the probability of a successful connection from S to T, can then be computed by the following steps i) Monte Carlo (MC)–sampling a large number M of random realizations of the states of the connecting arcs; ii) CA–computing, for each realization, if a path from S to T exists: the ratio of the number of successful S-T paths over the total number of realizations evaluated gives the network reliability.

The basic algorithm proceeds as follows:

1. $n = 0$
2. While MC – iteration $n < M$
   { $n = n + 1$
   Sample by MC a realization of the states of the connecting arcs w
   Apply the previously illustrated CA algorithm for S-T connectivity, to evaluate if there is a path from S to T
   If a path exists, then update the counter of successful system states }

3. Network reliability $p_{ST} = \dfrac{number\ of\ S-T\ successful\ paths}{M}$

5.4.2.1 Example

Fig. 4a shows the example network to be evaluated [29]: a path between the source node S and the terminal node T is required. Initially the source node S is activated. In iteration 1 nodes 1, 2, 3 and 4 are activated. Fig. 4b shows the network status for this iteration: dotted circles represent activated nodes. In the next iteration, nodes 5, 7, 8 and 9 are activated since they all have an activated node in their respective neighborhood (see Fig. 4c). In the next iteration, node 5 activates node 6, node 7 activates node T (Fig. 4d). Since the terminal node T has been activated, a path from node S to node T exists.

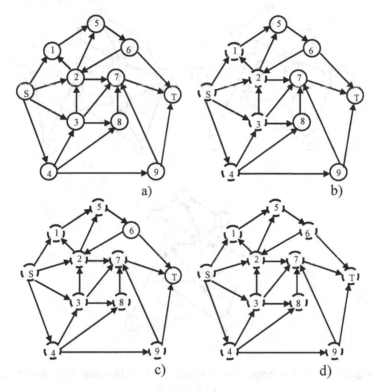

Fig. 4. Network for example 4.2.2 [29]: a) Iteration 0; b) Iteration 1; c) Iteration 2; d) Iteration 3

5.4.2.2 Connectivity Changes

Assume that a path is found in the network previously analyzed and a change occurs in the connectivity. For example arcs are eliminated or included. In these conditions it is not necessary to reinitiate the CA evaluation from the beginning.

For instance, in the previous example of Section 4.2.1 assume that the CA reached the status shown in Fig. 4d and arcs (4,9), (7,T) and (6,T) are eliminated. The network status for this case is shown in Fig. 5a.

Updating the current network configuration by means of (4.2), it can be seen that node 9 changes to the quiescent state since its neighborhood is empty. In the next iteration node T changes to its quiescent state since the only node in its neighborhood is also quiescent and the process stagnates. In this case the network has no S-T path (Fig. 5b).

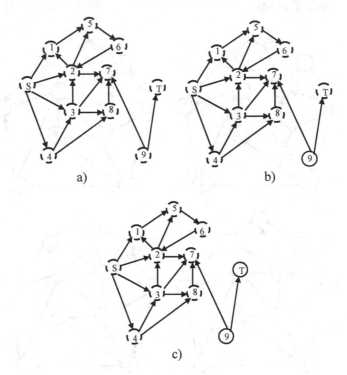

a)   b)

c)

Fig. 5. Network changes for example 4.2.2: a) Iteration 0; b) Iteration 1; c) Iteration 2

5.4.2.3 Steady-state Reliability Assessment

Assume the following reliability values of the arcs in the network of Figure 4a [29]: $r_{2,7} = 0.81$, $r_{S,4} = r_{4,8} = r_{4,9} = r_{9,7} = 0.981$, others $r_{i,j} = 0.9$. By the proposed evaluation approach with $M=10^4$ trials, the system reliability is estimated to be 0.997, which compares well with the value reported in [29].

As mentioned in the previous Section 4.2, step 4 is performed considering the quiescent status of the network topology before any changes. The network evaluation fro scratch requires on average 3.5 iterations whereas using the available information from the previous evaluated topology requires on average 1.01 steps. The number of removed/added arcs among consecutive trials varies between 1 and 14, with an average of 7.07.

5.4.3 The All-Terminal Evaluation Problem

5.4.3.1 The CA Model

The *All-Terminal* reliability of a network is defined as the probability that every node can communicate with every other node through some path. This means that the network forms at least a spanning tree (A *spanning tree* of a graph is a set of m-1 arcs that connect all vertices of the graph) [9].

To solve the problem using Cellular Automata computing, the procedure for the *S-T connectivity evaluation problem* of Section 4.1 needs to be properly modified. An additional cell is introduced whose neighborhood contains all other m cells, i.e. $N_{m+1} = \{ j\ j = 1, 2, .., m \}$.

As defined in Section 4.1, the transition rule governing the evolution of cell $i \neq m+1$ consists of the application of (4.1) to the states of the nodes in its neighborhood. At the same iteration step cell $m+1$ is subject to a different transition rule which consists in the application of the logic operator AND ($\wedge$) to the states of all m cells in its neighborhood:

$$s_{m+1}(t+1) = s_1(t) \wedge s_2(t) \wedge ... \wedge s_m(t), \tag{4.3}$$

According to this rule, cell $m+1$ is activated if all of the cells in its neighborhood are activated.

The computation ends when cell $m+1$ is activated or the process stagnates. The complete connectivity of all the m cells in the network can be computed in $O(m)$ time.

The basic algorithm proceeds as follows:
1. $t = 0$
2. Set all the cells state values to 0 (passive)
3. Set $s_j(0) = 1$ (a generic cell $j \neq m+1$ is activated)
4. While $t < m$
    $\{ t = t + 1$
    Update cells:

a. Update all cells states, excluding cell m+1, by means of rule (4.1)

b. Update the state of cell m+1, by means of rule (4.3)

If $s_T(m+1) = 1$, stop (all cells are connected) }

5. $s_T(m+1) = 0$: there is no spanning tree

5.4.3.2 Example

Consider the 11-nodes network shown in Fig. 6 [29]. Fig. 7a shows the network configuration at $t=0$: Cell 11 is initially activated. Note that we have included the additional cell $m+1=12$ (arcs from the other cells into cell 12 are not shown for preserving the clarity of the picture).

Fig. 7b shows the state of the network at $t=1$. Cells 1 to 4 are activated. At the next time iteration, $t=2$, cells 5,7,8, and 9 are activated (Fig. 7c). Fig. 7d shows the cells states at $t=3$: the remaining cells 6 and 10 are activated. At $t=4$, all the network cells have been activated, so that cell 12 is also activated, which means that at least a spanning tree exists.

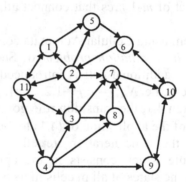

Fig. 6. Network used to illustrate the All-Terminal CA [29]

5.4.3.3 All-terminal Reliability Assessment: Application

Consider the network of 52 nodes and 72 links of the Belgian telephone inter-zones network shown in Fig. 8 [18]. The reliability of each arc is 0.90. Applying the MC approach described in Section 4.3, with $M=10^4$ samples, combined with the CA presented in Section 4.3.1, the *all-terminal* reliability estimate turns out to be $(65.28 \pm 0.47) \cdot 10^{-2}$ which compares well with the value of $(65.45 \pm 0.01) \cdot 10^{-2}$ obtained in [18] with $M=10^5$ samples.

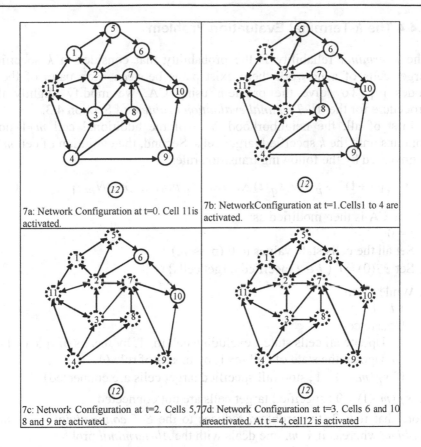

7a: Network Configuration at t=0. Cell 11is activated.

7b: NetworkConfiguration at t=1.Cells1 to 4 are activated.

7c: Network Configuration at t=2. Cells 5,7 8 and 9 are activated.

7d: Network Configuration at t=3. Cells 6 and 10 areactivated. At t = 4, cell12 is activated

Fig. 7. Network configuration at different times for example 4.3.2

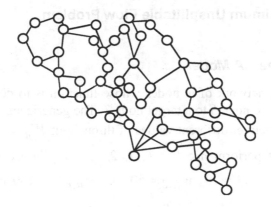

Fig. 8. Telephone network considered for example 4.3.3 [18].

5.4.4 The *k*-Terminal Evaluation Problem

The *k-terminal* reliability is the probability that considering *k* specified target cells of a network there exist paths between each pairs of the k nodes [9]. To solve the problem using CA, we modify slightly the procedure for the *All-Terminal evaluation problem* of Section 4.3.

First of all, the neighborhood N_{m+1} of the additional cell *m*+1 now contains only the *k* specified target cells. Second, the evolution of cell *m*+1 is governed by the following transition rule:

$$s_{m+1}(t+1) = s_p(t) \wedge s_q(t) \wedge ... \wedge s_r(t), p, q, ..., r \in N_{m+1} \qquad (4.4)$$

The CA is then modified as:
1. $t = 0$
2. Set all the cells state values to 0 (passive)
3. Set $s_j(0) = 1$ ($j \in$ {specified target cells})

4. While t < m
   { $t = t + 1$
   Update cells:
      a. Update all cells states, excluding cell m+1, by means of rule (4.1)
      b. Update the state of cell m+1, by means of rule (4.4)
   If $s_T(m+1) = 1$, stop (all specified target cells are connected) }
5. $s_T(m+1) = 0$: specified target cells are not connected

Note that if *k*=2, the problem reduces to the *S-T connectivity evaluation problem* whereas if *k*=*m*, one deals with the *All-terminal* problem.

5.4.5 Maximum Unsplittable Flow Problem

5.4.5.1 The CA Model

Consider a network of *m* nodes whose function is to deliver a throughput from a source node *S* to a target node *T*. The generic arc *ji* connecting node *j* to *i* may deliver different values of throughput $W_{ji} = w_{ji,k}$ depending on its level of performance $k = 0, 1, 2, ... , m_{ji}$ in ascending order (e.g. $w_{ji,0} = 0$, $w_{ji,1} = 20$, $w_{ji,2} = 50, ... , w_{ji,m_{ji}} = 100$ in arbitrary units).

The generic node i can handle different values of throughput, $X_i = x_{i,l}$, $l = 0, 1, 2, \ldots, m_i$ in ascending order (e.g. $x_{i,0} = 0$, $x_{i,1} = 30$, $x_{i,2} = 60$, $\ldots, x_{i,m_i} = 100$ in arbitrary units).

Physically, the generic node i receives as input from its generic neighbor $j \in N_i$ a throughput $\zeta_j(t)$ which is the minimum of node j throughput $_j(t)$ $= X_j(t)$ and of arc ji capacity $W_{ji}(t)$:

$$\zeta_j(t) = \min[s_j(t), w_{ji}(t)] \qquad (4.5)$$

Among the inputs $\zeta_j(t)$ received from all the neighbors $j \in N_i$, node i processes the maximum possible, but with the constraint of its own intrinsic throughput limitation X_i. Hence, the transition rule governing the evolution of the generic cell i consists of the application of the minimum operator to the current throughput level $X_i(t)$ of node i itself and to the maximum of the input throughput $\zeta_j(t)$ received from the neighboring nodes $j \in N_i$:

$$s_i(t+1) = \min\left\{ X_i(t), \max_{j \in N_i}\left[\zeta_j(t)\right]\right\} \qquad (4.6)$$

According to this rule a cell (node) is activated at the lowest level of throughput provided by the maximum input it receives and its own intrinsic capacity.

At the beginning of the evolution, the source S is the only active cell at a level X_S. The successive applications of the above transition rule to each node of the network, generates the paths of throughput transmission through the network. The computation ends when the target node is activated or the process stagnates.

In order to find the S-Tpath of maximum allowable throughput, for a fixed node – capacity and arc – throughput configuration, the following algorithm of CA evolution may be applied:
1. $t = 0$
2. Set all cells state values to 0
3. Set $s_S(0) = X_S$, the source capacity
4. While $t < m$-1
    { $t = t + 1$
    Update all cells states by means of rule (4.6)

If $s_T(t) = X_T \neq 0$, stop (target activated at the throughput level X_T) }

5. $s_T(m-1) = 0$ (target passive: there is no connection path from S to T).

5.4.5.2 Example

Consider the network in Figure 9 [8] and assume that all the nodes have a throughput of 100, in arbitrary units.

Let $s_S(0) = 71$. The computation of the network throughput proceeds as follows. At $t=1$ we have:

$s_2(1) = \min\{(100, \max(\min(71,12))\} = 12$;

$s_3(1) = \min\{(100, \max(\min(71,35))\} = 35$;

$s_4(1) = \min\{(100, \max(\min(71,24))\} = 24$

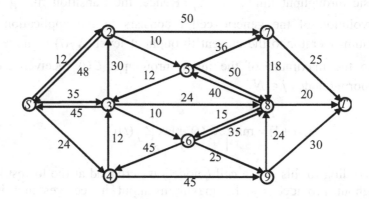

Fig. 9. Network considered for example 4.5.2 [8]. The numbers on the arcs represent their capacities

At $t=2$ we have:

$s_2(2) = \min\{(100, \max(12, \min(35,30))\} = 30$;

$s_3(2) = \min\{(100, \max(35, \min(24,12))\} = 35$;

$s_4(2) = \min\{(100,24\} = 24$;

$s_5(2) = \min\{(100, \max(\min(12,10)\} = 10$;

$s_6(2) = \min\{(100, \max(\min(35,10)\} = 10$;

$s_7(2) = \min\{(100, \max(\min(12,50)\} = 12$;

$s_8(2) = \min\{(100, \max(\min(35,24)\} = 24$;

$s_9(2) = \min\{(100, \max(\min(24,45)\} = 24$

At $t=3$ we have:

$s_2(3) = s_2(2)$; $s_3(3) = \min\{(100, \max(35, \min(12,12)))\} = 35$;

$s_4(3) = \min\{(100, \max(24, \min(10,45)))\} = 24$

$s_5(3) = \min\{(100, \max(\min(30,10), \min(24,40)))\} = 24$;

$s_6(3) = \min\{(100, \max(\min(35,10), \min(24,15)))\} = 15$;

$s_7(3) = \min\{(100, \max(\min(30,50), \min(10,36)))\} = 30$;

$s_8(3) = \min\{100, \max(\min(12,18), \min(10,50), \min(35,24),$
$\min(10,15), \min(24,24)) = 24$

Finally, at $t=4$ we have for the target node:

$s_T(4) = \min\{(100, \max(\min(30,25), \min(24,20), \min(24,30)))\} = 25$

Thus, even if the S-maximum splittable flow is potentially 71 units (the source throughput), the maximum unsplittable flow that can be transferred from S to T is only 25 units.

Finally, let us evaluate the reliability of the network of Figure 9 within the framework of unsplittable flow. We assume that starting from a source of 71, in arbitrary units, a maximum throughput of x units is required at the target and that the reliability of each arc is 0.90. Hence, a network configuration is successful if it provides at the target node a throughput $s_T \geq x$. Figure 10 shows the network reliability point estimate as a function of the throughput required, calculated by the combined procedure MC-CA with $M=10^4$ samples.

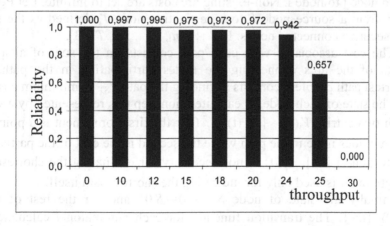

Fig. 10. Network reliability as a function the minimum throughput required

5.4.6 Maximum Reliability Path [23]

5.4.6.1 Shortest Path

The problem of finding the shortest path (SP) from a single source to a single destination in a graph arises as a sub-problem to many broader problems [11, 25]. In general, different path metrics are used for different application. For example, in communication systems, if each link cost is 1 then the minimum number of hubs is found. But cost can also represent the propagation delay, the link congestion or the reliability of each link. In the latter case, if the individual communication links operate independently, then the problem can be stated as to find what path has the maximum reliability [25].

This problem has some well-known polynomial algorithmic solutions, namely Bellman-Ford's or Dijkstra's [11, 25]. Recently neural networks have also been proposed to find the SP [3, 21]. Problems that require multiple fast computation of shortest path can benefit from more efficient methods, such as the application of a CA approach. In [1] an algorithm based on CA consisting of a discrete d-dimensional lattice of n cells with a neighborhood of variable radius is presented. The CA model to be considered is similar but rely on a faster and simpler implementation.

As described in Section 4.1, each node i is mapped into a spatial cell whose neighborhood N_i is the set of cells (i.e. network elements) which provide their input to it. Associated with each arc $ji \in \{1,2,...,m\}$ x $\{1,2,...,m\}$, $i \neq j$, is a nonnegative number C_{ji} that stands for the cost of arc from node j to node i. Non-existing arc costs are set to infinite. Let P_{ST} be a path from a source node S to a destination node T, defined as the set of consecutive connected nodes: $P_{ST} = \{S, n_1, n_2, ... n_i, T\}$ [3].

The cost associated with each path consists on the sum of all partial costs of the arcs connecting the nodes participating in the path. The shortest path problem consists in finding the path P_{ST} of minimum cost.

The state of each node, at each iteration step t, is represented by a vector with two entries $V_i(t) = \{V_i^1(t), V_i^2(t)\}$: the first component is a pointer to the previous node in the path while the second is the cost of the partial path up to node i [21]. $V_i^1(t)$ is not necessary for evaluating the shortest path length but it is used only for indicating the shortest path itself.

Initially the state of node S $V_S(0) = \{S,0\}$ and for the rest of nodes $V_i(0) = \{i,\infty\}$. The transition function for each automaton i calculates the minimum of the quantity $U_{i,j}(t_s) = (V_j^2(t) + C_{j,i})$ among the members in N_i. The node producing that minimum, will be denoted as node k. Then, the value of the second component V_i^2 is set to $V_k^2(t) + C_{k,i}$ [21]. In case of conflicts when two or more nodes the neighborhood have the same $U_{i,j}(t)$ quantity, then the k node is selected as the one with the minimum $V_k^2(t)$. The effect of this rule is the incremental calculation of the cost for all paths starting at node S.

At $t=0$, node$S$ is in a stable state and it will not change during the evaluations. During each iteration step at least a new node will become stable [21]. Since the longest path in the network consists of n-1 nodes, it can be computed in $O(n)$ time. This means that the computation ends after n-1 iteration steps or the process stagnates. The algorithm presented in [1] requires $O(\upsilon n)$upper time, where $\upsilon = \max(C_{pq})$ and $C_{pq} \in N$.

The basic algorithm is:

1. Initialize the state of the cells as $V_S(0)=\{S,0\}$ and$V_i(0)=\{i,\infty\},i\neq S$, $t = 0$
2. While$t\neq n$-1 or the process stagnates
   { Synchronously update each automaton i, $i\neq S$ finding $k \in N_i$ that minimize
   $U_{i,k}(t) = (V_k^2(t)+C_{k,i})$ or, in case of conflict, minimize $V_k^2(t)$ and set
   $V_i(t+1)=\{k, V_k^2(t)+C_{k,i}\}$
   $t = t+1$ }
3. Stop

The minimum S-Tpath cost is the second component of the state vector of node$T$. In case that this value is infinity then there exists no S-Tpath.

5.4.6.2 Example

Fig. 11 shows the network to be evaluated [22]: the shortest path between the source node S and the terminal node T is required.

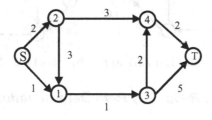

Fig. 11. Network for example 4.6.2 [22]

Fig. 12 presents the status of the network and $V_i(t)$ for each successive iteration step. After 4 time steps, the CA model gives the SP required.

Note that the model also provides the shortest path from the source node S to every node in the network. Fig. 12f) presents the shortest path between S and T (cost = 6).

5.4.6.3 Example

When the connectivity and/or the cost of the arcs in the network change, a CA evaluation from scratch is not required (Section 4.2.2). Let us consider the previous example and assume that the costs of the links (3, T) and (1,3) change to 1 and 10 respectively. Fig. 13 presents the status of the network for each successive iteration step. After 3 iteration steps, the new SP is obtained with a total cost of 3.

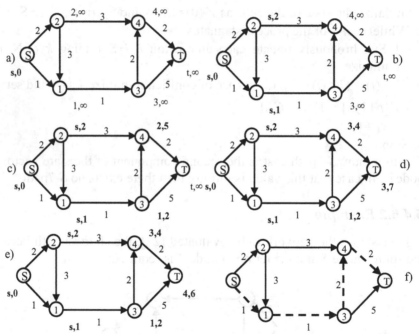

Fig. 12. Network status and $V_i(t)$: a) t=0; b) t=1; c) t=2; d) t=3 ; e) t=4; f) SP

5.4.6.4 Maximum Reliability Path Determination

If the individual links operate independently, then the reliability of a path $P_{ST} = \{S, n_1, n_2, ... , T\}$ is determined as: $R_{ST} = R_{S,n_2} R_{n_1,n_2} ..... R_{n_i,T}$

where R_{n_1,n_2} is the reliability of the link between n_1 and n_2. Taking the logarithm, $\log(R_{ST}) = \log(R_{S,n_2}) + \log(R_{n_1,n_2}) + ... + \log(R_{n_i,T})$.

Finding the maximum reliability R_{ST} is equivalent to maximize log (R_{ST}). Since $0 < R_{i,j} \leq 1$, then $\log(R_{i,j}) < 0$ and the problem is equivalent to minimize $-\log(R_{S,n_2}) - \log(R_{n_1,n_2}) - ..... - \log(R_{n_1,T})$.

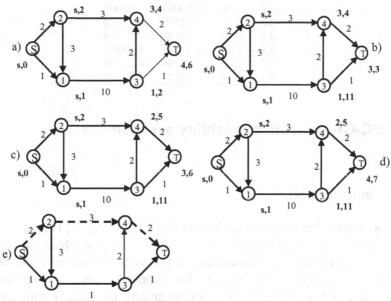

Fig. 13. Network changes for example 4.6.2 a) $t=0$; b) $t=1$; c) $t=2$; d) $t=3$; e) SP

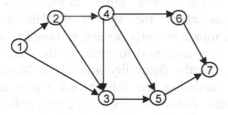

Fig. 14. Network configuration [25] considered for Maximum Reliability Path determination

As an example consider the network in Fig. 14 [25]. Table 1 presents the reliability of each arc and the corresponding log values. The maximum reliability path between nodes 1 and 7 corresponds to the shortest path evaluation, using $-\log(R_{i,j})$ as cost, and is given by 1-3-5-7.

Table 1. Data for network in Figure 14

(i,j)	$R_{i,j}$	log ($R_{i,j}$)	- log ($R_{i,j}$)
1,2	0.20	-0.69897	0.69897
1,3	0.90	-0.04576	0.04576
2,3	0.60	-0.22185	0.22185
2,4	0.80	-0.09691	0.09691
3,4	0.10	-1.00000	1.00000
3,5	0.30	-0.52288	0.52288
4,5	0.40	-0.39794	0.39794
4,6	0.35	-0.45593	0.45593
5,7	0.25	-0.60206	0.60206
6,7	0.50	-0.30103	0.30103

5.5 MC-CA network availability assessment

5.5.1 Introduction

Let us assume that the generic element $ji \in \{1,2,...,m\} \times \{1,2,...,m\}$ of the network system (arc ji if $i \neq j$; node i if $i=j$) can fail at stochastic times with constant failure rate λ_{ji}. Restoration is immediately started upon failure, lasting a stochastic time which, for simplicity and without loss of generality, is also assumed to be exponentially distributed with constant repair rates μ_{ji}. The outcome of the restoration action is a full renovation of the functionality of the network element to an *as good as new* condition. These assumptions allow the system availability to be computed analytically, for comparison purposes and without loss of generality.

The approach here proposed to evaluate the time-dependent network availability combines the flexibility offered by Monte Carlo simulation (Section 3), for modeling the failure and repair dynamics, with the efficiency of Cellular Automata (Section 2), for verifying the source-target connectivity in correspondence of the different system configurations (Section 4.1). As recalled in Section 3, the MC approach amounts to simulating a large number of realizations of the system life up to its mission time T_M. Each of these simulated histories corresponds to a virtual experiment in which the system is followed in its evolution throughout the mission time. During the simulation, in correspondence to each system stochastic transition, we check the *S-T* connectivity by running the CA solver (Section 4.1) and collect, in appropriately devised counters, the quantities of interest for the estimation of the system instantaneous availability (Section 3).

To simulate the network evolution we use the direct Monte Carlo approach according to which the transition times of all network elements are individually sampled and the minimum of them identifies the time at which the next transition occurs [16]. A single MC trial then corresponds to a realization of a random walk of the system:

$$(s_{t_0}, t_0), (s_{t_1}, t_1), \ldots, (s_{t_e}, t_e), (s_{t_{e+1}}, t_{e+1}), \ldots$$

After the e-th transition occurring at time t_e, the system state is identified by the binary state s_{te} of the terminal node T (active or non active) as resulting from the CA-based S-T connectivity evaluation procedure, and remains unchanged until the next transition at t_{e+1}. Correspondingly, we update the discrete counters for the estimation of the quantities of interest associated to the times $t \in [t_e, t_{e+1}]$. By performing several Monte Carlo histories, we obtain many independent realizations of the system random walks and by ensemble averaging the quantities in the counters we estimate the time-dependent system availability A(t). The basic algorithm is presented in Appendix.

If there are dependences among the failure and repair behaviors of different elements, the procedure must be modified to allow re-sampling of the dependent elements transitions when one of them has changed state [16].

Notice that in correspondence of each system transition, i.e. the failure or repair of an arc or node, the CA solver must be applied to the reconfigured network. This does not require repeating the propagation of the CA activation pulse starting from the initial network configuration, i.e. with the only active cell in the CA being the source. It is indeed more efficient to proceed inversely to activate or deactivate those cells whose neighbors turn out, as a result of the system transition, to contain newly active cells or to be empty of any active cell, respectively (Section 4.2.2).

Note that, when combining the CA and MC approaches, the CA parallel computation strategy can still be applied at all times that the network connectivity evaluation is required. Instead, the general direct MC approach is based on a series computation procedure. Indeed, the evaluation of the next system transition time t_{e+1} requires sampling of the transition time for only the element which made the last transition at time t_e. In the meantime, in order to compute t_{e+1}, all other network elements should "wait" for the outcome of this sampling operation so that they are computationally inactive. However, it still seems attractive to exploit a parallel computational strategy even if effective only for the evaluation of the network connectivity: for large-sized networks this is indeed where computation burden resides the most, as opposed to the evaluation of the next transition time.

5.5.2 A Case Study of Literature

Consider the network of Fig. 14 [25]. Table 2 reports the failure rates of each arc, in arbitrary units of inverse of time. Repair rates are assumed an order of magnitude larger than the corresponding failure rates. On the contrary, the nodes are assumed to be perfect, i.e. non-failing.

Table 2. Failure rates of the arcs in the network of Fig. 14 (in arbitrary units)

(j,i)	λ_{ji}
2,1	$1.609 \cdot 10^{-2}$
3,1	$1.054 \cdot 10^{-3}$
3,2	$5.108 \cdot 10^{-3}$
4,2	$2.231 \cdot 10^{-3}$
4,3	$2.303 \cdot 10^{-2}$
5,3	$1.204 \cdot 10^{-2}$
5,4	$9.163 \cdot 10^{-3}$
6,4	$1.049 \cdot 10^{-2}$
7,5	$1.386 \cdot 10^{-2}$
7,6	$6.931 \cdot 10^{-3}$

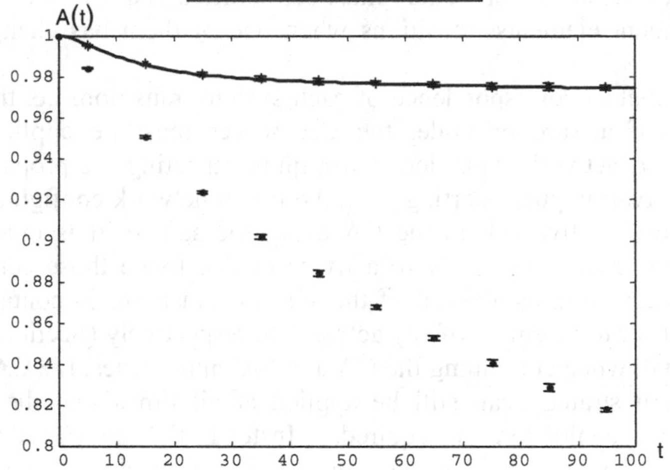

Fig. 15. Instantaneous availability A(t) of the network of Fig. 14. Solid line = analytical solution from minimal cut-sets; symbol *, with error bar of one standard deviation = MC-CA simulation with 10^5 MC trials of the network with only arcs failing; symbol •, with error bar of one standard deviation = MC-CA simulation with 10^5 MC trials of the network with also failing nodes

The analytical expression of the instantaneous availability A(t) can be obtained from the minimal cut-sets of the network. Fig. 15 shows the agreement of the analytical solution (solid line) with the results of the MC-CA approach with 10^5 MC trials (symbol *) with error bar of one standard deviation. The CPU time required was approximately 4 minutes in an Athlon 1400 MHz computer.

The significance of the use of the MC-CA procedure for the computation of the availability of a network system increases with the system complexity. When only connection arcs undergo stochastic transitions, in principle one can use the cut or path sets of the network to compute the analytical availability, with possibly intensive and computational burdensome methods. However, if the number of nodes and connections is large and/or if also the nodes can make transitions, as in realistic networks, the analytical procedures become impractical. It is in these cases that the MC-CA procedure can become a very powerful modeling and computational method.

The circles in Fig. 15 show, as an example, the instantaneous availability of the network of Fig. 14 when also the nodes can fail and be repaired, with constant rates equal to those of connection arc 31 (the more available one). The CPU time required was approximately 4min 30sec in an Athlon 1400 MHz computer.

5.6 Conclusions

CA are a general class of mathematical models, which are appealingly simple and yet capture a rich complexity of behavior of dynamical systems. They offer a significant computational potential due to their spatially and temporally discrete nature characterized by local interaction and an inherently parallel form of evolution.

MC sampling and simulation is a useful computational tool for reproducing several realizations of a system by sampling from known probability distributions the state of its components.

The combination of CA and MC is an attractive road for efficiently solving advanced network reliability problems. Besides the intuitive, physical simplicity of CA modeling and MC computing, their combination can enhance the performance of classical algorithms with respect to the reliability assessment of complex network systems and allow a straightforward parallel computing implementation.

The algorithms presented extend the simple CA approach previously developed for the *S-T connectivity* evaluation to the assessment of a network, with respect to the *all-terminal, k-terminal* and *maximum unsplittable flow* reliability and to its availability.

References

1. Adamatzky A.: "Computation of shortest path in cellular automata". Mathematical and Computer Modelling 23, 1996
2. Aggarwal K.K.: A simple method for reliability evaluation of a communication system. IEEE Trans Communication 1975; COM-23: 563-5.
3. Araújo F., Ribeiro B., Rodrígues L.: "A Neural Network for Shortest Path Computation", IEEE Transaction on Neural Networks, Vol. 12, N. 5, pp 1067-1072, Sep 2001
4. Avent T.: Availability evaluation of oil/gas production and transportation systems. Reliability Engineering and System Safety 1987;18:35-44.
5. Avent T.: Some considerations on reliability theory and its applications. Reliability Engineering and System Safety 1988;21:215-23.
6. Bar-Yam, Y, *Dynamics of Complex Systems*, Addison Wesley, 1997.
7. Billinton, R., Allan, R.N.: *Reliability evaluation of engineering systems, concepts and techniques,* 2nd ed. New York, Plenum Press, (1992).
8. Bulteau S., El Khadiri M. A Monte Carlo Simulation of the Flow Network Reliability using Importance and Stratified Sampling, Rapport de recherche No. 3122, Institut National de Recherche en Informatique et en Automatique, Rennes, France, 1997
9. Colbourn Ch.: *The Combinatorics of Network Reliability*, Oxford University Press, 1987
10. Dubi A, Monte Carlo Applications in Systems Engineering, Wiley, 1999
11. Evans J., Minieka E.: "Optimization Algorithms for Network and Graphs", Marcel Dekker Inc., New York, 1992
12. Fishman, G.: A Comparison of four Monte Carlo methods for estimating the probability of s-t connectedness. IEEE Trans Reliability 1986; R-35.
13. Henley EJ, Kumamoto H, Probabilistic Risk Assessment, IEEE Press, 1991
14. Jane CC, Lin JS, Yuan J.: Reliability evaluation of a limited-flow network in terms of MC sets. IEEE Trans Reliability 1993;R-42:354-61.
15. Kubat P.: Estimation of reliability for communication/computer networks simulation/analytical approach. IEEE Trans Communication 1989; 37:927-33.
16. Labeau P.E., Zio E., "*Procedures of Monte Carlo transport simulation for applications in system engineering*", Reliability Engineering and System Safety,**77** , 2002, pp 217-228
17. Lux I, Koblinger L, Monte Carlo particle transport methods: neutron and photon calculations, CRC Press, 1991
18. Manzi E., Labbé M., Latouche G., Maffioli F., Fishman's Sampling Plan for Computing Network Reliability, IEEE Trans Reliability 2001; R-50: 41-46.
19. Marseguerra M, Zio E, Basics of the Monte Carlo Method with Application to System Reliability. LiLoLe- Verlag GmbH (Publ. Co. Ltd.), 2002
20. Marseguerra M, Zio E, System Unavailability Calculations in Biased Monte Carlo Simulation: a Possible Pitfall, Annals of Nuclear Energy 2000, 27:1589-1605

21. Mérida E., Muñoz J., Dominguéz E.: "Implementación Neuronal del Algoritmo de Dijkstra", CAEPIA 2001, IX Conferencia de la Asociación Española para la Inteligencia Artificial, Gijon, Spain
22. Reingold, E., Nievergelt, J., Deo, N.: *Combinatorial algorithms: theory and practice*. New Jersey, Prentice-Hall (1982).
23. Rocco S., C.M., Moreno, J.A., Network reliability assessment using a cellular automata approach Reliability Engineering and System Safety 2002;78: 289 – 295.
24. Samad MA.: An efficient algorithm for simultaneously deducing MPs as well as cuts of a communication network. Microelectronic Reliability 1987; 27:437-41
25. Taha H.: Operations Research, An Introduction, Macmillan Publishing Co., Inc., New York, 1982
26. Wolfram, S, "Origins of randomness in physical systems", Phy. Rev. Lett. **55** (1985), 449-452.
27. Wolfram, S, ed., *Theory and Applications of Cellular Automata*, World Scientific (1986).
28. Yeh WC, Revised A.: Layered-network algorithm to search for all d-minpaths of a limited-flow acyclic network. IEEE Trans Reliability 1998; R-46:436-42
29. Yoo, YB, Deo, N. A comparison algorithm for terminal-pair reliability IEEE Trans Reliab 1988; R-37:

Appendix. Computation of the Time-dependent Availability of Network Systems with the MC-CA Approach

The basic algorithm for simulating M Monte Carlo trials proceeds as follows:

1. n = 0
2. While MC – iteration $n < M$,

    {

       $n = n + 1$ (cycle on MC trials)

       Sample by direct MC the transition times of the elements of the network

       Order the times of transitions which are smaller than the mission time in a schedule of system transitions, e.g. $t_1, t_2, ..., t_e, t_{e+1}$ on Figure A1

       While the time of transition is $\leq$ the mission time

       {

       Move the clock time of the system life to the next occurring transition in the schedule, for example identifying the generic t_e of Figure A1

Re-configure the system as a result of the occurred transition

Apply the CA algorithm for S-T connectivity to evaluate if there is a path from S to T, e.g. determining s_{te} (Figure A1)

If a path exists, then collect in the availability counters of the corresponding time channels the portion of time the system is available

}

}

3. Network availability at time t:

$$A(t) = \frac{\text{Content of the counter of the corresponding time channel}}{M * \text{time channel length}}$$

Note that, as said at step 8, in each time channel the portion of time during which the system is available is collected. This is done to account for the fact that, in general, the system transitions occur at time-points within the channels, e.g. t_e in Fig. A1, so that the unavailability contribution covers only a portion of the time channel.

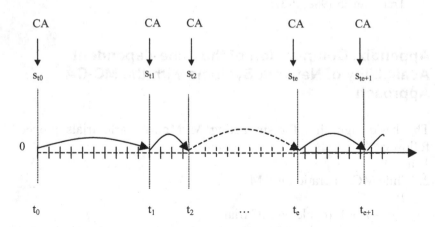

Fig. A1: A realization of the system life. The mission time is divided in time channels. The system configuration changes after each transition time of one of its elements and CA is used to evaluate the state of the re-configured system (i.e. available or unavailable).

Network Reliability Assessment through Empirical Models using a Machine Learning Approach

Claudio M. Rocco S.

Facultad de Ingeniería, Universidad Central Venezuela, Caracas

Marco Muselli

Istituto di Elettronica e di Ingegneria dell'Informazione e delle Telecomunicazioni, Consiglio Nazionale delle Ricerche, Genova, Italy

6.1 Introduction: Machine Learning (ML) Approach to Reliability Assessment

The reliability assessment of a system requires knowledge of how the system can fail, failure consequences and modeling, as well as selection of the evaluation technique [4].

For a reliability evaluation, almost all the systems are modeled using a Reliability Block Diagram (RBD), that is, a set of components that interact in some way to comply with the system purpose. System components are represented by blocks connected together either in series, in parallel, meshed or through a combination of these. For example, if the system failure occurs when all the components are failed, they are represented in a reliability network as a parallel set.

An RBD can be considered as an undirected or a directed connected graph. For example, in a communication system each node represents a communication center and each edge a transmission link between two such centers. It is assumed that each edge (link) functions independently of all other edges and that the edge operation probability is known.

The literature offers two main categor of techniques to evaluate the reliability of a system: analytical and simulation methods. The first one analyzes the topology of the equivalent graph to obtain a symbolic expres-

M.R.S. Claudio and M. Muselli: *Network Reliability Assessment through Empirical Models using a Machine Learning Approach*, Computational Intelligence in Reliability Engineering (SCI) **40**, 145–174 (2007)
www.springerlink.com

sion for the occurrence of a failure. On the other hand, the simulation approach allows a practical way of evaluating the reliability of the network in specific situations. In particular, when complex operating conditions are considered or the number of events is relatively large, Monte Carlo (MC) techniques offer a valuable way of evaluating reliability [4]. This approach is widely used when real engineering systems are to be analyzed.

In general, any reliability index can be obtained as the *expected value* of a System Function (*SF*) [14] or of an Evaluation Function (*EF*) [25] applied to a system state x (vector representing the state of each element in the network). This function determines whether a specific configuration corresponds to an operating state or a failed one [24], according to a specific criterion. For example, if connectivity between two particular nodes, s (the source) and t (the terminal) must be ensured, the system is operating if there exists at least a working path from the source node s to the terminal node t.

The corresponding reliability measure (*s-t* reliability) has been widely studied in the literature; in this case a depth-first procedure [23, 29] can be employed as an *EF*.

In other systems, for example in communication networks, the success criterion assumes that a network performs well if and only if it is possible to transmit successfully a specific required capacity. For these systems, the connectivity is not a sufficient condition for success, as it is also required that an adequate flow is guaranteed between s and t, taking into account the capacity of the links involved. In this case, the max-flow min-cut algorithm [23,29] can be adopted to evaluate if a given state is capable or not of transporting a required flow; alternatively, procedures based on the concept of composite paths [1, 28] can be used as the *EF*.

In general, the reliability assessment of a specific system requires the computation of performance metrics using special *EF*. An important characteristic of these metrics and their extensions is that the solution of an NP-hard problem [35] is needed for their evaluation in almost all the contexts of interest. In this situation MC techniques are used to estimate performance metrics. However, an MC simulation requires a large number of *EF* evaluations to establish any reliability indices with high computational effort. For this reason, it is convenient to approximate the *EF* using a Machine Learning (ML) method.

Two different situations can be identified: ML predictive methods reconstruct the desired SF through a black box device, whose functioning is not directly comprehensible. On the contrary, ML descriptive methods provide a set of intelligible rules describing the behavior of the *SF* for the system at hand. Support Vector Machines (SVM) is a widely used predictive method, successfully adopted in reliability assessment [30], whereas

Decision Trees (DT) [31] and Shadow Clustering (SC) [20] are two descriptive methods that are able to discover relevant properties for reliability analysis.

The chapter is organized as follows: In Sec. 2 some definitions are presented. Section 3 introduces the three machine learning methods considered for approximating the reliability of a network, while Sec. 4 compares the results obtained by each method for a specific network. Finally, Sec. 5 contains the conclusions.

Acronyms:

ARE	Approximate Reliability Expression
DT	Decision Tree
EF	Evaluation Function
HC	Hamming Clustering
ML	Machine Learning
RBD	Reliability Block Diagram
RE	Reliability Expression
SC	Shadow Clustering
SF	Structure Function
SVM	Support Vector Machine

6.2 Definitions

Consider a system S composed by several units interconnected by d links; the functioning of S directly depends on the state x_i, $i = 1, \ldots, d$, of each connection, which is viewed as an independent random variable assuming two possible values 1 and 0, associated with the operating and the failed condition, respectively. In particular, we have [3]:

$$x_i = \begin{cases} 1 \ (\text{operating state}) & \text{with probability } P_i \\ 0 \ (\text{failed state}) & \text{with probability } Q_i = 1 - P_i \end{cases} \quad (1)$$

where P_i is the probability of success of component (link) i.

The state of a system S containing d components is then expressed by a random vector $x = (x_1, x_2, \ldots, x_d)$, which uniquely identifies the functioning of S. Again, it can be operating (coded by the value 1) or failed (coded by 0). To establish if x leads to an operating or a failed state for S, we adopt a proper Evaluation Function (EF):

$$y = EF(x) = \begin{cases} 1 & \text{if the system is operating in state } x \\ 0 & \text{if the system is failed in state } x \end{cases} \quad (2)$$

If the criterion to be used for establishing reliability is simple connectivity, a depth-first procedure [23,29] can be employed as an *EF*. In the
case of capacity requirements, the *EF* could be given by the max-flow
min-cut algorithm [23, 29]. For other metrics, special *EF* may be used.
Since x and y include only binary values, the functional dependence between y and x is given by a Boolean function, called Structure Function
(*SF*) [10], which can be written as a logical sum-of-products involving the
component states x_i or their complements $\overline{x}_i$. Consider the system in Fig. 1
that contains four links.

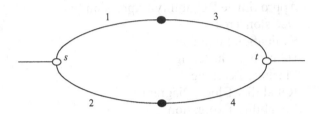

Fig. 1. A 4-components network

If the connectivity between the source node s and the terminal node t
must be ensured in an operating state for the network, the following *SF* is
readily obtained:

$$y = SF(x) = x_1x_3 + x_2x_4 \tag{3}$$

where the OR operation is denoted by '+' and the AND operation is denoted by '·'. Like for standard product among real numbers, when no con-
fusion arises the AND operator can be omitted.

The *reliability* of a system is defined as the expected value of its structure function [15]. When the *SF(x)* has the form of a logical sum-of-
products, a closed-form expression for the system reliability, called *Reliability Expression (RE)*, can be directly obtained by substituting in the
logical sum-of-products, according to (1) and to the independence of the
random variables x_i, every term x_i with P_i and every $\overline{x}_i$ with Q_i. After this

substitution logical sums and products must be changed into standard sums
and products among real numbers. For example, the *RE* deriving from the
SF(x) in (3) is $P_1P_3 + P_2P_4$, which gives the value of the system reliability
when substituting the actual values of P_i into this expression.

Since the Boolean expression for the *SF* of a system can be derived only
for very simple situations, it is important to develop methods that are able
to produce an estimate of the system reliability by examining a reduced
number of different system states x_j, $j = 1, ..., N$, obtained by as many ap-

plications of the *EF*. A possible approach consists in employing machine learning techniques, which are able to reconstruct an estimate of the system function $SF(x)$ starting from the collection of N states x_j, called in this case *training set*. Some of these techniques generate the estimate of the *SF* as a logical sum-of-products, which can be used to produce (through the simple procedure described above) an *Approximate Reliability Expression* (*ARE*) that is (hopefully) close to the actual *RE*.

In this case, the estimation of the system reliability can be easily performed, by substituting into the *ARE* the actual values of the P_i. On the other hand, when an approximation to the *SF* is available and cannot be written in the form of a logical sum-of-products, a standard Monte Carlo approach can be adopted to estimate the system reliability. By using the approximate *SF* to establish the state y instead of the *EF*, the computational cost is reduced.

6.3 Machine Learning Predictive Methods

In this section three different machine learning techniques are described: Support Vector Machines (SVM), Decision Trees (DT) and Shadow Clustering (SC). All these methods are able to solve two-class classification problems, where a decision function g: $\Re^d \rightarrow \{0,1\}$ have to be reconstructed starting from a collection of examples (x_1,y_1), (x_2,y_2), ..., (x_N,y_N). Every y_j is a (possibly noisy) evaluation of $g(x_j)$ for every $j = 1, ..., N$. Thus, they can be employed to reconstruct the *SF* of a system S when a set of N pairs (x_j,y_j), where x_j is a system state and $y_j = EF(x_j)$, is available.

In this case, DT and SC are able to generate a logical sum-of-products that approximates the *SF*. On the other hand, SVM produces a linear combination of real functions (Gaussians, polynomial, or others) that can be used to establish the state y associated to a given vector x. Consequently, SVM cannot be directly used to generate an ARE for the system at hand.

6.3.1 Support Vector Machines

In the last ten years Support Vector Machines (SVM) have become one of the most promising approach for solving classification problems [11,36]. Their application in a variety of fields, ranging from particle identification, face identification and text categorization to engine detection, bioinformatics and data base marketing, has produced interesting and reliable results, outperforming other widely used paradigms, like multilayer perceptrons and radial basis function networks.

For symmetry reasons, SVM generates decision functions $g : \Re^d \rightarrow$ 1,+1}; like other classification techniques, such as multilayer perceptrons or radial basis function networks, SVM constructs a real function $f : \Re^d \rightarrow \Re$ starting from the collection of N samples (x_j, y_j), then writing $g(x) = \text{sign}(f(x))$, being $\text{sign}(z) = +1$ if $z \geq 0$ and $\text{sign}(z) = -1$ otherwise.

The learning algorithm for SVM stems from specific results obtained in statistical learning theory and is based on the following consideration [36]: every classification problem can always be mapped in a high-dimensional input domain $\Re^D$ with $D \gg d$, where a linear decision function performs very well. Consequently, the solving procedure adopted by SVM amounts to selecting a proper mapping $\Phi : \Re^d \rightarrow \Re^D$ and a linear function $f(x) = w_0 + w \cdot \Phi(x)$, such that the classifier $g(x) = \text{sign}(w_0 + w \cdot \Phi(x))$ gives the correct output y when a new pattern x not included in the training set has to be classified.

Some theoretical results ensure that, once chosen the mapping Φ, the optimal linear function $f(x)$ is obtained by solving the following quadratic programming problem:

$$\min_{w_0, w, \xi} \frac{1}{2} w \cdot w + C \sum_{j=1}^{N} \xi_j$$
$$\text{subject to} \quad y_j(w_0 + w \cdot \Phi(x_j)) \geq 1 - \xi_j \tag{4}$$
$$\xi_j \geq 0 \quad \text{for every } j = 1, ..., N$$

where the variables ξ_j take account of possible misclassifications of the patterns in the training set. The term C is a regularization constant that controls the trade-off between the training error $\sum_j \xi_j$ and the regularization factor $w \cdot w$.

The parameters (w_0, w) for the linear decision function $f(x)$ can also be retrieved by solving the Lagrange dual problem:

$$\min_{\alpha} \frac{1}{2} \sum_{j=1}^{N} \sum_{k=1}^{N} \alpha_j \alpha_k y_j y_k \Phi(x_j) \cdot \Phi(x_k) - \sum_{j=1}^{N} \alpha_j$$
$$\text{subject to} \quad \sum_{j=1}^{N} \alpha_j y_j = 0 \tag{5}$$
$$0 \leq \alpha_j \leq C \quad \text{for every } j = 1, ..., N$$

which is again a quadratic programming problem where the unknowns α_j are the Lagrange multipliers for the original (primal) problem. The directional vector w can then be obtained by the solution α through the equation

$$w = \sum_{j=1}^{N} \alpha_j y_j \Phi(x_j) \tag{6}$$

It can be easily seen that there is a 1-1 correspondence between the scalars α_j and the examples (x_j, y_j) in the training set.

Now, the Karush-Kuhn-Tucker (KKT) conditions for optimization problems with inequality constraints [34], assert that in the minimum point of the problem at hand it must be

$$\alpha_j (y_j (w_0 + w \cdot \Phi(x_j)) - 1 + \xi_j) = 0 \tag{7}$$

Thus, for every $= 1, \ldots, N$ either the Lagrange multiplier α_j is null or the constraint $y_j (w_0 + w \cdot \Phi(x_j)) \geq 1 - \xi_j$ is satisfied with equality. It should be noted that only the points $_j$ with $\alpha_j > 0$ gives a contribution to the sum in (6); these points are called *support vectors*. If x_{j^-} and x_{j^+} are two support vectors with output +1 and −1, respectively, the bias w_0 for the function $f(x)$ is given by

$$w_0 = \frac{w \cdot \Phi(x_{j^+}) + w \cdot \Phi(x_{j^-})}{2} = \frac{\sum_{j=1}^{N} \alpha_j y_j \Phi(x_j) \cdot \Phi(x_{j^+}) + \sum_{j=1}^{N} \alpha_j y_j \Phi(x_j) \cdot \Phi(x_{j^-})}{2}$$

The real function $f(x)$ is then completely determined if the mapping $\Phi: \Re^d \rightarrow \Re^D$ is properly defined according to the peculiarities of the classification problem at hand. However, if the dimension D of the projected space is very high, solving the quadratic programming problem (4) or (5) requires a prohibitive computational cost.

A possible way of getting around this problem derives form the observation that both in the optimization problem (5) and in the expression for w_0 always appears the inner product between two instances of the function Φ. Therefore, it is sufficient to define a *kernel* function $K: \Re^d \times \Re^d \rightarrow \Re^+$ that implements the inner product, i.e. $K(u,v) = \Phi(u) \cdot \Phi(v)$. This allows control of the computational cost of the solving procedure, since in this way the dimension D of the projected space is not explicitly considered. As the kernel function K gives the result of an inner product, it must be always non negative and symmetric. In addition, specific technical constraints, described by the Mercer's theorem [38], have to be satisfied to guarantee consistency.

Three typical choices for (u,v) are [36]:

the linear kernel $K(u,v) = u \cdot v$

the Gaussian radial basis kernel (GRBF) $K(u,v) = \exp\left(-\|u-v\|^2/2\sigma^2\right)$

the polynomial kernel $K(\boldsymbol{u},\boldsymbol{v}) = (\boldsymbol{u}\cdot\boldsymbol{v}+1)^p$

where the parameters σ and p are to be chosen properly.

By substituting in (5) the kernel function K we obtain the following quadratic programming problem:

$$\min_{\alpha} \frac{1}{2}\sum_{j=1}^{N}\sum_{k=1}^{N}\alpha_j\alpha_k y_j y_k K(\boldsymbol{x}_j,\boldsymbol{x}_k) - \sum_{j=1}^{N}\alpha_j$$

$$\text{subject to } \sum_{j=1}^{N}\alpha_j y_j = 0$$

$$0 \le \alpha_j \le C \qquad \text{for every } j = 1,...,N$$

which leads to the decision function $g(\boldsymbol{x}) = \text{sign}(f(\boldsymbol{x}))$, being

$$f(\boldsymbol{x}) = w_0 + \sum_{j=1}^{N}\alpha_j y_j K(\boldsymbol{x}_j,\boldsymbol{x}) \tag{8}$$

Again, only support vectors with $\alpha_j > 0$ gives a contribution to the summation above; for this reason classifiers adopting (8) are called *Support Vector Machines* (*SVM*). The average number of support vectors for a given classification problem is strictly related to the generalization ability of the corresponding SVM: the lower is the average number of support vectors, the higher is the accuracy of the classifier $g(\boldsymbol{x})$.

The bias w_0 in (8) is given by:

$$w_0 = \frac{1}{2}(\sum_{j=1}^{N}\alpha_j y_j K(\boldsymbol{x}_j,\boldsymbol{x}_{j^+}) + \sum_{j=1}^{N}\alpha_j y_j K(\boldsymbol{x}_j,\boldsymbol{x}_{j^-}))$$

where again $\boldsymbol{x}_{j^-}$ and $\boldsymbol{x}_{j^+}$ are two support vectors with output (+1) and (−1), respectively.

The application of SVM to the problem of estimating the system function $SF(\boldsymbol{x})$ starting from a subset of possible system states can be directly performed by employing the above general procedure to obtain a good approximation for the SF. Since SF is a Boolean function we must substitute the output value $y = -1$ in place of $y = 0$ to use the standard training procedure for SVM.

The choice of the kernel is a limitation of the SVM approach. Some work has been done on selecting kernels using prior knowledge [7]. In any case, the SVM with lower complexity should be preferred. Our experience in the reliability field has confirmed the good quality of the GRBF kernel having parameter $(1/2\sigma^2) = 1\mathcal{U}$, as suggested in [9].

For example, consider the system shown in Fig. 1, whose component and system states are listed in Tab. 1, if a continuity criterion is adopted. As it is usually the case in a practical application, suppose that only a sub-set of the whole collection of possible states (shown in Tab. 2) is available.

Table 1. Component and system states for the network shown in Fig. 1

x_1	x_2	x_3	x_4	$y = EF\ (x)$
0	0	0	0	0
0	0	0	1	0
0	0	1	0	0
0	0	1	1	0
0	1	0	0	0
0	1	0	1	1
0	1	1	0	0
0	1	1	1	1
1	0	0	0	0
1	0	0	1	0
1	0	1	0	1
1	0	1	1	1
1	1	0	0	0
1	1	0	1	1
1	1	1	0	1
1	1	1	1	1

Table 3 shows the support vectors obtained using the linear kernel and the LIBSVM software [9]. Note that in this case only 6 support vectors are derived. These support vectors are able to completely separate the training set. However, when the model is applie the test set (states from Tab. 1 not included in Tab. 2), only 6 out of 8 states are correctly classified, as shown in Table 4.

From Tab. 3 it is clear that support vectors can not be easily interpreted, since the expression generated does not correspond to a logical sum-of-products.

Table 2. Available subset of states for the network shown in Fig. 1 (training set)

x_1	x_2	x_3	x_4	$y = EF(x)$
0	0	0	1	0
0	0	1	1	0
0	1	1	1	1
1	0	0	1	0
1	0	1	1	1
1	1	0	0	0
1	1	0	1	1
1	1	1	1	1

Table 3. Support vectors for the available subset of states shown in Tab. 2

x_1	x_2	x_3	x_4	$y = EF(x)$
0	0	0	1	0
0	0	1	1	0
0	1	1	1	1
1	0	0	1	0
1	0	1	1	1
1	1	0	0	0
1	1	0	1	1
1	1	1	1	1

Table 4. SVM estimation for the test set

x_1	x_2	x_3	x_4	$y = EF(x)$	SVM estimate
0	0	0	0	0	0
0	0	1	0	0	0
0	1	0	0	0	0
0	1	0	1	1	0
0	1	1	0	0	0
1	0	0	0	0	0
1	0	1	0	1	0
1	1	1	0	1	1

6.3.2 Decision Trees

Decision tree based methods represent a non-parametric approach that turns out to be useful in the analysis of large data sets for which complex data structures may be present [2,5,27]. A DT solves a complex problem by dividing it into simpler sub-problems. The same strategy is recursively applied to each of these sub-problems.

A DT is composed of nodes, branches and terminal nodes (leaves). For our network problem, every node is associated with a component of the network. From each node start two branches, corresponding to the operating or failed state of that component. Finally, every terminal node represents the network state: operating or failed. Many authors use the convention to draw the false branch on the left side of the node and the true branch on the right.

Consider for example the DT shown in Fig. 2 and suppose a new system state is presented for classification. At the root node the state of the component x_2 is checked: if it is failed, the left branch is chosen and a new test on component x_1 is performed. Again, if x_1 is failed, the left branch is chosen and $y = 0$ is concluded.

Even if it may seem reasonable to search for the smallest tree (in terms of numbers of nodes) that perfectly classifies training data, there are two problems:
1) its generation requires the solution of an NP-hard problem and
2) it is not guaranteed that this tree performs well on a new test sample.

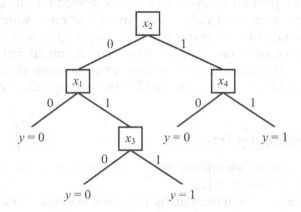

Fig. 2. Example of a decision tree

For this reason DT methods usually exploit heuristics that locally perform a one-step look-ahead search, that is, once a decision is taken it is never reconsidered. However, this heuristic search (hill-climbing without backtracking) may be stuck in a local optimal solution. On the other hand, this strategy allows building decision trees in a computation time that increases linearly with the number of examples [26].

A DT can be used to derive a collection of intelligible rules in the form **if-then**. It is sufficient to follow the different paths that connect the root to the leaves: every node encountered is converted into a condition to be added to the **if** part of the rule. The **then** part corresponds to the final leaf: its output value is selected when all the conditions in the **if** part are satisfied.

Since the tree is a directed acyclic graph, the number of rules that can be extracted from a DT is equal to the number of terminal nodes. As an example, the tree in Fig. 2 leads to three rules for the output $y = 0$ and two rules for $y = 1$.

All the conditions in the **if** part of a rule are connected by a logical AND operation; different rules are considered as forming an **if-then-else** structure. For example, the problem in Fig. 1 is described by the following set of rules:

if $x_1 = 0$ AND $x_2 = 0$ **then** $y = 0$
else if $x_1 = 0$ AND $x_4 = 0$ **then** $y = 0$
else if $x_2 = 0$ AND $x_3 = 0$ **then** $y = 0$
else if $x_3 = 0$ AND $x_4 = 0$ **then** $y = 0$
else $y = 1$

which is equivalent to the *SF* for this network $SF(x) = x_1x_3 + x_2x_4$.

Since all the possible output values are considered for rule generation, a complex decision tree can yield a very large set of rules, which is difficult to be understood. To recover this problem, proper optimization procedures have been proposed in the literature, which aim at simplifying the final set of rules. Several different tests have shown that in many situations the resulting set of rules is more accurate than the corresponding decision tree [26].

6.3.2.1 Building the Tree

In general, different algorithms use a *top-down induction* approach for constructing decision trees [22]:
1. If all the examples in the training set T belong to one class, then halt.
2. Consider all the possible tests that divide T into two or more subsets. Employ a proper measure to score how well each test splits up the examples in T.
3. Select the test that achieves the highest score.
4. Divide T into subsets according to the selected test. Run this procedure recursively by considering each subset as the training set T.

For the problem at hand, a test on a component state with two possible values will produce at most two child nodes, each of which corresponds to a different value. The algorithm considers all the possible tests and chooses the one that optimizes a pre-defined goodness measure.

Since small trees lead to simpler set of rules and to an increase in performance, the procedure above is performed by searching for tests that best separates the training set T. To achieve this goal, the most predictive components are considered at Step 3 [5,27].

6.3.2.2 Splitting Rules

Several methods have been described in the literature to measure how effective is a split, that is how good is a component attribute (operating or failed) to discriminate the system state. The most used are:

1. Measures depending on the difference between the training set T and the subsets obtained after the splitting; a function of the class proportion, e.g. the entropy, is typically employed.
2. Measures related to the difference between the subsets generated by the splitting; a distance or an angle that takes into account the class proportions is normally used.
3. Statistical measures of independence (typically a χ^2) between the subsets after the splitting and the class proportions.

In this paper the method used by C4.5 [27] is considered; it adopts the information gain as a measure of the difference between the training set T and the subsets generated by the splitting. Let p be the number of operating states and n the number of failed states included in the training set T. The entropy $E(p,n)$ of T is defined as:

$$E(p,n) = -\frac{p}{p+n}\log(\frac{p}{p+n}) - \frac{n}{p+n}\log(\frac{n}{p+n}) \tag{9}$$

Suppose the component x_j is selected for adding a new node to the DT under construction; if the test on attribute x_j leads to a splitting of T in k disjoint subsets, the average entropy $E_j(p,n)$ after the splitting is given by

$$E_j(p,n) = \sum_{i=1}^{k} \frac{p_i + n_i}{p+n} E(p_i,n_i) \tag{10}$$

where p_i and n_i are the number of instances from each class in the ith subset. Note that in our case $k = 2$ since every component has only two possible states (operating or failed).

The information gain $I_j(p,n)$ is then given by the difference between the values of the entropy before and after the splitting produced by the attribute x_j:

$$I_j(p,n) = E(p,n) - E_j(p,n) \tag{11}$$

At Step 3 of the DT procedure the component x_j that scores the maximum information gain is selected; a test on that component will divide the training set into $k = 2$ subsets.

For example, consider the system shown in Fig. 1 and the training set shown in Tab. 2. There are $p = 4$ operating states and $n = 4$ failed states; thus, the entropy $E(4,4)$ assumes the value:

$$E(4,4) = -4/(4+4)\log[4/(4+4)] - 4/(4+4)\log[4/(4+4)] = 1$$

Looking at the class proportion after the splitting we note that for $x_1 = 0$ there are one system operating state and two failed states, whereas for $x_1 = 1$ there are three system operating states and two failed states. Thus:

$$E(1,2) = 0.918296, \ E(3,2) = 0.970951$$
$$E_1(4,4) = 3/8 \cdot E(1,2) + 5/8 \cdot E(3,2) = 0.951205$$
$$I_1(4,4) = E(4,4) - E_1(4,4) = 1 - 0.951205 = 0.048795$$

Now, for $x_2 = 0$ and $x_3 = 0$ there are one system operating state and three failed states, whereas for $x_2 = 1$ and $x_3 = 1$ there are three system operating states and one failed state. Consequently, we have:

$$E(1,3) = E(3,1) = 0.811278$$
$$E_2(4,4) = E_3(4,4) = 4/8 \cdot E(1,3) + 4/8 \cdot E(3,1) = 0.811278$$
$$I_2(4,4) = I_3(4,4) = 1 - 0.811278 = 0.188722$$

Finally, the fourth component x_4 presents only one system failed state for $x_4 = 0$, whereas for $x_4 = 1$ we detect four system operating states and three failed states. Consequently, we obtain:

$$E(0,1) = 0 \ , \ E(4,3) = 0.985228$$
$$E_4(4,4) = 1/8 \cdot E(0,1) + 7/8 \cdot E(4,3) = 0.862075$$
$$I_4(4,4) = 1 - 0.862075 = 0.137925$$

Since both x_2 and x_3 score the maximum information gain, one of them must be considered for the first node of the DT. Suppose that the second component x_2 is selected as the root node. It is important to note that, in general, the component chosen for the first node holds a primary importance [2].

After this choice the training set in Tab. 2 is split into the following two subsets:

x_1	x_2	x_3	x_4	$y = EF(x)$	x_1	x_2	x_3	x_4	$y = EF(x)$
0	0	0	1	0	0	1	1	1	1
0	0	1	1	0	1	1	0	0	0
1	0	0	1	0	1	1	0	1	1
1	0	1	1	1	1	1	1	1	1

The former contains the examples with $x_2 = 0$ and the latter those with $x_2 = 1$. If we repeat the procedure for the addition of a new node by considering the first subset as T, we obtain:

$$I_1(1,3) = I_3(3,1) = 0.311278 \ , \quad I_4(3,1) = 0$$

Consequently, the highest information gain is achieved with the choice of x_1 or x_3; the same procedure allows to select the component x_4 for the second subset, which yields, after a further splitting, the DT in Fig. 2. A direct inspection of the DT allows generating the following set of rules:

if $x_2 = 1$ AND $x_4 = 0$ **then** $y = 0$
else if $x_1 = 0$ AND $x_2 = 0$ **then** $y = 0$
else if $x_1 = 1$ AND $x_2 = 0$ AND $x_3 = 0$ **then** $y = 0$
else $y = 1$

Although this set of rules correctly classifies all the system configurations in Tab. 2, it is not equivalent to the desired system function $SF(x) = x_1x_3 + x_2x_4$. This means that the DT procedure is not able to recover the lack of information deriving from the absence of eight feasible system configurations, reported in Tab. 1 and not included in the training set.

6.3.2.3 Shrinking the Tree

The splitting strategy previously presented relies on a measure of the information gain based on the examples included in the available training set. However, the size of the subset analyzed to add a new node to the DT decreases with the depth of the tree. Unfortunately, estimates based on small samples will not produce good results for unseen cases, thus leading to models with poor predictive accuracy, which is usually known as overfitting problem [13]. As a consequence, small decision trees consistent with the training set tend to perform better than large trees, according to the Occam's Razor principle [13].

The standard approach followed to take into account these considerations amounts to *pruning* branches off the DT. Two general groups of pruning techniques have been introduced in the literature: 1) pre-pruning methods that stop building the tree when some criteria is satisfied, and 2) post-pruning methods that first build a complete tree and then prune it back. All these techniques decide if a branch is to be pruned by analyzing the size of the tree and an estimate of the generalization error; for implementation details, the reader can refer to [26,27].

It is interesting to note that in the example presented in section 3.2.2 neither nodes nor branches can be removed from the final decision tree in Fig. 2 without degrading significantly the accuracy on the examples of the training set. In this case the pruning phase has no effect.

6.3.3 Shadow Clustering (SC)

As one can note, every system state x can be associated with a binary string of length d: it is sufficient to write the component states in the same order as they appear within the vector x.

For example, the system state $x = (0, 1, 1, 0, 1)$ for $d = 5$ will correspond to the binary string 01101. Since also the variable y, denoting if the considered system is operating or failed, is Boolean, at least in principle, any technique for the synthesis of digital circuits can be adopted to reconstruct the desired SF from a sufficiently large training set $\{(x_j, y_j), j = 1, \ldots, N\}$. Unfortunately, the target of classical techniques for Boolean function reconstruction, such as MINI [17], ESPRESSO [6], or the Quine-McCluskey method [12], is to obtain the simplest logical sum-of-products that correctly classifies all the examples provided. As a consequence, they do not generalize well, i.e. the output assigned to a binary string not included in the given training set can be often incorrect.

To recover this drawback a new logical synthesis technique, called Hamming Clustering (HC) [18,19] has been introduced. In several application problems HC is able to achieve accuracy values comparable to those of best classification methods, in terms of both efficiency and efficacy. In addition, when we are facing with classification problem the Boolean function generated by HC can be directly converted into a set of intelligible rules underlying the problem at hand.

Nevertheless, the top-down approach adopted by HC to generate logical products can require an excessive computational cost when the dimension d of the input vector is very high, as it can be the case in the analysis of system reliability. Furthermore, HC is not well suited for classification problems that cannot be easily coded in a binary form. To overcome this shortcoming, an alternative method, named Shadow Clustering (SC) [20], has been proposed. It is essentially a technique for the synthesis of monotone Boolean functions, writable as a logical sum-of-products not containing the complement (NOT) operator.

The application of a proper binary coding allows the treatment of general classification problems; the approach followed by SC, which resembles the procedure adopted by HC, leads to the generation of a set of intelligible rules underlying the given classification problem. Preliminary tests [21] show that the accuracy obtained by SC is significantly better than that achieved by HC in real world situations.

Since the system function may not be a monotone Boolean function, an initial coding β is needed to transform the training set, so as a logical sum-of-products not including the complement operator can be adopted for realizing the $SF(x)$. A possible choice consists in using the coding $\beta(x)$ that produces a binary string z with length $2d$, where every component x_i in x gives rise to two bits z_{2i-1} and z_{2i} according to the following rule:

$$z_{2i-1} = 1, z_{2i} = 0 \quad \text{if } x_i = 0, \text{ whereas } z_{2i-1} = 0, z_{2i} = 1 \quad \text{if } x_i = 1$$

It can be easily seen that this coding maps any binary training set into a portion of the truth table for a monotone Boolean function, which can then be reconstructed through SC. A basic concept in the procedure followed by

SC is the notion of *cluster*, sharing the same definition *iofplicant*   in classic theory of logical synthesis.

A cluster is the collection of all the binary strings having the value 1 in a same fixed subset of components. As an example, the eight binary strings 01001, 01011, 01101, 11001, 01111, 11011, 11101, 11111 form a cluster since all of them only have the value 1 in the second and in the fifth component. This cluster is usually written as 01001, since in the synthesis of monotone Boolean functions the value 0 serves as a don't care symbol and is put in the positions that are not fixed. Usually the cluster 01001 is said *to becovered* by the eight binary strings mentioned above.

Every cluster can be associated with a logical product among the components of x, which gives output 1 for all and only the binary strings which cover that cluster. For example, the cluster 01001 corresponds to the logical product x_2x_5, obtained by considering only the components having the value 1 in the given cluster. The desired monotone Boolean function can then be constructed by generating a valid collection of clusters for the binary strings in the training set with output 1. This collection is consistent, if none of its elements is covered by binary strings of the training set having output 0.

After the application of the binary coding β on the examples (x_j, y_j) of the training set, we have obtained a new collection of input-output pairs (z_j, y_j) with $z_j = \beta(x_j)$, which can be viewed as a portion of the truth table of a monotone Boolean function. If T and F contain the binary strings z_j with corresponding output $y_j = 1$ and $y_j = 0$, respectively, the procedure employed by SC to reconstruct the sum-of-products expression for the desired monotone Boolean function $f(z)$ consists of the following four steps:

1. Set $S = T$ and $C = \varnothing$.
2. Starting from the implicant $000\cdots0$, turn some 0 in 1 to obtain a cluster c that is covered by the greatest number of binary strings in   and by no element of F.
3. Add the cluster c to the set C. Remove from S all the binary strings that cover c. If S is not empty go to Step 2.
4. Simplify the collection C of clusters and build the corresponding monotone Boolean function.

As one can note, SC generates the sum-of-products expression for the desired monotone Boolean function f (Step 4) by examining a collection C of clusters incrementally built through the iteration of Steps 2–3. To this aim, SC employs an auxiliary set S to maintain the binary strings of T that do not cover any cluster in the current collection C.

The following subsections describe the solutions adopted by SC to construct the clusters to be added in C (Step 2) and to simplify the final collection C (Step 4), thus improving the generalization ability of the resulting monotone Boolean function.

6.3.3.1 Building Clusters

Starting from the largest cluster $0\cdots00$, containing only 0 values, an implicant c has to be generated for its inclusion in the collection C. The only prescription to be satisfied in constructing this cluster is that it cannot be covered by any binary string in F.

As suggested by the Occam's Razor principle, smaller sum-of-products expressions for the monotone Boolean function to be retrieved perform better; this leads to prefer clusters that are covered by as many as possible training examples in S and contain more don't care values 0 inside them.

However, searching for the optimal cluster in this sense leads to an NP-hard problem; consequently, greedy alternatives must be employed to avoid an excessive computing time. In these approaches an iterative procedure changes one at a time the components with value 0 in the cluster under construction, until no elements of F cover the resulting implicant c.

Every time a bit in c is changed from 0 to 1 a (possibly empty) subset R_i of binary strings in S do not cover anymore the new implicant. The same happens for a (possibly empty) subset $G_i \subset F$.

It can be easily seen that the subset R_i contains all the elements in S that cover c and has a value 0 in the ith component; likewise, the subset G_i includes all and only the binary strings in F covering c and having a 0 as the ith bit.

It follows that a greedy procedure for SC must minimize the cardinality $|R_i|$ of the subset R_i, while maximizing the number of elements in G_i, when the ith bit of c is set. In general, it is impossible to satisfy both these prescription; thus, it is necessary to privilege one of them over the other.

Trials on artificial and real-world classification problems suggest that the most promising choice consists in privileging the minimization of $|R_i|$, which leads to the Maximum-covering version of SC (MSC) [21]. Here, at every iteration the cardinality of the sets R_i and G_i is computed for every component i of c having value $c_i = 0$. Then the index i^* that minimizes $|R_i|$ is selected; ties are broken by taking the maximum of $|G_i|$ under the same value of $|R_i|$.

As an example, consider the network in Fig. 1, having system function $SF(x) = x_1x_3 + x_2x_4$, and the training set shown in Tab. 2. It can be noted that in this case the SF is a monotone Boolean function and therefore can be reconstructed by SC without recurring to the preliminary coding β. However, to illustrate the general procedure followed by SC, we apply anyway the mapping β, thus obtaining the binary training set (z_j, y_j) in Tab. 5.

Table 5. Binary training set for the system in Fig. 1 obtained by applying the coding β.

z_1	z_2	z_3	z_4	z_5	z_6	z_7	z_8	$y = SF(x)$
1	0	1	0	1	0	0	1	0
1	0	1	0	0	1	0	1	0
1	0	0	1	0	1	0	1	1
0	1	1	0	1	0	0	1	0
0	1	1	0	0	1	0	1	1
0	1	0	1	1	0	1	0	0
0	1	0	1	1	0	0	1	1
0	1	0	1	0	1	0	1	1

It can be directly obtained that the set T contains the strings 10010101, 01100101, 01011001, and 01010101, whereas F includes 10101001, 10100101, 01101001, and 01011010. Starting at Step 2 of SC with the generic implicant 00000000, we compute the cardinalities $|R_i|$ and $|G_i|$ for $i = 1, \ldots, 8$, thus obtaining:

$$|R_1| = 3, |R_2| = 1, |R_3| = 3, |R_4| = 1, |R_5| = 3, |R_6| = 1, |R_7| = 4, |R_8| = 0$$
$$|G_1| = 2, |G_2| = 2, |G_3| = 1, |G_4| = 3, |G_5| = 1, |G_6| = 3, |G_7| = 3, |G_8| = 1 \quad (12)$$

Then, $|R_i|$ is maximized for $i = 8$; by changing the eight bit from 0 to 1, we obtain the cluster 00000001. At this time the subsets R_i for $i = 1, \ldots, 7$ remain unchanged, whereas the cardinalities of the subsets G_i are

$$|G_1| = 1, |G_2| = 2, |G_3| = 0, |G_4| = 3, |G_5| = 1, |G_6| = 2, |G_7| = 3$$

Now, the minimum value of $|R_i| = 1$ is obtained for $i = 2, 4, 6$, but the maximization of $|G_i|$ suggests to change from 0 to 1 the fourth bit, thus obtaining the cluster $c = 00010001$. Since this implicant is not covered by any element of F, it can be inserted into C (Step 3). Then the set S is reduced by removing from it the binary strings that cover c, namely 10010101, 01011001, and 01010101; it follows that $= \{01100101\}$.

The procedure is then repeated at Step 2, by considering again the generic implicant 00000000 and by computing the cardinalities $|R_i|$ for $i = 1, \ldots, 8$. We obtain:

$$|R_1| = 1, |R_2| = 0, |R_3| = 0, |R_4| = 1, |R_5| = 1, |R_6| = 0, |R_7| = 1, |R_8| = 0$$

Note that the subsets G_i in (12) are not changed since F was not altered. It is immediately seen that the best choice corresponds to $i = 6$; changing the corresponding bit from 0 to 1 yields the cluster 00000100. The cardinalities $|G_i|$ now become

$$|G_1| - 0, |G_2| - 1, |G_3| - 0, |G_4| = 1, |G_5| = 1, |G_7| = 1, |G_8| = 0$$

Consequently, the index $i = 2$ is selected, thus leading to the implicant c = 01000100 to be inserted into C.

Removing at Step 3 the last element from S, which covers c, we obtain that S becomes empty and the execution of SC follows at Step 4 with the simplification of the resulting collection $C = \{00010001, 01000100\}$.

6.3.3.2 Simplifying the Collection of Clusters

Usually, the repeated execution of Steps 2-3 leads to a redundant set of clusters, whose simplification can improve the prediction accuracy of the corresponding monotone Boolean function. In analogy with methods for decision trees, the techniques employed to reduce the complexity of the resulting sum-of-products expressions are frequently called *pruning algorithms*.

The easiest effective way of simplifying the set of clusters produced by SC is to apply the *minimal pruning* [19,21]: According to this greedy technique the clusters that is covered by the maximum number of elements in T are extracted one at a time. At each extraction, only the binary strings not included in the clusters already selected are considered. Breaks are tied by examining the whole covering.

The application of minimal pruning to the example analyzed in the previous subsection begins with the computation of the covering associated with each of the two clusters generated in the training phase. It can be readily observed that 00010001 covers three examples of Tab. 3 (precisely the binary strings 10010101, 01011001 and 01010101), whereas the covering of 01000100 is equal to 2. Consequently, the cluster 00010001 is firstly selected.

After this choice only the binary string 01100101 does not cover any implicant, which leads to the selection of the second cluster 01000100. No simplification is then possible in the collection C, which leads to the monotone Boolean function $z_4 z_8 + z_2 z_6$. By applying in the opposite way the coding β we obtain the desired expression for the system function $SF(x) = x_2 x_4 + x_1 x_3$, i.e. the correct SF for the system in Fig. 1.

6.4 Example

To evaluate the performance of thethods presented in the previous sections, the network shown in Fig. 3 has been considered [39]. It is assumed that each link has reliability P_i and capacity of 100 units. A system failure occurs when the flow at the terminal node t falls below 200 units. Consequently, a max-flow min-cut algorithm is used to establish the value of the *EF* [23,29].

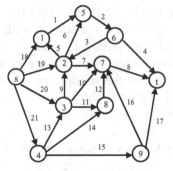

Fig. 3. Network to be evaluated [39]

In order to apply a classification method it is first necessary to collect a set of examples (x,y), where $y = EF(x)$, to be used in the training phase and in the subsequent performance evaluation of the resulting set of rules. To this aim, 50000 system states have been randomly selected without replacement and for each of them the corresponding value of the *EF* has been retrieved.

To analyze how the size of the training set influences the quality of the solution provided by each method, 13 different cases were analyzed with 1000, 2000, 3000, 4000, 5000, 6000, 7000, 8000, 9000, 10000, 15000, 20000 and 25000 examples in the training set. These examples were randomly extracted with uniform probability from the whole collection of 50000 system states; the remaining examples were then used to test the accuracy of the model produced by the machine learning technique. An average over 30 different choices of the training set for each size value was then performed to obtain statistically relevant results.

The performance of each model is evaluated using standard measures of *sensitivity, specificity* and *accuracy* [37]:

$$\text{sensitivity} = TP/(TP+FN); \quad \text{specificity} = TN/(TN+FP);$$

$$\text{accuracy} = (TP+TN)/(TP+TN+FP+FN)$$

where

- TP (resp. TN) is the number of examples belonging to the class $y = 1$ (resp. $y = 0$) for which the classifier gives the correct output,
- FP (resp. FN) is the number of examples belonging to the class $y = 1$ (resp. $y = 0$) for which the classifier gives the wrong output.

For reliability evaluation, *sensitivity* gives the percentage of correctly classified operating states and *specificity* provides the percentage of correctly classified failed states.

6.4.1 Performance Results

Different kernels were tried when generating the SVM model and it was found that the best performance is achieved with a Gaussian radial basis function (GRBF) kernel with parameter $(1/2\sigma^2) = 0.05$. All SVM models obtained are able to completely separate the corresponding training set. The optimization required in the training phase was performed using the LIBSVM software [9]. Table 6 shows the average performance indices during the testing phase.

Table 6. Average performance indices for SVM (test)

Training Set Size	Accuracy %	Sensitivity %	Specificity %
1000	95.12	93.93	95.61
2000	96.90	96.96	96.87
3000	97.63	97.93	97.51
4000	98.09	98.50	97.92
5000	98.42	98.84	98.25
6000	98.66	99.09	98.48
7000	98.83	99.27	98.64
8000	98.97	99.41	98.79
9000	99.07	99.50	98.90
10000	99.17	99.59	98.99
15000	99.48	99.81	99.35
20000	99.64	99.88	99.53
25000	99.73	99.93	99.64

Table 7. Average performance indices for DT

Training Set Size	Accuracy (%)		Sensitivity (%)		Specificity (%)	
	Training	Test	Training	Test	Training	Test
1000	98.71	95.65	98.26	92.67	98.90	96.91
2000	99.25	97.30	98.89	95.48	99.40	98.08
3000	99.42	98.06	99.15	96.52	99.54	98.71
4000	99.54	98.63	99.44	97.72	99.59	99.01
5000	99.63	98.94	99.51	98.17	99.68	99.27
6000	99.69	99.10	99.58	98.45	99.73	99.38
7000	99.74	99.23	99.67	98.70	99.77	99.45
8000	99.80	99.37	99.71	98.89	99.83	99.57
9000	99.81	99.42	99.75	98.94	99.84	99.63
10000	99.83	99.47	99.75	99.02	99.86	99.66
15000	99.90	99.69	99.85	99.42	99.92	99.80
20000	99.93	99.80	99.88	99.57	99.96	99.89
25000	99.95	99.85	99.92	99.71	99.97	99.92

As for DT, the resulting models do not classify correctly all the examples in the training set. Table 7 presents the average performance indices during training and testing. Finally, Table 8 shows the average performance indices obtained by SC only during testing, since also SC does not commit errors on the system states of the training set.

Table 8. Average performance indices for SC (test)

Training Set Size	Accuracy %	Sensitivity %	Specificity %
1000	96.30	94.52	97.05
2000	98.07	97.34	98.38
3000	98.67	98.11	98.91
4000	99.01	98.64	99.16
5000	99.26	99.00	99.37
6000	99.45	99.32	99.50
7000	99.54	99.43	99.58
8000	99.60	99.49	99.64
9000	99.65	99.54	99.69
10000	99.69	99.59	99.73
15000	99.84	99.80	99.85
20000	99.89	99.89	99.90
25000	99.92	99.92	99.92

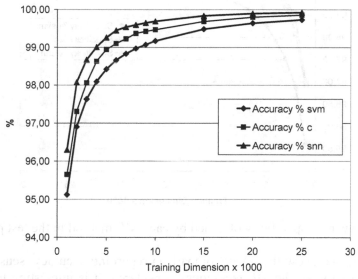

Fig. 4. Average accuracy obtained by each ML method in the test phase

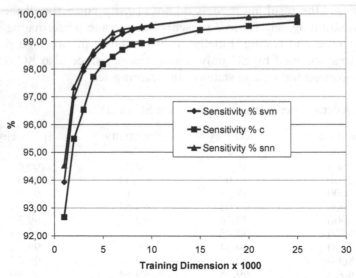

Fig. 5. Average sensitivity obtained by each ML method in the test phase

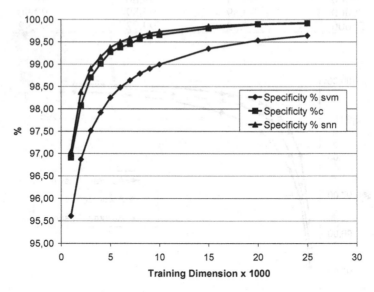

Fig. 6. Average specificity obtained by each ML method in the test phase

Figures 4–6 show the result comparison regarding accuracy, sensitivity and specificity for the ML techniques considered. It is interesting to note that the index under study for each model increases with the size of the training set. SC has the best behavior for all the indices. However, for the sensitivity index, the performances of SC and SVM are almost equal. For the specificity index, the performance of SC and DT are almost equal.

This means that SC seems to be more stable when considering the three indices simultaneously. In [33] different networks are evaluated using different *EF*: the behavior obtained is similar to the one observed in the network analyzed in this chapter.

6.4.2 Rule Extraction Evaluation

As previously mentioned, DT and SC are able to extract rules that explain the behavior of the systems in the form of a logical sum-of-products approximating the *SF*. DT rules are in disjoint form, so the *ARE* can be easily determined. Rules generated by SC are not disjoint; thus, an additional procedure, such as the algorithm KDH88 [16], has to be used to perform this task.

Table 9 shows the average number of paths and cuts generated by both procedures. As can be seen in Fig. 7, both techniques are able to extract more and more path and cut sets as long as the training set is increased (the system under study has 43 minimal paths and 110 minimal cuts). However, for a given training set size, SC can produce more path and cut sets than DT.

Table 9. Average number of paths and cuts extracted by DT and SC

Training Set Size	PATHS		CUTS	
	DT	SC	DT	SC
1000	2.2	3.5	17.5	21.2
2000	5.3	8.5	24.9	27.9
3000	7.9	12.5	29.1	34.3
4000	9.9	15.5	32.9	38.2
5000	11.6	18.6	36.1	41.4
6000	13.5	21.1	39.0	45.4
7000	14.7	23.0	41.0	48.5
8000	16.6	24.5	43.6	51.6
9000	18.2	26.1	45.5	52.7
10000	19.1	27.4	47.7	55.0
15000	23.0	31.6	55.8	64.9
20000	26.0	34.8	61.4	72.3
25000	27.9	36.3	66.2	77.7

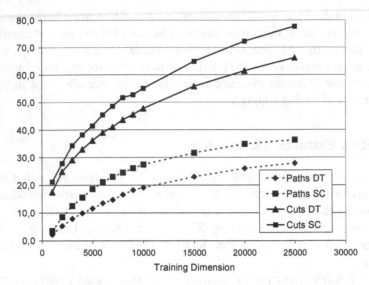

Fig. 7. Average number of paths and cuts extracted by DT and SNN

Once DT and SC are trained, their resulting *ARE* is used to evaluate the network reliability. Table 10 shows the network reliability evaluated using the correct *RE* and the *ARE* obtained by both models, for $P_i = 0.90$; the relative errors are also included for completeness. Both models produce excellent results, but SC errors are significantly lower. On the other hand, for a specific relative error, SC requires a training set with lower size.

Table 10. Average network reliability and relative error using the path sets extracted by DT and SC

Training Set Size	ARE Evaluation		Rel. Error (%)	
	DT	SC	DT	SC
1000	0.64168	0.73433	28.85	18.57
2000	0.79388	0.87961	11.98	2.47
3000	0.85864	0.89315	4.80	0.97
4000	0.88066	0.89637	2.35	0.61
5000	0.88980	0.89830	1.34	0.39
6000	0.89234	0.89972	1.06	0.24
7000	0.89401	0.90034	0.87	0.17
8000	0.89701	0.90078	0.54	0.12
9000	0.89784	0.90087	0.45	0.11
10000	0.89812	0.90115	0.42	0.08
15000	0.89992	0.90180	0.22	0.01
20000	0.90084	0.90194	0.12	0.00
25000	0.90119	0.90200	0.08	0.00

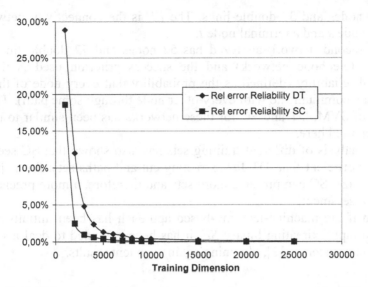

Fig. 8. Reliability relative error using the path sets extracted by DT and SC

6.5 Conclusions

This chapter has evaluated the excellent capability of three machine learning techniques (SVM, DT and SC) in performing reliability assessment, in generating the *Approximate Reliability Expression* (*ARE*) of a system and in determining cut and path sets for a network.

SVM produce an approximation to the *SF*, which cannot be written in the form of a logical sum-of-products. However, the model generated can be used within a standard Monte Carlo approach to replace the *EF*.

On the other hand, DT and SC are able to generate an approximation to the *SF* in the form of a logical sum-of-products expression, even from a small training set. The expression generated using DT is in disjoint form, which allows to easily obtain the corresponding *ARE*. Expressions generated by SC need to be converted in disjoint form, so as to produce the desired *ARE*. Both DT and SC provide information about minimum paths and cuts.

The analysis of the results on a 21-link network has shown that SC is more stable for all the performance indices evaluated, followed by DT and SVM.

The same set of experiments has been used to evaluate the performance of three ML techniques in two additional networks [33]. The first network has 20 nodes and 30 double-links. The *EF* is the connectivity between a source node *s* and a terminal node *t*.

The second network analyzed has 52 nodes and 72 double links (the Belgian telephone network) and the success criterion used is the all-terminal reliability (defined as the probability that every node of the network can communicate with every other node through some path). The behavior of SVM, DT and SC for these networks has been similar to the results reported here.

The analysis of different training sets has also shown that SC seems to be more efficient than DT for extracting cut and paths sets: for a specific data set size, SC can produce more sets and therefore, a more precise reliability assessment.

Even if the machine-learning-based approach has been initially developed for approximating binary *SF*, it has been extended to deal also with multi-state systems [32], obtaining again excellent results.

References

1. Aggarwal K.K., Chopra Y.C., Bajwa J.S.: Capacity consideration in reliability analysis of communication systems, *IEEE Transactions on Reliability*, 31, 1982, pp. 177–181.
2. Bevilacqua M., Braglia M., Montanari R.: The classification and regression tree approach to pump failure rate analysis, *Reliability Engineering and System Safety*, 79, 2002, pp. 59–67.
3. Billinton, R. Allan R.N: *Reliability Evaluation of Engineering Systems, Concepts and Techniques (second edition)*, Plenum Press, 1992.
4. Billinton, R. Li W.: *Reliability Assessment of Electric Power System Using Monte Carlo Methods*, Plenum Press, 1994.
5. Breiman L., Friedman J. H., Olshen R. A., Stone C. J.: *Classification and Regression Trees*, Belmont: Wadsworth, 1994.
6. Brayton R. K., Hachtel G. D., McMullen C. T., Sangiovanni-Vincentelli A. L.: *Logic Minimization Algorithms for VLSI Synthesis*, Hingham, MA: Kluwer Academic Publishers, 1984.
7. Campbell C.: An introduction to kernel methods, In Howlett R.J. and Jain L.C., editors, *Radial Basis Function Networks: Design and Applications*, p. 31. Springer Verlag, Berlin, 2000.
8. Cappelli C., Mola F., Siciliano R.: A statistical approach to growing a reliable honest tree, *Computational Statistics & Data Analysis*, 38, 2002, pp. 285–299.
9. Chang C.-C., Lin C.-J.: *LIBSVM: A Library for Support Vector Machines*, software available at http://www.csie.ntu.edu.tw/~cjlin/libsvm/index.html, 2001.

10. Colbourn Ch.: *The Combinatorics of Network Reliability*, Oxford University Press, 1987.
11. Cristianini N., Shawe-Taylor J.: *An Introduction to Support Vector Machines*, Cambridge University Press, 2000.
12. Dietmeyer D. L., *Logical Design of Digital Systems (third edition)*, Boston, MA: Allyn and Bacon, 1988.
13. Duda R. O., Hart P. E., Stork D. G.: *Pattern Classification*, John Wiley & Sons, 2001.
14. Dubi A.: Modeling of realistic system with the Monte Carlo method: A unified system engineering approach, *Proceedings of the Annual Reliability and Maintainability Symposium*, Tutorial Notes, 2001.
15. Grosh D.L.: *Primer of Reliability Theory*, John Wiley & Sons, New York, 1989
16. Heidtmann K. D.: Smaller sums of disjoint products by subproducts inversion, *IEEE Transactions on Reliability*, 38, 1989, pp. 305–311.
17. Hong S. J., Cain R. G., Ostapko D. L.: MINI: A heuristic approach for logic minimization, *IBM Journal of Research and Development*, 18, 1974, pp. 443–458.
18. Muselli M., Liberati D.: Training digital circuits with Hamming Clustering, *IEEE Transactions on Circuits and Systems*, 47, 2000, pp. 513–527.
19. Muselli M., Liberati D.: Binary rule generation via Hamming Clustering, *IEEE Transactions on Knowledge and Data Engineering*, 14, 2002, pp. 1258–1268.
20. Muselli M., Quarati A.: Reconstructing positive Boolean functions with Shadow Clustering, *ECCTD 2005 – European Conference on Circuit Theory and Design*, Cork, Ireland, 2005.
21. Muselli M.: Switching Neural Networks: A new connectionist model for classification, *WIRN '05 – XVI Italian Workshop on Neural Networks*, Vietri sul Mare, Italy, 2005.
22. Murthy S., Kasif S., Salzberg S.: A system for induction of oblique decision tree, *Journal of Artificial Intelligence Research*, 2, 1994, pp. 1–32.
23. Papadimitriou C. H., Steiglitz K.: *Combinatorial Optimisation: Algorithms and Complexity*, Prentice Hall, New Jersey, 1982.
24. Pereira M. V. F., Pinto L. M. V. G.: A new computational tool for composite reliability evaluation, *IEEE Power System Engineering Society Summer Meeting*, 1991, 91SM443-2.
25. Pohl E. A., Mykyta E. F.: Simulation modeling for reliability analysis, *Proceedings of the Annual Reliability and Maintainability Symposium*, Tutorial Notes, 2000.
26. Portela da Gama J. M.: *Combining Classification Algorithms*, PhD. Thesis, Faculdade de Ciências da Universidade do Porto, 1999.
27. Quinlan J. R.: *C4.5: Programs for Machine Learning*, Morgan Kaufmann Publishers, 1993.
28. Rai S, Soh S.: A computer approach for reliability evaluation of telecommunication networks with heterogeneous link-capacities, *IEEE Transactions on Reliability*, 40, 1991, pp. 441–451.

29. Reingold E., Nievergelt J., Deo N.: *Combinatorial Algorithms: Theory and Practice*, Prentice Hall, New Jersey, 1977.
30. Rocco C. M., Moreno J. M.: Fast Monte Carlo reliability evaluation using Support Vector Machine, *Reliability Engineering and System Safety*, 76, 2002, pp. 239–245.
31. Rocco C. M.: A rule induction approach to improve Monte Carlo system reliability assessment, *Reliability Engineering and System Safety*, 82, 2003, pp. 87–94.
32. Rocco C. M., Muselli M.: Approximate multi-state reliability expressions using a new machine learning technique, *Reliability Engineering and System Safety*, 89, 2005, pp. 261–270.
33. Rocco C. M., Muselli M.: Machine learning models for reliability assessment of communication networks, submitted to *IEEE Transactions on Neural Networks*.
34. Shawe-Taylor J., Cristianini N.: *Kernel Methods for Pattern Analysis*, Cambridge University Press, 2004.
35. Stivaros C., Sutner K.: Optimal link assignments for all-terminal network reliability, *Discrete Applied Mathematics*, 75, 1997, pp 285–295.
36. Vapnik V.: *Statistical Learning Theory*, John Wiley & Sons, 1998.
37. Veropoulos K., Campbell C., Cristianini N.: Controlling the sensitivity of Support Vector Machines, *Proceedings of the International Joint Conference on Artificial Intelligence*, Stockholm, Sweden, 1999, pp. 55–60.
38. Wahba G.: *Spline Models for Observational Data*, SIAM, 1990.
39. Yoo Y. B., Deo N.: A comparison of algorithm for terminal-pair reliability, *IEEE Transaction on Reliability*, 37, 1988, pp. 210–215.

Neural Networks for Reliability-Based Optimal Design

Ming J Zuo, Zhigang Tian

Department of Mechanical Engineering, University of Alberta, Canada

Hong-Zhong Huang

School of Mechanical and Electronic Engineering, University of Electronic Science and Technology of China, P.R. China

7.1 Introduction

7.1.1 Reliability-based Optimal Design

Today's engineering systems are sophisticated in design and powerful in function. Examples of such systems include airplanes, space shuttles, telecommunication networks, robots, and manufacturing facilities. Critical measures of performance of these systems include reliability, cost, and weight. Optimal system design aims to optimize such performance measures.

The traditional system reliability theory assumes that a system and its components may only experience one of two possible states: working or failed. As a result, we call it binary reliability theory. Under the binary assumption, the reliability of a system is defined to be the probability that the system will perform its functions satisfactorily for a certain period of time under specified conditions. The reliability of a system depends on the reliabilities of the constituent components and the configuration of the system. A design of a system provides a specification of the reliabilities of the components and the system configuration. In optimal system design, one aims to find the best design that optimizes various measures of performance of the system.

M.J. Zuo et al.: *Neural Networks for Reliability-Based Optimal Design*, Computational Intelligence in Reliability Engineering (SCI) **40**, 175–196 (2007)
www.springerlink.com   © Springer-Verlag Berlin Heidelberg 2007

One of the most studied system configurations in the literature is the series-parallel system configuration. A series-parallel system consists of N subsystems connected in series such that the system works if and only if all the subsystems work wherein subsystem i ($1 \leqslant i \leqslant N$) consists of n_i components connected in parallel such that the subsystem fails if and only if all the components in this subsystem fail. Fig. 1 shows such a series-parallel configuration. The reliability of such a series-parallel system is expressed as:

$$R_s = \prod_{i=1}^{N}\left(1 - \prod_{j=1}^{n_i}(1 - p_{ij})\right),$$ (1)

where p_{ij} is the reliability of component j in subsystem i. For such a system, a typical optimization problem involves finding the number of parallel components in each subsystem to maximize system reliability subject to constraints on budget, volume, and/or weight. Requirement on system reliability may be treated as a constraint while one of the constraints may be treated as the objective function to be maximized or minimized. It is a nonlinear programming problem involving integer variables. Since the focus is on finding the optimal redundancy level in each subsystem, such an optimal design problem is also referred to as a redundancy allocation problem.

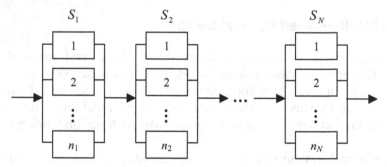

Fig. 1. Structure of a series-parallel system

Many variations of the redundancy allocation problem have been studied in the literature. The design variables may include the number of redundant components in each subsystem, the reliability value of each component, and the selection of component versions that are available on the market. The redundancy structure in each subsystem may be in the form of k-out-of-n (Coit and Smith 1996) or in the form of standby (Zhao and Liu

2004). The constituent components have also been modeled as being multi-state (Liu et al 2003) or having fuzzy lifetimes (Zhao and Liu 2004).

Another well studied system configuration in the literature is a general network configuration. A network consists of nodes and links. One is interested in determining what links should be present between pairs of nodes. The measure of performance of the network to be optimized may be cost, two-terminal reliability, or all-terminal reliability (AboElFotoh and Al-Sumait 2001 and Srivaree-ratana et al 2002). Such optimization problems are non-linear integer programming problems.

7.1.2 Challenges in Reliability-based Optimal Design

A major challenge in reliability based optimal design problems is the evaluation of system reliability given a system design. This is a time-consuming task for large systems. In optimal system design, system reliability has to be evaluated frequently for each candidate design. Thus, efficient algorithms for system reliability evaluation are essential for solving these problems.

To search for optimal solutions of reliability-based optimal design problems, efficient optimization algorithms are needed. Kuo et al. (2001) surveyed and classified optimization techniques for solving redundancy allocation problems. They compared the pros and cons of the following classical optimization techniques: integer programming, transforming non-linear to linear functions, dynamic programming, the sequential unconstrained minimization technique (SUMT), the generalized reduced gradient method (GRG), the modified sequential simplex pattern search, and the generalized Lagrangian function method. Other examples of integer programming solutions to the redundancy allocation problems are presented by Misra and Sharma (1991), Gen et al. (1990), and Gen et al. (1993). In the process of searching for more efficient optimization algorithms, researchers have used artificial neural networks (ANN) as a function approximator and as an optimizer for solving all kinds of reliability based design problems.

7.1.3 Neural Networks

Neural networks consist of simple elements called neurons operating in parallel. The structure of neural networks is inspired by biological neurological systems. According to Rojas (1996), McCulloch and Pitts introduced the first abstract model of neurons by mimicking biological neurons and Hebb presented a learning law so that a network of neurons can be trained. The research on neural networks achieved significant progress in the 1980s. Neural network models have found many applications in the past fifteen years.

Neural networks have been trained to perform complex functions in various fields of application including pattern recognition, identification, classification, speech, vision and control systems (Rojas, 1996). Neural networks can be trained to solve problems that are difficult for conventional computers or human beings. The advantages of neural networks include: (1) Adaptive learning: an ability to learn how to do tasks based on the data given for training or initial experience. (2) Self-organization: an neural network can create its own organization or representation of the information it receives during learning time. (3) Real time operation: neural network computations may be carried out in parallel, and special hardware devices are being designed and manufactured which take advantage of this capability. (4) Fault tolerance via redundant information coding: partial destruction of a network leads to the corresponding degradation of performance. However, some network capabilities may be retained even with major network damage.

Two types of neural networks are most widely used in reliability-based optimal design: feed-forward neural networks as a function approximator, and Hopfield networks as an optimizer. These two types of neural networks and their applications will be discussed in details in the following sections.

7.1.4 Content of this Chapter

In this chapter, we explore the applications of artificial neural networks for solving reliability-based optimal design problems. The remaining part of this chapter is organized follows. In Section 2, we summarize the advantages of artificial neural networks that are specifically useful for solving reliability based optimal design problems. The use of ANN as a function approximator is presented in Section 3 while the use of ANN as an optimizer is given in Section 4. Section 5 provides a summary and points out future research topics in application of ANN for solving reliability based design problems.

7.2 Feed-forward Neural Networks as a Function Approximator

7.2.1 Feed-forward Neural Networks

The most widely used type of neural networks is the feed-forward neural network. The structure of a feed-forward neural network with three layers is shown in Fig. 2. It has one input layer, one hidden layer, and one output layer. A feed-forward neural network is used for nonlinear mapping. That is, based on the available data sets of input and output pairs, a neural network can be trained to model the mapping relationship between inputs and outputs.

For example, from function $y = x^2$, we generate five input/output pairs: [1, 1], [2, 4], [3, 9], [4, 16], and [5, 25]. The neural network we are using should have one neuron in the input layer and one neuron in the output layer, since there are only one input and one output. In this example, we can simply use one hidden layer with 3 hidden neurons. After training with the provided training pairs, the neural network can pretty much model the hidden mapping relationship $y = x^2$, and thus we can calculate what is the output value when the input is say 1.5.

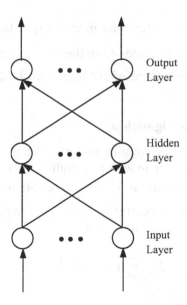

Fig. 2. Structure of a feed-forward neural network

A three-layer feed-forward neural network is capable of modeling any nonlinear mapping (Rojas, 1996). A feed-forward neural network may have more than three layers. However, too many hidden layers make the model more complex and the generalization capability of the network will become worse. Thus, feed-forward neural networks with one or two hidden layers are the most widely used ones in practical applications. Building nonlinear mapping relationship is a major advantage of feed-forward neural networks. We do not have to know the interior mechanism of the system to be modeled. As long as we have a set of input and output pairs, the feed-forward neural network can be trained to approximate the relationship between the output and the input to any specified degree of accuracy.

The function represented by a neural network model is determined largely by the connections between neurons. We can train a neural network to perform a particular function by adjusting the values of the connections (weights) between neurons (Rojas, 1996). Typically many input/output pairs are used in the process called supervised learning to train a network. The back-propagation (BP) algorithm is a widely used training algorithm for feed-forward neural networks. The BP algorithm aims at minimizing the following error function:

$$E = \frac{1}{2}\sum_j \left(T_j - v_j\right)^2 \tag{2}$$

where j represents all the neurons in the output layer, T_j is the desired output in the training pair, and v_j is the actual output from the current neural network. The procedure of BP algorithm is shown as follows (Fu, 1994).

The Backpropagation Algorithm

A. *Initialization*. Set all weights and node thresholds to small random numbers. Give each neuron an index (including the input neurons).

B. *Feed-forward calculation*. Use v_j to denote the output of neuron j. (1) The output of an input neuron is equal to the input value. (2) The output of a hidden neuron or output neuron is:

$$v_j = F\left(\sum_i w_{ji} v_i - \theta_j\right) \tag{3}$$

where w_{ji} is the weight from neuron i to j, θ_j is the threshold, and F is the so-called activation function. In a feed-forward neural network, a neuron only gets inputs from the immediately preceding layer. A commonly used activation is the sigmoid function given by: $F(x) = 1/(1 + e^{-x})$.

C. *Backpropagation weight training.* (1) Start from the output layer and work backward to the hidden layers recursively to calculate error δ_j. For output neurons:

$$\delta_j = v_j(1 - v_j)(T_j - v_j) \tag{4}$$

For hidden neurons:

$$\delta_j = v_j(1 - v_j)\sum_k \delta_k w_{kj} \tag{5}$$

where δ_k is the error at neuron k to which a connection points from hidden neuron j. (2) Adjust the weights as follows:

$$w_{ji}(t+1) = w_{ji}(t) + \Delta w_{ji}$$
$$\Delta w_{ji} = \eta \delta_j v_i \tag{6}$$

where η is the positive scalar and called learning rate.

D. Repeat the feed-forward calculation and backpropagation weight training until the stopping criterion in terms of output errors is met.

Other techniques can be applied to improve the BP algorithm, like using the momentum terms. There are also other training algorithms based on other optimization methods, such as quasi-Newton methods and conjugate gradient methods (Rojas, 1996).

The ability of ANN to approximate a function of many variables to any degree of accuracy has been put into good use in solving reliability based optimal design problems. Coit and Smith (1996) used ANN to estimate the reliability of a series-parallel system wherein each subsystem has a k-out-of-n configuration in order to solve the optimal redundancy allocation problem. Zhao and Liu (2004) considered a series-parallel system wherein the redundancy configuration may be either parallel or standby and the lifetime of the system and that of each component is modeled as a fuzzy random variable. They used an ANN model to approximate the expected system lifetime and system reliability as a function of the redundancy levels and component lifetimes. Liu et al (2003) used ANN to approximate the expected system utility of a series-parallel system wherein the state of the system and that of each component is modeled as a continuous multi-state random variable. Huang et al (2005) used an ANN model to represent the relationship between a designer's preference score and the performance measures such as cost, reliability, and weight of the system given a specific design as a part in their integrated interactive multi-objective optimization approach. Srivaree-ratana et al (2002) used an ANN model to represent the relationship between the all-terminal reliability of a network and the failure-prone links to be installed in the network. Papadrakakis and Lagaros (2002) used an ANN model to represent the relationship between performance measures such as stress and failure probability and the design variables in optimal design of large-scale 3-D frame structures. These uses of ANN as a function approximator will be discussed in details in this section.

In a feed-forward ANN, the number of layers, the number of neurons in each layer, and the connection weights between neurons define the structure of the ANN. Training data specify the desired relationship between output and input. Through training, a feed-forward ANN can be used to approximate any continuous function to any degree of accuracy (Cybenko 1989). This capability has been used in many reliability based optimization problems. In this section, we summarize applications of ANN as a function approximator.

7.2.2 Evaluation of System Utility of a Continuous-state Series-parallel System

Liu et al (2003) report a study on optimal redundancy allocation for a continuous-state series-parallel system. The structure of the considered multi-state series-parallel system can also be represented by Fig. 1. It consists of N subsystems, S_1 to S_N, connected in series. Each subsystem, say S_i, has n_i identical components connected in parallel. The state of each component and the system may be modeled as a continuous random variable taking values in the range of [0, 1]. The definition of a multi-state series-parallel system provided by Barlow and Wu (1978) is used here. That is, the state of a parallel system is the state of the best component in the system while the state of a series system is the state of the worst component in the system. Let x_{ij} denote the state of component j in subsystem S_i. Then, the system state can be expressed as

$$\varphi(x) = \min_{1 \le i \le N} \max_{1 \le j \le n_i} x_{ij}$$

When the system is in state s, the utility of the system is denoted by $\mu(s)$. Given the state density function of each component, namely $f_{ij}(s)$, we can evaluate the state distribution of the system. With the system state distribution obtained, we can then find the expected utility of the system.

The design problem concerned is maximization of the expected system utility subject to cost constraints through determination of the optimal redundancy level in each subsystem. The optimization model is as follows:

Maximize:

$$U(n_1,n_2,\cdots,n_N) = -\int_0^1 \mu(s) \frac{\mathrm{d}}{\mathrm{d}s}\left(\prod_{i=1}^{N}\left(1- \prod_{j=1}^{n_i}\left(\int_0^s f_{ij}(t)\mathrm{d}t \right) \right) \right) \mathrm{d}s \tag{7}$$

Subject to:

$$\sum_{i=1}^{N}\sum_{j=1}^{n_i} C_{ij} \le C_T$$

where C_{ij} is the cost of component j in subsystem i and C_T is the total budget allowed.

This optimization model includes both integration and differentiation in the objective function. The objective function is very complicated and the problem is very difficult to solve using a classical optimization algorithm. This situation arises when (1) the number of subsystems, N, is large; (2) the state distributions of components in the same subsystem are not identical; and/or (3) component state distribution is not a simple distribution. In addition, in some cases, the component state distribution function may have to be expressed in an empirical form and, as a result, no analytical expression of $U(n_1,n_2,\cdots,n_N)$ is available. The critical problem is that evaluating this utility function directly is very time-consuming, and thus evaluating it repetitively in the optimization process is very hard and sometimes impossible.

The system state distribution $g(s,n_1,n_2,\cdots,n_N)$ is defined as:

$$g(s,n_1,n_2,\cdots,n_N) = \prod_{i=1}^{N}\left(1- \prod_{j=1}^{n_i}\left(\int_0^s f_{ij}(t)\mathrm{d}t \right) \right) \tag{8}$$

Liu et al (2003) use feed-forward neural networks to approximate this system state distribution. They use a three-layer neural network, with sigmoidal activation functions used in the hidden layer, and linear activation functions used in the output layer. Training and testing (validation) data sets are generated as follows: first, a suitable number of input vectors, $(s,n_1,n_2,\cdots,n_N)$, are chosen or generated randomly from the allowed value ranges of s, n_1, n_2, ..., and n_N; next, the input vectors are normalized and for each input vector, the desired output, $g(s, n_1, n_2, \cdots, n_N)$, is calculated with equation (8). The training and testing data sets consist of different pairs of the input vector and its corresponding output.

No matter how complicated $g(s, n_1, n_2, \cdots, n_N)$ might be, the approximate analytical expression of system distribution function, $\hat{g}(s, n_1, n_2, \cdots, n_N)$, is always a linear combination of a finite number of sigmoidal functions. The approximate objective function $\hat{U}(n_1, n_2, \cdots, n_N)$ constructed from $\hat{g}(s, n_1, n_2, \cdots, n_N)$ usually has analytical expression (Liu et al 2003), and thus much less time-consuming to calculate. And the redundancy allocation of continuous-state series-parallel systems can be implemented more efficiently. The following example is given to illustrate this approach (Liu et al 2003).

Example 1

In this example, we use a 4-stage series-parallel system, in which $N = 4$; $\mu(s)=10s$; $C_{1i}=3200$; $C_{2i}=1700$; $C_{3i}=830$; $C_{4i}=2500$; and $C_T=160,000$. The probability density functions of components in the four subsystems are three commonly used distributions: unit distribution, triangular distribution and Beta distribution.

$f_{1i}=1$: unit distribution;
$f_{2i}=2s$: triangular distribution;

$$f_{3i}(s) = \frac{\Gamma(\alpha+\beta)}{\Gamma(\alpha)\Gamma(\beta)} s^{\alpha-1}(1-s)^{\beta-1}: \text{Beta distribution where } \alpha=2, \beta=3.5;$$

$$f_{4i}(s) = \frac{\Gamma(\alpha+\beta)}{\Gamma(\alpha)\Gamma(\beta)} s^{\alpha-1}(1-s)^{\beta-1}: \text{Beta distribution where } \alpha=5, \beta=2.$$

The density functions of the components in the 4 subsystems are plotted in Fig. 3.

The desired output targets for ANN training, i.e., actual system distribution function, are calculated with

$$g(s, n_1, n_2, n_3, n_4) = \prod_{i=1}^{4} (1 - \prod_{j=1}^{n_i} (\int_0^s f_{ij}(t)dt)) \tag{9}$$

$$= (1-s)^{n_1}(1-s^2)^{n_2}(1-(3.5(1-s^{4.5})-4.5(1-s^{3.5}))$$

$$+ 0.0635)^{n_3}(1-6s^5+5s^6)^{n_4}$$

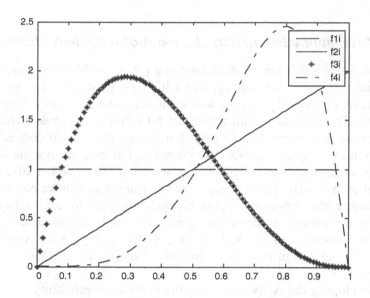

Fig. 3. Component state distribution functions for Example 1

We can see that although equation (9) is also complicated, it is calculable. But the objective function $U(n_1,n_2,n_3,n_4)$ as in Equation (7) is too complex and seems impossible to calculate analytically.

On the contrary, it is straightforward to solve this optimization problem using the ANN approximation. After training the neural network and obtaining the approximate system distribution function $\hat{g}(n_1,n_2,n_3,n_4)$, we can construct the approximate objective function $\hat{U}(n_1,n_2,n_3,n_4)$, which is an analytical expression (Liu et al 2003).

The training data set containing 1600 training pairs is generated randomly. The size of the hidden layer is 19. After 2500 epochs of training iterations, the actual percent training error is 0.9859%. The maximum validation error is 2.4532%. After an exhaustive search, the final optimal solution is

$$M_1=14; \quad M_2=9; \quad M_3=36; \quad M_4=15$$

and the corresponding approximate utility is $\hat{U}(12,7,35,32) = 8.4384$.

7.2.3 Other Applications of Neural Networks as a Function Approximator

7.2.3.1 Reliability Evaluation of a k-out-of-n System Structure

Coit and Smith (1996) consider the optimal design problem of a series system with k-out-of-n redundancy. In such a system structure, there are N subsystems connected in series and each subsystem adopts a k-out-of-n structure. The optimal design model aims to find the optimal number of components n_i in subsystem i ($n_i \geq k$) such that the system cost is minimized subject to system reliability constraints. For each subsystem, several types of components are available for selection. The number of components of each available type needs to be determined so that we can find the total number of components of possibly different types for each subsystem. The decision variable values are determined through an optimization process using genetic algorithms. However, the reliability of the system for each candidate design must be evaluated. This may be a time-consuming process.

In developing the ANN model for the purpose of reliability estimation of a k-out-of-n system, Coit and Smith (1996) use full factorial design of the critical parameters k, n, and three underlying distributions (uniform, quadratic skewed-left and quadratic skewed-right) in equal proportions for component reliabilities. The skewed distributions are used to make sure that the developed ANN model is accurate for high component reliabilities or low component failure probabilities. Analytical methods are used to find the training data for the neural networks.

Since the ANN model is developed for estimation of the reliability of a k-out-of-n system structure, it actually includes the parallel redundancy as a special case. This makes the ANN model more general than for a simple parallel structure. However, caution has to be taken in assessing the error in the estimate of the reliability of the k-out-of-n subsystem. When these reliability values of the subsystems are multiplied together to get the system reliability, these errors may be magnified (Coit and Smith 1996). One needs also to take into consideration the optimizer to be used in the training of the ANN model.

7.2.3.2 Performance Evaluation of a Series-parallel System Under Fuzzy Environment

Zhao and Liu (2004) consider the problem of redundancy allocation of a series system with parallel redundancy or standby redundancy. The system has n subsystems connected in series. Subsystem i ($1 \le i \le n$) has n_i components either connected in parallel or connected in standby. The lifetime of each component can be represented by a fuzzy random variable. The measures of performance of the system may be the expected lifetime of the system, system reliability, or the so-called (α,β)-system lifetime (Zhao and Liu 2004).

The key for solving this optimal allocation problem is to evaluate the expected lifetime of the system, system reliability, and the so-called (α,β)-system lifetime. For a given design, a random fuzzy simulation approach was proposed by Zhao and Liu (2004) to evaluate these performance measures. Since this approach is very time-consuming, they used it to generate training data to train a feedforward ANN which will then be used to approximate these measures of performance during the design optimization process. The used ANN model has one input layer, one hidden layer, and one output layer. The number of neurons in the input layer is equal to the number of decision variables. The number of output neurons is equal to the number of performance measures of interest. The number of neurons in the hidden layer is determined by the pruning algorithm of Castellano et al (1997).

Once the ANN model is trained, it is used as a function approximator in the optimization model. Genetic algorithms are used to solve the optimization problems. Examples are used to illustrate this approach for solving redundancy allocation problems including parallel redundancy and standby redundancy.

7.2.3.3 Evaluation of All-terminal Reliability in Network Design

Srivaree-ratana et al. (2002) consider a network design problem in which the nodes are fixed and perfect while the links are failure prone. The question to be answered is what links should be installed to minimize the total cost of installing these selected links subject to requirement on all-terminal reliability of the network.

The most time-consuming task in this network design problem is the evaluation of all-terminal reliability given a network design. Approaches such as enumeration and Monte Carlo simulation can be used for evaluation or approximation of network reliability, but they are very time-consuming. Srivaree-ratana et al (2002) decide to use ANN to estimate the all-terminal reliability as a function of a selected design. A feedforward ANN model is adopted and the backpropagation training algorithm is used. A hyperbolic activation function is used for all neurons and a learning rate of 0.3 is used for hidden neurons and of 0.15 is used for the output neurons. The total number of hidden neurons is chosen to be identical to the number of input neurons.

Experiment results provided by Srivaree-ratana et al (2002) show that the ANN model works very well for estimating all-terminal reliability. Future research topics may include fine tuning the ANN model and imbed occasional evaluation of the exact all-terminal reliability of preferred designs.

7.2.3.4 Evaluation of Stress and Failure Probability in Large-scale Structural Design

Reliability based optimal design of large-scale structural systems is extremely computation intensive. Papadrakakis and Lagaros (2002) address the optimal design of multi-storey 3-D frames. The goal is to minimize the weight of the structure subject to constraints on allowed stress, displacement, and failure probability. Due to the randomness in loads to be applied, material properties, and member geometry, evaluation of the stress, displacement, and failure probability given a structure designed is very time-consuming.

The measures such as stress, displacement, and failure probability can be evaluated using finite element method, the limit elasto-plastic method, and Monte Carlo simulations given certain design parameters. These calculated values and the corresponding design parameters can then be used as the training data set for training of a feedforward ANN which can then be used for approximation of these measures during the optimization process. The actual optimization algorithm used is the genetic algorithm. Numerical results are presented to illustrate the effectiveness of the proposed approaches.

There are more application examples of neural networks as function approximator in the literature, like the representation of the preference structure of the designer in multi-objective design optimisation (Huang et al 2005).

7.3 Hopfield Networks as an Optimizer

7.3.1 Hopfield Networks

Another widely used structure of ANN is the Hopfield network (Hopfield 1982, Hopfield and Tank 1985). In a Hopfield network, all neurons are in a single layer (Fig. 4). Every pair of neurons are connected with the same connection weights. That is

$$w_{ji} = w_{ij}, \text{ for } i \neq j$$
$$w_{ii} = 0 \tag{10}$$

Each neuron may represent a binary variable because its output may take values of 0 or 1 only. An activation function is used to map the total input to a neuron to a 0-1 output value.

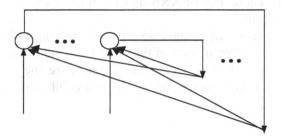

Fig. 4. Structure of a Hopfield network

Through a dynamic update equation of the input value of each neuron, the Hopfield ANN converges to a state that minimizes an energy function of the neural network. Consider a Hopfield ANN with n neurons wherein w_{ij} denotes the connection weight between neurons i and j, u_i the input to neuron i, $v_i = F(u_i)$ the output of neuron i through activation function $F(.)$, and θ_i the bias of neuron i. Then, the energy function of the ANN is given by Hopfield and Tank (1985) as:

$$E = -\frac{1}{2}\sum_{i=1}^{n}\sum_{j=1}^{n}w_{ij}v_i v_j - \sum_{i=1}^{n}\theta_i v_i \tag{11}$$

The dynamics of the Hopfield ANN is defined as

$$\frac{du_i}{dt} = \sum_{j=1}^{n} w_{ij} v_j + \theta_i \qquad (12)$$

Equation (12) actually represents the steepest descent direction of the energy function given in equation (11). If the input to neuron i, u_i, is updated following the direction given by equation (12), the energy function will converge to a local minimum. This is why the Hopfield ANN can be used to solve optimization problems.

A major advantage of Hopfield networks is their efficiency in solving optimization problems (Nourelfath and Nahas, 2003). Such an ANN as an optimizer was first introduced by Hopfield and Tank (1985). The concept of quantized neurons was introduced by Matsuda (1999). AboElFotoh and Al-Sumait (2001) used Hopfield networks for solving a network design problem. Nourelfath and Nahas (2003) used quantized Hopfield networks for selection of the components in a series system for system reliability maximization. These uses of ANN as an optimizer will be discussed in details in this section.

The key in the use of a Hopfield ANN for solving reliability optimization problems is in formulation of the energy function and definition of decision variables v_i. In this section, we summarize the work reported by AboElFotoh and Al-Sumait (2001) and Nourelfath and Nahas (2003) for this purpose.

7.3.2 Network Design with Hopfield ANN

AboElFotoh and Al-Sumait (2001) considered a network design problem. There are n perfect nodes in the network. The question to be answered is what links should be installed to minimize the total cost of the network subject to all-terminal reliability requirement. The reliability and the cost of each possible link are given as data.

Notation:

n Number of nodes in the network
i, j Network nodes
(i, j) The link between node i and node j
$p_{i,j}$ Reliability of link (i, j)
$c_{i,j}$ Cost of link (i, j)
R_0 All-terminal reliability requirement of the network
R_S All-terminal reliability of the network
$v_{i,j}$ Takes the value of 1 if link (i, j) is selected and 0 otherwise

A, B, C   Positive constants that may be adjusted in the optimization process

The optimization model for the network design problem is

$$\text{Minimize} \sum_{i=1}^{n} \sum_{j=1}^{n} c_{i,j} v_{i,j}$$

(13)

$$\text{Subject to:} \qquad R_S \geq R_0$$

To use a Hopfield ANN to solve this optimization problem, AboElFotoh and Al-Sumait (2001) use (i, j) to denote a neuron and the following to present the energy function:

$$E = -A \cdot R_S + B \sum_{i=1}^{n} \sum_{j=i+1}^{n} c_{i,j} v_{i,j} + C \cdot \left| R_S - R_0 \right|$$

(14)

This energy function consists of three terms added together. Since we are to minimize this energy function, the first encourages network reliability maximization, the second term encourages cost minimization, and the third term discourages the ANN from adding new links to increase network reliability unnecessarily over R_0. The negative derivative of the energy function given in equation (14) with respect to each decision variable $v_{i,j}$ is given by

$$-\frac{\partial E}{\partial v_{i,j}} = \begin{cases} (A-C) \cdot \dfrac{\partial R_S}{\partial v_{i,j}} - B c_{i,j}, & \text{if } R_S > R_0 \\[2ex] (A+C) \cdot \dfrac{\partial R_S}{\partial v_{i,j}} - B c_{i,j} & \text{if } R_S < R_0 \end{cases}, $$

(15)

The update equation for the input $u_{i,j}$ is then given by

$$u_{i,j}(t+1) = u_{i,j}(t) + \frac{\partial E}{\partial v_{i,j}}$$

(16)

To use the above equations, one has to have an algorithm to calculate the all-terminal reliability for each given network design. Since this is an NP-hard problem, AboElFotoh and Al-Sumait (2001) provide a lower bound and upper bound on this network reliability. Either bound may be used in equation (14) to approximate the reliability of the network. The activation function used is a simple threshold function, namely

$$v_{i,j} = \begin{cases} 1, & \text{if } u_i > \text{UTP} \\ 0, & \text{if } u_i < \text{LTP}, \\ \text{unchanged}, & \text{otherwise} \end{cases} \qquad (17)$$

where UTP and LTP are the threshold values.

Many network cases were generated to test the Hopfield ANN approach for this network design problem. AboElFotoh and Al-Sumait (2001) conclude that this approach is very efficient for design of large networks but does not guarantee global optimal solutions. Possible future research work include consideration of node failures, more efficient algorithm for evaluation of all-terminal network reliability, and better rules for selection of parameters of the energy function.

7.3.3 Series System Design with Quantized Hopfield ANN

Nourelfath and Nahas (2003) considered a series system with N components. For component j ($1 \leq j \leq N$), there are M_j choices available. These choices correspond to different costs, reliabilities, weights, and possibly other characteristics. We are interested in making a choice for each of the N components such that system reliability is maximized subject to cost and other constraints. The optimization model for this problem can be expressed as

$$\text{Maximize} \qquad R_S = \prod_{j=1}^{N} \left(\sum_{i=1}^{M_j} x_i^j R_i^j \right)$$

$$\text{Subject to:} \qquad \sum_{j=1}^{N} \sum_{i=1}^{M_j} C_i^j x_i^j \leq B \qquad (18)$$

$$\sum_{i=1}^{M_j} x_i^j = 1, \quad \forall j = 1, 2, \cdots, N$$

where R_S is the system reliability, R_i^j is the reliability of choice i for component j, C_i^j is the cost of choice i for component j, B is the budget for the system, and x_i^j is a 0-1 variable that takes the value of 1 if choice i is selected for component j. This is a 0-1 non-linear programming problem. However, the objective function can be transformed into a linear function as follows (Nourelfath and Nahas 2003):

$$\text{Minimize} \qquad \psi = -\ln R_S = \sum_{j=1}^{N} \sum_{i=1}^{M_j} x_i^j \left| \ln R_i^j \right|. \qquad (19)$$

With this transformation, the optimization problem becomes a 0-1 linear programming problem.

Without loss of generality, the budget amount B is assumed to be an integer value. After introducing a slack variable t to convert the inequality constraint into an equality constraint, the optimization model becomes:

$$\text{Minimize} \quad \psi = -\ln R_S = \sum_{j=1}^{N} \sum_{i=1}^{M_j} x_i^j \left| \ln R_i^j \right|$$

$$\text{Subject to} \quad \sum_{j=1}^{N} \sum_{i=1}^{M_j} C_i^j x_i^j + t = B$$

$$\sum_{i=1}^{M_j} x_i^j = 1, \quad \forall j = 1, 2, \cdots, N$$

(20)

x_i^j are 0-1 variable and t is integer.

Though the Hopfield ANN in its originally proposed form allows only 0-1 variables, the quantized Hopfield ANN developed by Matsuda (1999) can be used to deal with integer variables too. Applying this model, Nourelfath and Nahas (2003) use the following energy function of the quantized Hopfield ANN for solving the series system optimization problem:

$$E = \frac{A_1}{2} \left(\sum_{j=1}^{N} \sum_{i=1}^{M_j} x_i^j \left| \ln R_i^j \right| \right)^2 + \frac{A_2}{2} \left(\sum_{j=1}^{N} \sum_{i=1}^{M_j} x_i^j C_i^j + t - B \right)^2$$

$$+ \frac{A_3}{2} \sum_{j=1}^{N} \left(\sum_{i=1}^{M_j} x_i^j - 1 \right)^2, \qquad (21)$$

where A_1, A_2, and A_3 are positive parameters.

Simulation studies are conducted to test the quantized Hopfield ANN model for the series system optimization problem. The following two forms of optimization objectives other than that in equation (20) were tested as well

$$\text{Minimize} \quad \psi = \sum_{j=1}^{N} \sum_{i=1}^{M_j} x_i^j \frac{1}{\ln R_i^j}, \text{ and}$$

$$\text{Minimize} \quad \psi = \sum_{j=1}^{N} \sum_{i=1}^{M_j} x_i^j \left(1 - \ln R_i^j\right),$$

(22)

and simulation results showed that no effects occur when considering these forms of objective functions.

Nourelfath and Nahas (2003) conclude that the quantized Hopfield ANN reduces the number of neurons needed to represent the series system optimization model and as a result reduces the computation time in finding optimal solutions. Unfortunately, the quantized Hopfield ANN does not guarantee global optimal solutions either. Other possible future research topics include application of this model to solving other reliability based optimization problems.

7.4 Conclusions

Reliability based optimal design presents challenging optimization problems. These problems often involve time-consuming tasks of evaluation of various system performance measures such as reliability, expected utility, lifetime, stress, displacement, and failure probability. Neural network models have been used for the purpose of function approximation to significantly reduce the computation needs in the on-line optimization process because ANN models can be trained off-line. The Hopfield ANN model has also been used as a local optimization routine in search for optimal solutions.

When ANN is used for function approximation, the main concern is its accuracy. Usually the ANN approximation can not be 100% accurate. To have a better accuracy, a larger training sample size is required, which leads to more computation efforts. The users need to verify the accuracy of ANN approximation, and find out whether or not the accuracy is acceptable. When ANN is used as an optimizer, the main concern is its local optimisation characteristic. It is possible that the global optimum can never be reached.

Future research directions for application of ANN models in reliability based design includes improvement of the global search ability of the Hopfield neural networks, systematic selection of the parameters of the energy function of the Hopfield neural networks, combination of ANN function approximator with occasional evaluation of the exact values of the functions being approximated, the issue of error propagation in the function approximators, and the interaction between the function approximator and the actual optimization routine used.

References

H.M.F. AboElFotoh and L.S. Al-Sumait, (2001) A neural approach to topological optimization of communication networks, with reliability constraints. IEEE Transactions on Reliability, Vol. 50, No. 4, pp. 397-408.

R.E. Barlow and A.S. Wu, (1978), Coherent systems with multi-state Components, Mathematics of Operations Research, Vol. 3, No. 4, pp. 275-281.

G. Castellano, A.M. Fanelli and M. Pelillo, (1997) An iterative pruning algorithm for feedforward neural networks, IEEE Transactions on Neural Network, Vol. 8, pp. 519–537.

G. Cybenko, (1989), Approximation by superpositions of a sigmoidal function. Mathematics of control, signals, and systems, Vol. 2, No. 4, pp. 303-314

D.W. Coit and A.E. Smith. (1996) Solving the redundancy allocation problem using a combined neural network/genetic algorithm approach. Computers & Operations Research, Vol. 23, No. 6, pp. 515-526.

L. Fu, (1994), Neural Networks in Computer Intelligence, McGraw-Hill, Inc., New York.

M. Gen, K. Ida and J. U. Lee, (1990), A computational algorithm for solving 0-1 goal programming with GUB structures and its applications for optimization problems in system reliability, Electronics and Communication in Japan: Part 3, Vol. 73, pp. 88-96.

M. Gen, K. Ida, Y. Tsujimura and C. E. Kim, (1993), Large scale 0-1 fuzzy goal programming and its application to reliability optimization problem, Computers and Industrial Engineering, Vol. 24, pp. 539-549.

J.J. Hopfield. (1982) Neural networks and physical systems with emergent collective computational abilities. Proceedings of the National Academy of Sciences of the United States of America-Biological Sciences, Vol. 79, No. 8, pp. 2554-2558.

J. J. Hopfield and D.W. Tank, (1985) Neural computation of decisions in optimization problems, Biological Cybernetics, vol. 52, pp. 141–152.

H. Huang, Z. Tian, and M.J. Zuo, (2005) Intelligent interactive multi-objective optimization method and its application to reliability optimization". IIE Transactions, Vol. 37, No. 11, pp. 983-993.

W. Kuo, V. R. Prasad, F. A. Tillman and C. L. Huang, (2001), Optimal Reliability Design, Cambridge University Press, New York.

P.X. Liu, M.J. Zuo and M. Q-H Meng, (2003) A neural network approach to optimal design of continuous-state parallel-series systems. Computers and Operations Research, Vol. 30, pp. 339–352.

S. Matsuda, (1999) Quantized Hopfield networks for integer programming. Systems and computers in Japan, pp. 1354–64.

K. B. Misra and U. Sharma, (1991), An efficient approach for multiple criteria redundancy optimization problems, Microelectronics and Reliability, Vol. 31, No. 2, pp. 303-321.

M. Nourelfath and N. Nahas. (2003) Quantized hopfield networks for reliability optimization. Reliability Engineering & System Safety, Vol. 81, No. 2, pp. 191-196.

M. Papadrakakis and N.D. Lagaros. (2002) Reliability-based structural optimization using neural networks and Monte Carlo simulation. Computer Methods in Applied Mechanics and Engineering, Vol. 191, No. 32, pp. 3491-3507.

R. Rojas, (1996). Neural networks: a system introduction. Berlin: Springer.

C. Srivaree-Ratana, A. Konak and A.E. Smith. (2002) Estimation of all-terminal network reliability using an artificial neural network. Computers & Operations Research, Vol. 29, No. 7, pp.: 849-868.

R.Q. Zhao and B.D. Liu. (2004) Redundancy optimization problems with uncertainty of combining randomness and fuzziness. European Journal of Operational Research, Vol. 157, No. 3, pp. 716-735.

Software Reliability Predictions using Artificial Neural Networks

Q.P. Hu, M. Xie and S.H. Ng

Department of Industrial and Systems Engineering
National University of Singapore, Singapore

8.1 Introduction

Computer-based artificial systems have been widely applied in nearly every field of human activities. Whenever people rely heavily on some product/technique, they want to make sure that it is reliable. However, computer systems are not as reliable as expected, and software has always been a major cause of the problems. With the increasing reliability of hardware and growing complexity of software, the software reliability is a rising concern for both developer and users. Software reliability engineering (SRE) has attracted a lot of interests and research in the software community and software reliability modeling is one major part of SRE research.

Software reliability modeling describes the fault-related behaviors of the software testing process and is one of the important achievements in software reliability research activities. The information provided by the models is helpful in making management decisions on issues regarding the software reliability. They have been successfully applied in practical software projects, such as cost-analysis [17, 34], testing-resource allocation [6, 37], test-stopping decision [21, 32] and fault-tolerance system analysis [11, 19].

Generally, software reliability models can be grouped into two categories: analytical software reliability growth models (SRGMs) and data-driven models. Analytical SRGMs use stochastic models to describe the software failure process under several assumptions to provide mathematical tractability [18, 22, 23, 24, 31]. The major drawbacks of these models are their restrictive assumptions. On the other hand, most data-driven models follow the approach of time series analysis, including both traditional autoregressive methods [5] and modern artificial neural network (ANN) techniques [15]. These models are developed from past software failure history data. Specially, ANNs are universal functional approxima-

Q.P. Hu et al.: *Software Reliability Predictions using Artificial Neural Networks*, Computational Intelligence in Reliability Engineering (SCI) **40**, 197–222 (2007)
www.springerlink.com   © Springer-Verlag Berlin Heidelberg 2007

tors. They are generally nonlinear and have the capability for generalization [12]. ANN models have recently attracted more and more attention [3, 10, 13, 16, 25, 28, 29]. ANN models are developed with respect to software failure data under specific network architecture. Compared to SRGMs, ANN models are much less restrictive in assumptions. Besides these major kinds of models, there are also some models recently developed through Fuzzy theory [2], Bayesian networks [1], etc.

Currently, most software reliability models, both analytical and data-driven models, assume (explicitly or implicitly) that software faults can be removed immediately once they are detected. As a result, these models only describe the dynamic behavior of the software fault detection process (FDP). In real practice, after a fault is detected, it has to be reported, diagnosed, removed and verified before it can be noted as corrected. This related correction time is not that trivial to be ignored [30, 39]. Furthermore, the fault-correction time is an important factor in some management decision analysis, such as stopping time for testing, fault-correction control, and fault correction resource allocation. Therefore, utilizing only software fault detection data series can result in highly inaccurate predictions of the software reliability. When both fault detection and correction data series are available, they can be utilized by incorporating fault correction process (FCP) into software reliability models to make the software reliability models more realistic.

Some extensions on current software reliability models have been explored. For analytical models, Schneidewind (1975) proposed to model fault correction process as a separate process following the fault detection process with a constant time lag [26]. This idea was extended in several ways in [35]. Schneidewind (2001) further extended the original model by assuming the time lag is a random variable [27]. These works are based on non-homogeneous Poisson process (NHPP) models where the time delay is the critical aspect of the modeling. As this approach is based on the traditional software reliability models, much time on modeling is saved. In addition, with only one extra factor of correction time, this model provides a simple analysis approach. Within the Markov framework, a non-homogeneous continuous time Markov chain has been proposed [9]. Due to its complexity, analysis is not tractable and is often done trough simulation. Moreover, there is a state explosion problem with such models. Similarly, ANN models can also be extended to model both FDP and FCP. This can be done with a separate network for FCP in addition to the original one for FDP. As a general designation, all these modeling approaches will be called separate approaches, for they are developed through a separate way.

However, applying this separate approach to either analytical or ANN model still fails to describe the interactions between FDP and FCP. Specifically, the paired analytical models treat the software FDP as a NHPP independent from FCP and no influence from FCP is considered, while the influence of FDP on FCP is described by the time delay, which is not always the case for variant software testing processes. As for a separate ANN model, no interactions between these two processes are considered at all, and the two processes are treated as uncorrelated from each other. However, the feedback of fault correction on detection can not be ignored. Intuitively, slow fault correction should have negative effects on the fault detection process, and in extreme, it can make the successive detection process halt; while fast correction process would add pressure to the fault detection indirectly, through action on the testing personnel. In addition, as a following process to FDP, FCP can be described better by incorporating more information from FDP models. None of the above described approaches can meet this requirement.

Compared with the analytical approach, the ANN modeling framework is flexible in combining multiple processes together [4], and has the potential to overcome the deficiencies described above. This problem is explored comprehensively in this chapter under this framework. The combined ANN models with both the fault detection and correction processes are studied, focusing on the incorporation of the bi-directional influences between these two processes. In comparison with the former two schemes, more accurate prediction can be expected. Specifically, comparing with paired analytical models, the combined ANN model extracts the feedback from FCP on FDP, enabling better prediction for FDP. However, as the analytical models can describe the impact of FDP on FCP with time delay assumption, these two models would compete in predicting FCP. Compared to the separate ANN model, the combined model describes better the influences between the two processes, so they can be expected to perform better in both FDP and FCP prediction. Further in this chapter two kinds of framework for the combined ANN model are proposed and comparisons among these available models are made.

This chapter is organized as follows. In section 2, an overview on traditional software reliability models and their extensions to incorporate FCP is presented. In section 3, with the formulation of this problem, combined ANN models are described in detail. Two specific frameworks are introduced, one feedforward and one recurrent, which are modeled through different approaches. Section 4 applies these two combined ANN models to real software reliability data, presenting the comparisons within the two frameworks. Detailed comparisons with the paired analytical models and separate ANN model are given in section 5. Section 6 presents our conclusions and discussions on further studies on combined ANN model.

8.2 Overview of Software Reliability Models

8.2.1 Traditional Models for Fault Detection Process

In this section, we provide an overview of the major modeling approaches, adopting the classification approach similar to [31]. Generally, software reliability growth models (SRGMs) have both analytical and data-driven models. Analytical SRGMs have three major sub-categories: non-homogeneous Poisson process (NHPP) models, Markov models and Bayesian models. They are constructed by analyzing the dynamics of the software failure process, and their applications are developed by fitting them against software failure data.

8.2.1.1 NHPP Models

Denote $N(t)$ as the cumulative number of software failures occurred by time t. The process $\{N(t); t \geq 0\}$ is assumed to follow a Poisson distribution with characteristic MVF (Mean Value Function) $m(t)$. By assuming perfect and immediate fault-correction, the failure (fault-detection) process is also a fault-removal process.

Generally, different fault detection models can be obtained by using different nondecreasing MVF $m_d(t)$. For finite $m_d(t)$ models, there are two representative models as GO-model and S-shaped NHPP model. The GO-model [8] describes the fault detection process with exponential decreasing intensity with MVF as.

$$m_d(t) = a \cdot (1 - e^{-bt}), \ a, b > 0 \tag{2.1}$$

The S-shaped model [36] describes the fault detection process with an increasing-then-decreasing intensity, which can be interpreted as a learning process. The MVF is given as

$$m_d(t) = a \cdot [1 - (1 + bt)e^{-bt}], \ a, b > 0. \tag{2.2}$$

In both models, a is the final number of faults that can be detected by the testing process, and b can be interpreted as the failure occurrence rate per fault.

8.2.1.2 Markov Models

The best-known software reliability model, the JM-model, is a Markov model [14]. This model has the following underlying assumptions:
- the number of initial software faults is an unknown but fixed constant;
- a detected fault is removed immediately and no new fault is introduced;
- times between failures are independent, exponentially distributed random quantities;
- all remaining software faults contribute the same amount to the software failure intensity as ψ.

Denote N_0 as the number of software faults in the software before testing starts. From the assumptions, after the kth failure, there are (N_0-k) faults left, and the failure intensity decreases to $\psi(N_0-k)$. Then the time between failures T_i, $i = 1, \ldots, N_0$, are independent exponentially distributed random variables with respective parameter as $\lambda(i)=\psi[N_0-(i-1)]$, $i=1,\ldots, N_0$.

8.2.1.3 Bayesian Models

Bayesian analysis is a commonly accepted approach to incorporate previous knowledge in software testing. Most Bayesian formulations are based on the previous two kinds of models. One of the best-known Bayesian model is the LV-model [20]. It assumes that the time between failures are independent exponentially distributed with a parameter that is treated as random variable,

$$f(t_i \mid \lambda_i) = \lambda_i \cdot e^{\lambda_i t_i} , i = 1, 2, \ldots, n. \tag{2.3}$$

in which λ_i is assumed to have a Gamma prior distribution as

$$f(\lambda_i \mid \alpha, \psi(i)) = \frac{[\psi(i)]^\alpha \lambda_i^{\alpha-1}}{\Gamma(\alpha)} e^{\psi(i)\lambda_i} \tag{2.4}$$

where α is the shape parameter and $\psi(i)$ is the scale parameter depending on the number of detected faults.

8.2.1.4 ANN Models

ANN approach to model software is originally proposed in [15]. The reliability prediction here is regarded as an explanatory or causal forecasting problem [38]. The mapping between inputs and outputs of ANN can be written as follows: for generalization training $n_t = f(t_n)$; while for prediction training $n_t = f(t_{n-1})$, where t is the time when n failures occur. Most of the recent models [3, 13, 25, 29] take software reliability prediction as a time series forecasting problem [38]. The mapping of ANN can be written as $y_{t+1} = f(y_t, y_{t-1}, \ldots, y_{t-p})$, as illustrated in Fig. 1. Different failure data and network architectures are applied. y_t could be the inter/accumulated failure time/number. Both feedforward and recurrent neural networks have been applied. Usually, one-step predictions are developed for the measurement of failure time or number, and after that multi-step predictions can be obtained iteratively to show the trend of software failure behavior. ANN models have been successfully applied to solve software optimal release time problem with multi-step reliability prediction [7].

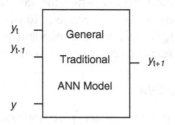

Fig. 1. General Traditional ANN Model

8.2.2 Models for Fault Detection and Correction Processes

8.2.2.1 Extensions on Analytical Models

Analytical model extensions use SRGMs to model the fault detection process, and describe the fault correction process as a time-delayed process due to time delay for correction. With FDP modeled as Schneidewind's SRGM, by assuming that fault correction has the same rate as detection, the FCP is modeled as a delayed FDP with a constant, random or time-dependent time-lag [26, 27, 35]. Extensions can be made to model FDP with other NHPP (Non-Homogeneous Poisson Process) SRGMs and the FCP can be modeled as a correspondingly delayed process. GO-Model and delayed S-shaped model are typical NHPP models, with the S-shaped model focusing on describing the learning-phenomenon along with software testing. Specifically, if FDP is modeled with GO-model, software FDP and FCP are described as two processes with the following paired characteristic MVFs (Mean Value Functions)

$$\begin{cases} m_d(t) = a(1 - e^{-bt}), & a > 0, b > 0 \\ m_c(t) = a\left[1 - e^{-b(t-\Delta_t)}\right], & t \geq \Delta_t \end{cases} \qquad (2.5)$$

If FDP is modeled as delayed S-shaped model to describe the learning phenomenon, the paired MVFs are given as

$$\begin{cases} m_d(t) = a\left[1 - (1 + bt)e^{-bt}\right], & a > 0, b > 0 \\ m_c(t) = a\left[1 - (1 + b(t - \Delta_t))e^{-b(t-\Delta_t)}\right], & t \geq \Delta_t \end{cases} \qquad (2.6)$$

where Δ_t denotes the time-delay of FCP with respect to FDP, which can be a constant or a time-dependent value.

8.2.2.2 Extensions on ANN Models

In parallel, similar to the paired analytical models describing these two processes separately, traditional ANN models can also be extended to model both FDP and FCP in a separate way. Originally, software reliability ANN models use the cumulative detected faults number data sequence $\{d_1, d_2, ..., d_n\}$ collected from FDP to establish the model presented in Fig. 2a. Separately, the FCP model can be incorporated with the cumulative corrected faults number data sequence $\{c_1, c_2, ..., c_n\}$ collected from FCP.

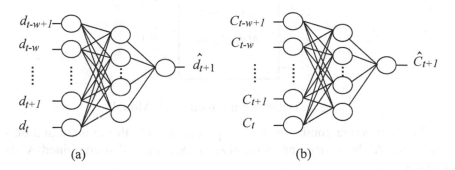

(a)   (b)

Fig. 2. Separate ANN Model Architecture, for FDP (a) and for FCP (b)

The corresponding framework is shown in Fig. 2b. As the models for FDP and FCP constitute two separate networks, they are further referred to as separate ANN models.

8.3 Combined ANN Models

In order to provide more accurate software reliability data prediction, it is essential to model the related dynamic phenomenon more realistically. Software testing (random testing) is a complicated and interactive process, and from the viewpoint of software reliability, there are both software fault detection and fault correction processes. These two processes are correlated. Once a fault is detected, it will be submitted for correction. This requires time for diagnosing, removal and verification. If the fault does not hamper the detection process, these two processes will proceed in parallel, but if it is so severe that the software is deemed inoperable, the detection process would wait until the fault is corrected; if the detection rate is very high, it will bring pressure to correction process, and vice versa.

Traditional SRGMs and ANN models only describe the fault detection process by assuming immediate and perfect correction. The practical extensions, paired analytical models and separate ANN models mentioned in the previous section, account for the fault correction process. However, they fail to model the interactions between these two processes. In this section, we propose the combined ANN models, as illustrated in Fig. 3, to model both FDP and FCP using the method of multivariate time series prediction [4] and to incorporate the interactions between these two processes.

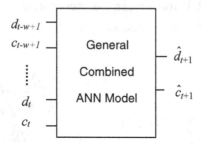

Fig. 3. General combined ANN Model

The architecture combines the two processes in both the input and output of the ANN model, and is the reason they are called combined ANN models.

Specifically, there are two major kinds of ANNs: feedforward and recurrent, and both have their advantages in time series predictions. Feedforward ANNs have been adopted by most researchers and there are some "rules of thumb" to follow in modeling network architecture. Construction with this well-studied framework is effortless and effective. Although recurrent ANNs are less studied, they have the ability to incorporate temporal information as they feedback inner states or outputs into the input layer. Both frameworks will be explored in this chapter.

8.3.1 Problem Formulation

Within the framework of neural networks modeling [4, 15], we formulate our problem as follows. By denoting $D(i), i = 1, 2, 3...$ and $C(i), i = 1, 2, 3...$, as the cumulative number of detected faults and corrected faults after testing period i respectively, we define software testing process as a bi-process combining both FDP and FCP, $S(i) = \begin{bmatrix} D(i) \\ C(i) \end{bmatrix}, i = 1, 2, 3....$ With ongoing testing process, software fault-related data can be collected as data sets of $\{s_1, s_2, ..., s_{t-1}, s_t\} = \{\begin{bmatrix} d_1 \\ c_1 \end{bmatrix}, \begin{bmatrix} d_2 \\ c_2 \end{bmatrix}, ..., \begin{bmatrix} d_{t-1} \\ c_{t-1} \end{bmatrix}, \begin{bmatrix} d_t \\ c_t \end{bmatrix}\}$, from the beginning of software system random testing until current testing period t. $D(i)$ and $C(i), i = 1, 2, 3...$ are two correlated processes. To make testing related decisions, at the end of every testing period, we are interested in knowing the possible outcomes of the following time period. In other words, we need to develop one-step predictions based on the historical data sequence $\{s_1, s_2, ..., s_{t-1}, s_t\}$ to get $\hat{S}_{t+1} = \begin{bmatrix} \hat{D}_{t+1} \\ \hat{C}_{t+1} \end{bmatrix}$, the predicted number of cumulative faults by the end of testing period $t+1$. Then with the updating of new data s_{t+1}, we can evaluate the performance of the previous prediction and develop the prediction for the fault number in the next interval $\hat{S}_{t+2} = \begin{bmatrix} \hat{D}_{t+2} \\ \hat{C}_{t+2} \end{bmatrix}$. The prediction process is continually updated as new testing data becomes available from ongoing testing.

8.3.2 General Prediction Procedure

With pre-set configurations of the network, the prediction is a sequentially updating process, with stepwise prediction utilizing each newly collected software faults data from the ongoing testing. At each point, with the latest and all past data, the network is retrained for new prediction in three specific steps: data normalization, network training, and prediction. The specific prediction procedure for any one point with the combined ANN model is described generally as follows.

8.3.2.1 Data Normalization

Collected cumulative software fault data $\{s_1, s_2, ..., s_{t-1}, s_t\}$ cannot be fed into networks directly as they need to be normalized between [0, 1]. Normalization functions varies, and for our case, the simple normalization scheme of $s_i^{norm} = s_i/s_{max}$ is adopted. As $D(i)$ and $C(i)$, $i = 1, 2, 3...$ are incremental processes, the collected or predicted data would show an increasing trend, so we need to estimate the upper limit of $\hat{D}(t+1)$ and $\hat{C}(t+1)$. With the available cumulative data, this value can be calculated by estimating the maximum possible increments of Δd and Δc. This number can be estimated from past experience of similar projects. Then s_{max} at the end of testing period t is calculated as

$$s_{max}^t = Max(d_t, c_t) + Max(\Delta d, \Delta c) \tag{3.1}$$

To simplify our notation further we assume that $\{s_1, s_2, ..., s_{t-1}, s_t\}$ is already normalized.

8.3.2.2 Network Training

With available normalized data, the neural networks with pre-defined configuration can be trained to model these two processes. The collected historic data sequence $\{s_1, s_2, ..., s_{t-1}, s_t\}$ should be grouped into as many as $t-w$ past-to-future mapping patterns denoted as $\{s_{k-w}, s_{k-w+1}, ..., s_{k-1} \mid s_k\}, k = w+1, ..., t$. These training patterns abstract the historic input-output relationships of the network. The patterns are used to train the network by adjusting its weights and bias, which are initially set randomly. Typically, backpropagation algorithms are used to train the networks and there are some variations in the algorithms. These algorithms usually look for ANN parameters (weights of internodes' connections and node biases) to fit the patterns by minimizing the deviation of the network outputs from the outputs of training patterns. To overcome the overfitting problem, usually the generalization technique is adopted.

8.3.2.3 Fault Prediction

With the trained network, which has "fit" the training patterns out of the collected data set $\{s_1, s_2, ..., s_{t-1}, s_t\}$, we can use the most recent w data set $\{s_{t-w+1}, s_{t-w+2}, ..., s_t\}$ to generate the next pattern as $\{s_{t-w+1}, s_{t-w+2}, ..., s_t \mid \hat{s}_{t+1}\}$. Then we can get our predictions for the next time point as $\hat{s}_{t+1} = \begin{bmatrix} \hat{d}_{t+1} \\ \hat{c}_{t+1} \end{bmatrix}$.

An initialization problem exists in the training algorithms. Different initial values for network weights and bias would generate different training results. For the generalized training algorithm adopted here, the initial values are assigned randomly. For each point predictions, m replicated runs are usually performed with different initializations, and the mean is used as the prediction outputs [15, 29] given as

$$\hat{s}_{t+1} = \frac{1}{m} \sum_{i=1}^{m} \hat{s}_{t+1}^{i} \tag{3.2}$$

8.3.3 Combined Feedforward ANN Model

8.3.3.1 ANN Framework

The framework of the combined feedforward ANN model is illustrated in Fig. 4.

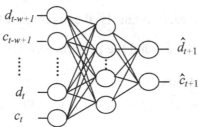

Fig. 4. Combined Feedforward ANN Model Architecture

It has inputs of both $\{d_{t-w+1}, ..., d_t\}$ and $\{c_{t-w+1}, ..., c_t\}$ and outputs of both $\hat{d}_{t+1}$ and $\hat{c}_{t+1}$. Specifically, the model is trained with the data from the bi-process $\{s_1, s_2, ..., s_{t-1}, s_t\}$ (both $\{d_{t-w+1}, ..., d_t\}$ and $\{c_{t-w+1}, ..., c_t\}$), combined with the past information of these two interactive processes together. Then with the well trained networks, the prediction can be generated from the latest w data points $S_{t-w+1},..,S_{t-1},S_t, 0 \le w \le t$ as the following function:

$$\hat{S}_{t+1} = F(S_{t-w+1},\ldots,S_{t-1},S_t) \tag{3.3}$$

8.3.3.2 Performance Evaluation

As an on-line prediction procedure, it starts tracking from the early stages of software testing. With the ongoing testing process, prediction is developed with the arrival of every updated data. For each single point prediction, the prediction is expected to be close to the collected data in the coming time period. Therefore, the prediction can not be evaluated until the next updated data is collected. As a whole, the prediction performance of the ANN model is evaluated with respect to all the past predictions with the data obtained from the whole testing process.

Specifically, suppose dataset $\{s_1, s_2, \ldots, s_t\}$ is used for network configuration. Within this data set, we simulate the sequential stepwise prediction process as in real software testing. Assume t_0 is the first point for prediction, and all the preceding data points $\{s_1, s_2, \ldots, s_{t_0-1}\}$ are used to train the network to get the prediction $\hat{s}_{t_0}$. With m different network initialization, m prediction repetitions are developed as $\hat{s}_{t_0,j}, j = 1,2,\ldots,m$. This procedure is carried on to get the following stepwise predictions $\hat{s}_{i,j}, i = t_0, \ldots, t, j = 1,2,\ldots,m$.

The prediction of each point is the average of the m repetitions $\hat{s}_i = \dfrac{1}{m}\sum_{j=1}^{m} \hat{s}_{i,j}, \quad i = t_0, \ldots, t$ and the performance is evaluated with its departure from the actual data as

$$SE_i = \left|\hat{s}_i - s_i\right|^2 = (\hat{d}_i - d_i)^2 + (\hat{c}_i - c_i)^2 = SE_i^d + SE_i^c, i = t_0, \ldots, t.$$

It is expected that the selected model works well through the whole testing process. The overall performance of the configuration is then determined by

$$MSE = \frac{1}{t-t_0+1}\sum_{i=t_0}^{t} SE_i^d + \frac{1}{t-t_0+1}\sum_{i=t_0}^{t} SE_i^c = MSE^d + MSE^c \quad (3.4)$$

8.3.3.3 Network Configuration

Feedforward networks are the most common networks and are widely studied. There are several "rules of thumb" to develop these networks that we adopt here. Within the context of this specific feedforward model, we pre-configure the network as follows. The architecture has three layers: an input layer, a hidden layer, and an output layer. Each ANN has $2*w$ inputs, which corresponds to w data sets [d, c]' presented to the network. In order to overcome the overfitting problem, the number of the hidden nodes should not be large. By comparing some practical recommendations, we chose this number as double the number of input nodes [38], i.e., $4*w$. The sigmoid function (logistic) is used as the activation function for each node in both the hidden and the output layers.

Using Eq. 3.4 as the performance criterion, the trial and error approach is used to determine the remaining parameters of the training algorithm.

8.3.4 Combined Recurrent ANN Model

8.3.4.1 ANN Framework

Similar to the combined feedforward ANN model, the proposed recurrent ANN model has the combined architecture as shown in Fig. 5, with feedback from the inner states to the input layer. Similarly, the model is trained with the data from the bi-process $\{s_1, s_2, ..., s_{t-1}, s_t\}$ (both $\{d_{t-w+1}, ..., d_t\}$ and $\{c_{t-w+1}, ..., c_t\}$), combined with the past information of these two interactive processes together. Then with the well trained networks, the prediction can be generated from the latest w data points $S_{t-w+1}, ..., S_{t-1}, S_t, 0 \leq w \leq t$ as the following function:

$$\hat{S}_{t+1} = F(S_{t-w+1}, ..., S_{t-1}, S_t; State_t\} \quad (3.5)$$

With respect to the model constraints, some parameters can be pre-configured as follows. Similar to the feedforward architecture, the basic Elman adopted here has architecture of three layers, one input layer, one hidden layer, and one output layer. Differently, Elman network has feed-back from the hidden states into the network inputs. This network has $2*w$ inputs, which corresponds to w data sets $[d, c]$' presented to the network. The sigmoid function (logistic) is used as the activation function for each node in both the hidden and the output layers.

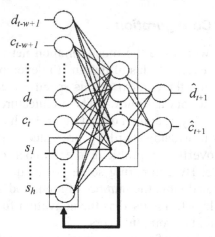

Fig. 5. Combined Recurrent ANN Model Architecture

8.3.4.2 Robust Configuration Evaluation

Unlike the evaluation on the combined feedforward network, the per-formance for the combined recurrent network is evaluated differently. Similarly, for each single point prediction, the prediction is expected to be close to the collected data in the coming time period. In addition, because the prediction is random, some repetitions are generated and small vari-ance is also expected. With a given dataset, which represents the history of a period of software testing, the configuration of network can be evaluated with its average performance in prediction through this period from the first prediction point. Different from the evaluation for the combined feed-forward ANN model, robustness criterion is adopted for combined recur-rent model.

The prediction of each point is the average of the m repetitions

$$\hat{s}_i = \frac{1}{m} \sum_{j=1}^{m} \hat{s}_{i,j} , \quad i = t_0, \dots, t$$ and the performance is evaluated by its depar-

ture from the actual data

$$SE_i = |\hat{s}_i - s_i|^2 = (\hat{d}_i - d_i)^2 + (\hat{c}_i - c_i)^2 = SE_i^d + SE_i^c , i = t_0, \dots, t$$

The dispersion of these m repetitions is given as

$$S_i^2 = \frac{1}{m} \sum_{j=1}^{m} (\hat{d}_{i,j} - \hat{d}_i)^2 + \frac{1}{m} \sum_{j=1}^{m} (\hat{c}_{i,j} - \hat{c}_i)^2 = S_{i,d}^2 + S_{i,c}^2, i = t_0, \ldots, t.$$

It is expected that the selected model works well through the whole testing process, and the overall performance of the configuration is evaluated by the following two criteria

$$
\begin{cases}
MSE = \dfrac{1}{t - t_0 + 1} \sum_{i=t_0}^{t} SE_i = MSE^d + MSE^c \\[3mm]
MS^2 = \dfrac{1}{t - t_0 + 1} \sum_{i=t_0}^{t} S^2 = MS_d^2 + MS_c^2
\end{cases}
\tag{3.6}
$$

However, as both criteria can be contradictory, in to balance both the prediction location and dispersion, their summation is used to evaluate the performance instead. Hence, for one specific network configuration θ, the performance function is expressed as

$$L(\theta) = MSE + MS^2 \tag{3.7}$$

8.3.4.3 Network Configuration through Evolution

There is less guidance for recurrent network configuration, and also the network training requires much more time than feedforward networks. Therefore some automatic "trial-and-error" approach is useful. Evolutionary programming provides an approach to optimize complex problems with specific fitness function, which suits our problem well, i.e., to search for an optimal configuration setting θ^* from the parameter space with respect to fitness function of $L(\theta)$ in Eq. 3.7. In fault detection prediction, genetic algorithm has been applied to optimize the network architecture parameters to determine the number of inputs and hidden nodes for feedforward architecture [29]. Specially, we also make the configuration setting include the algorithm parameters, for they are found to have great influence on the performance.

The configuration evolving process is described as following steps:
- *Step 1:* Encode the configuration setting θ into chromosome.
- *Step 2:* Generate an initial population of l individuals $\theta_1, \theta_2, \ldots, \theta_l$
- *Step 3:* Calculate the fitness function $L(\theta_i)$, $i=1,2,\ldots,l$ for each individual

– *Step 4:* Select parent settings for next generation according to the fitness values:
 • Crossover breeding operator
 • Mutation operator
 • Cull inferior solution
– *Step 5.* Repeat step 3 until stopping criteria are met and return optimal setting θ^*.

8.4 Numerical Analysis

To illustrate the application of combined ANN models we apply the two suggested models to real data collected from a middle sized application software testing process. The collected interval data set includes both fault detection and correction data, $\Delta D(t)$ and $\Delta C(t)$, as shown in Table 1.

Table 1. Fault detection and correction data (number per week)

Week	$\Delta d(t)$	$d(t)$	$\Delta c(t)$	$c(t)$
1	12	12	3	3
2	11	23	0	3
3	20	43	9	12
4	21	64	20	32
5	20	84	21	53
6	13	97	25	78
7	12	109	11	89
8	2	111	9	98
9	1	112	9	107
10	2	114	2	109
11	2	116	4	113
12	7	123	7	120
13	3	126	5	125
14	2	128	2	127
15	4	132	0	127
16	9	141	8	135
17	3	144	8	143

The proposed combined ANN models are used to develop one-step prediction for both $D(t)$ and $C(t)$, starting from some early point and tracking the software testing process till the end with the continuous updating of collected cumulative data .

With the combined ANN model, using either the feedforward or recurrent network architecture, some pre-configuration can be developed with respect to the constraints of the specific problem. At the beginning of this testing process, available data is scarce. Therefore, prediction needs to be developed as soon as possible, providing timely decision-making assistance for the testing procedure. However, for such data-driven modeling approaches like ANN, the model cannot be well adjusted without essential number of data points. As a result prediction cannot begin until enough data is collected. To compromise, the size of sliding window w cannot be large, and we set it to $w = 3$, and start prediction from the 6th testing period. Then we know our combined ANN model will have 6 inputs. As to the number of hidden nodes and some parameters related to algorithm, they are configured differently for feedforward and recurrent networks. Obviously, the network model has two outputs.

8.4.1 Feedforward ANN Application

As a common "rule of thumb", the number of hidden nodes is set to be double the number of input nodes. With different configurations on the training algorithm parameters, the following procedure is developed with trial-and-error to get a fully-configured network for further prediction out of sample. With the available data sequence as $\{s_1, s_2, ..., s_{t-1}, s_t\}$ the prediction can be developed as follows.

1. Data normalization:
Based on experience from similar past projects and current testing personnel allocation, the expected incremental number of $\Delta D(t)$ and $\Delta C(t)$ cannot exceed 25: $Max(\Delta d, \Delta c) = 25$. This number if set fixed for the whole prediction process with ongoing testing process. The data is normalized with the maximum number calculated from Eq. 3.1.

2. Network training:
With the normalized data set, the training patterns are generated for both frameworks respectively as $\{s_{k-w}, s_{k-w+1}, ..., s_{k-1} \mid s_k\}, k = w+1, ..., t$.

For our case, backpropagation algorithm is adopted to train the ANN with the generated patterns. To improve generalization of the training, the regularization method is implemented by adding the mean of the sum of squares of network weights and biases $g_i, i = 1, ..., l$, *MSW*, to the network performance function *MSE* in the following form

$$MSE_{reg} = \gamma \cdot MSE + (1-\gamma) \cdot MSW \tag{4.1}$$

where γ is the performance ratio, and $MSW = \dfrac{1}{l}\sum\limits_{j=1}^{l} g_j^2$. Such regulariza-

tion can force the network to have smaller weights and biases, which provides the smoother network response. It also reduces the chance of overfitting. The parameter γ is set by trial and error

3. Prediction:

With the well-trained network, the latest w data is fed and the prediction is generated as the network outputs. 50 runs of prediction for each point are performed, yielding 50 predictions for point i, $\hat{s}_{i,j}$, $i = 6, j = 1, ..., 50$.

Related variances are calculated to estimate the robustness of the model.

The prediction process is performed from the 6th testing period till the end of the testing. The prediction for each point is evaluated with the updated fault data. The prediction sequence is obtained in the form $\left\{\hat{d}_6, \hat{d}_7, ..., \hat{d}_{16}, \hat{d}_{17}\right\}$ and $\left\{\hat{c}_6, \hat{c}_7, ..., \hat{c}_{16}, \hat{c}_{17}\right\}$. Then the prediction performance of the model over the whole testing process is evaluated by comparing with the true data with mean squared errors calculated with Eq. 3.4. The corresponding prediction results are summarized in Table 2.

Table 2. One-step Predictions with Combined Feedforward ANN Model

Week	$\hat{d}(t)$ Mean	$\hat{d}(t)$ Var	SE_t^d	$\hat{c}(t)$ Mean	$\hat{c}(t)$ Var	SE_t^c
6	96.29	0.4324	0.50	73.43	0.7162	20.87
7	102.48	0.0895	42.55	94.53	0.9200	30.58
8	113.21	0.0032	4.88	91.66	0.0138	40.23
9	116.02	0.0024	16.12	97.76	1.0854	85.47
10	112.80	0.0838	1.43	110.17	0.9192	1.38
11	113.38	0.0550	6.86	111.63	0.0029	1.87
12	118.28	0.0024	22.31	115.25	0.0012	22.53
13	129.78	0.0062	14.30	123.50	0.0097	2.25
14	126.86	0.0682	1.29	127.18	0.0906	0.03
15	128.68	0.1775	11.03	129.45	0.0036	5.99
16	134.82	0.0019	38.21	130.10	0.0153	23.99
17	144.54	0.0066	0.29	140.42	0.0011	6.66
Ave.		0.0774	13.31		0.3149	20.15

Var in the table denotes the variance of the repeated predictions at each point. From Table 2, we can see that under this network configuration, the predictions along the period of this dataset can fit the observed value well with small variances.

8.4.2 Recurrent ANN Application

Adopting similar preset architecture parameters as the feedforward model, the remaining architecture parameter for the recurrent model is the number of hidden nodes n_h. Besides the architecture parameter, the network configuration θ should also include critical training algorithm parameters. The specific prediction procedure for this dataset is similar to that in section 3.1. From the former analysis, we have found that the performance ratio γ is a critical parameter. Here, back-propagation algorithm with learning rate and momentum is adopted. These two parameters are important to the algorithm performance. As the learning rate is adaptive, it is important for network training to set a proper value for momentum m_o.

The network parameters to be configured can be determined as $\theta = \begin{bmatrix} n_h & m_o & \gamma \end{bmatrix}$, i.e. the hidden nodes number, the momentum, and the performance ratio. For each specific configuration, such a prediction process is performed from the 6th testing period till the last one, obtaining predictions $\hat{s}_{i,j}$, $i = 6,7,\ldots,17, j =1,2,\ldots,5$. The corresponding fitness function value, i.e., the network performance value, can be calculated through

$$L(n_h, m_o, \gamma) = MSE + MS^2 .$$

This way, the evolving procedure in the former section is developed to find the proper value of $\theta^* = \begin{bmatrix} n_h^* & m_o^* & \gamma^* \end{bmatrix}$.

Table 3. One-step Predictions with Combined Recurrent ANN Model

Week	$\hat{d}(t)$		SE_t^d	$\hat{c}(t)$		SE_t^c
	Mean	Var		Mean	Var	
6	95.68	0.3919	1.76	72.64	1.4984	28.71
7	102.60	0.5183	40.96	86.78	1.9346	4.94
8	112.93	0.1001	3.72	98.83	0.2321	0.69
9	114.82	0.0887	7.95	102.81	0.1440	17.56
10	116.62	0.0113	6.86	108.82	0.0476	0.03
11	117.85	0.0376	3.42	110.86	0.1338	4.58
12	119.31	0.0517	13.62	113.52	0.1996	41.99
13	124.07	0.1142	3.72	121.06	0.3803	15.52
14	127.34	0.0099	0.44	126.34	0.0332	0.44
15	129.51	0.0708	6.20	129.85	0.1820	8.12
16	131.65	0.0000	87.42	132.73	0.0000	5.15
17	136.02	0.0389	63.68	137.77	0.0815	27.35
Ave.		0.1194	19.98		0.4056	12.92

With respect to this dataset, a configuration of network has been evolved with genetic algorithm as the hidden nodes number = 14; performance ratio = 0.9450; momentum = 0.9711. With this configuration, 20 more repeated predictions are obtained again for each time point. The prediction results are shown in the Table 3.

From these results, we can see that under this network configuration, the predictions along the period of this dataset can fit the observed value very well with small variances.

8.4.3 Comparison of Combined Feedforward & Recurrent Model

Both these two types of ANN models have been applied to model software reliability prediction. These two architectures have been also compared through different criteria with respect to different dataset [7, 13, 15, 29]. Although Elman architecture is advocated to incorporate the temporal patterns, there is no consistent advantage from these experimental results. As far as our dataset is concerned, we compare these two architectures with their predictive performance using both location (MSE) and dispersion (MS^2). This is summarized in Table 4.

Table 4. Comparison: Combined Feedforward *VS* Recurrent ANN

	MSE	MS2	L
Combined Feedforward ANN models	33.46	0.3923	33.8523
Combined Recurrent ANN models	32.90	0.5250	33.4250

From this table, we can see that there is slight advantage of combined recurrent ANN model over feedforward model, with respect to the "robust" performance L. However, if we take the criteria of either *MSE* or MS^2, contradictory conclusions will be drawn, although the differences are small. With respect to this data set, these two models are nearly the same. Therefore, both configured models can be set to develop predictions for the coming data points. Comparatively, combined feedforward ANN model would be more effective.

8.5 Comparisons with Separate Models

In this section, we proceed to verify that the proposed combined model would perform better than separate models. Accordingly, the comparisons of the combined ANN models with those two separate models mentioned in section 1 are developed.

8.5.1 Combined ANN Models vs Separate ANN Model

In the separate ANN model (Fig. 1.) two separate networks are modeled: one for $D(t)$ and the other for $C(t)$. The comparison between combined and separate ANN models is of interest because both of them are data-driven ANN models and their differences would focus on the effectiveness of the incorporating the correlations between these two processes. For the data set in Table 1, the "online" prediction process is developed with separate ANN model as follows. Feedforward network is adopted and the configuration is as follows. In the combined ANN models, the size of sliding window is set at $w=3$ and the prediction starts from the 6th point. The training, prediction and evaluation procedures are also the same as combined models. The prediction results are listed in Table 5.

Table 5. One-step Predictions with Separate ANN Model

Week	$\hat{d}(t)$		SE_t^d	$\hat{c}(t)$		SE_t^c
	Mean	Var		Mean	Var	
6	95.64	0.2296	1.84	66.17	0.3204	139.85
7	102.35	0.0263	44.23	89.78	1.4164	0.61
8	114.27	3.4484	10.69	91.49	0.0002	42.33
9	116.49	0.0007	20.18	100.13	0.0306	47.22
10	108.13	0.0039	34.41	109.78	0.0025	0.61
11	113.69	0.0014	5.35	112.23	0.0003	0.59
12	114.90	0.0000	65.68	114.89	0.0001	26.09
13	122.67	5.6761	11.12	117.44	0.0002	57.08
14	129.32	0.9281	1.75	125.94	0.0002	1.13
15	130.79	0.3720	1.47	129.84	0.0010	8.08
16	134.54	0.0515	41.71	129.80	0.0006	27.01
17	144.07	0.0002	0.00	135.14	0.0004	61.85
Ave.		0.8949	19.87		0.1477	34.37

From the results shown in Tables 2 - 5, we can compare these two kinds of models in two ways. With respect to the overall performance of MSE, the combined models outperform the separate one, which verifies the advantages of modeling the two processes together. In addition, from the prediction performance for each point, SE_t^d and SE_t^c, we observe an interesting phenomenon: prediction of the first point for $\hat{C}(t)$ is not acceptable, however the first prediction for $\hat{D}(t)$ performs well. This reflects the delay of FCP over FDP, which results in some data shortage for prediction of $\hat{C}(t)$ at the initial phase. Fortunately, prediction is reinforced by combining outputs of $C(t)$ and $D(t)$ in the combined ANN model.

8.5.2 Combined ANN Models vs Paired Analytical Model

When applying paired analytical models to the fault data, one faces the problem of model selection for FDP from many available NHPP SRGMs. As far as our case (Table 1) is concerned, the interval detected faults show an increasing trend in the early phase of software testing. Delayed S-shaped NHPP model is designed to describe such learning phenomenon. In addition, as this project takes relatively short testing period and is common application software, detected faults should be common ones and they are handed to available correctors that are stable though testing process. Therefore, instead of using Schneidewind's model directly, the slight extension as described in Eq. 2.6 is adopted, assuming constant time-delay between FDP and FCP.

In a similar way, "on-line" prediction is developed by fitting the model against historical data collected with the ongoing testing process. As a model-driven method, the prediction can be started from earlier points. However, in order to compare with the combined ANN model in the same time horizon, the prediction is also developed from the 6th point. The application results of the actual data with analytical models are presented in Table 6.

From the results shown in Tables 2, 3 and 6, we can see that combined ANN model performs over analytical model in prediction of both fault detection and correction. Further observations show that large prediction errors happen in the 8th, 9th, 16th and 17th points. Referring to Table 1, we see that these are the points where some unusual changes happen. Comparatively, ANN models work better on these points, showing more flexibility and sensitivity to the abnormal change. This difference can also be regarded as the difference in prediction approaches. The analytical model develops the prediction through fitting the historical data with respect to time; however, the ANN models develop networks to fit input-output patterns which incorporate the trend of data inside. More importantly, the simple time-delay assumption between the relationship of fault detection and correction does not fit this dataset well. The ANN models perform better in capturing the correlated relationships between these two processes.

Table 6. One-step Prediction with Paired Analytical Models

Week	$\hat{d}(t)$	SE_t^d	$\hat{c}(t)$	SE_t^c
6	100.97	15.74	72.71	27.95
7	113.36	18.99	91.17	4.71
8	120.71	94.28	103.71	32.64
9	121.48	89.79	110.07	9.45
10	121.55	57.00	114.10	26.05
11	120.97	24.72	116.19	10.18
12	120.93	4.29	117.80	4.82
13	122.80	10.22	120.56	19.73
14	124.78	10.36	123.13	14.98
15	126.34	32.04	125.13	3.49
16	127.70	176.84	126.81	67.01
17	130.37	185.82	129.65	178.27
Ave.		**55.15**		**28.95**

As a short conclusion, the software fault detection and correction processes are two correlated processes, and to develop accurate predictions, information about both of them should be incorporated into the model. Combined ANN models are a flexible way to implement this. Paired analytical model can describe one-directional effects, and in some cases it can perform better. However, the combined ANN models provide a unified approach to model the two processes together, which is more favorable than the analytical approach since more effort is needed on model selection in the analytical approach.

8.6 Conclusions and Discussions

In this chapter we have studied the use of neural networks to model both the software fault detection and correction processes together (referred to as combined ANN model), focusing on describing the interactions between these two correlated processes. This approach is regarded as an extension of separate ANN model under the same modeling framework, and is a complement to analytical models which only describe the influence of FDP on FCP as a time delay. With practical software testing data, this approach shows its advantage in incorporating more information than the separate ANN model and paired analytical model. Also, within the combined ANN models, both feedforward and recurrent frameworks perform well with the given dataset.

The combined ANN models are beneficial in incorporating the correlation between FDP and FCP. They model the software debugging process more realistically, with more accurate predictions. However, this model still has some aspects for further investigation. First, faults number is one important measure of software reliability, and predictions on some other measure such as detection rate would be interesting. [33] showed detection rate can be assumed to be the same as earlier projects/versions, and ANN models would help abstract this information when datasets from previous projects are available. Second, FCP is different from FDP, where some fault-correction factors (such as personnel) can be controlled. With more understanding of the interactions between FDP and FCP, some useful software fault correction policies can be proposed for more effective testing resource allocation. Third, software reliability prediction is just an initial step of reliability analysis. The prediction results need to provide assistance on decision-making for testing management. With potential to provide more accurate and multi-step predictions, and with the modeling of both fault-detection and correction processes, the combined ANN models are expected to be more helpful in testing management, such as decisions on stopping time. Application of this software reliability approach to decision problems with larger datasets will be useful in further understanding its potential.

Acknowledgement

This research is partly supported by the National University of Singapore under the research grant R-266-000-020-112, "Modelling and Analysis of Firmware Reliability".

References

[1] Bai CG, Hu QP, Xie M (2005) Software failure prediction based on a Markov Bayesian network model. Journal of Systems and Software 74(3) pp 275-282

[2] Cai KY (1996) Fuzzy Methods in Software Reliability Modeling. In: Introduction to Fuzzy Reliability. Kluwer Academic Publishers, pp. 243–276

[3] Cai KY, Cai L, Wang WD, Yu ZY, Zhang D (2001) On the neural network approach in software reliability modeling. Journal of Systems and Software 58(1) pp. 47-62

[4] Chakraborty K, Mehrotra K, Mohan CK, Ranka S (1992) Forecasting the Behavior of Multivariate Time-Series Using Neural Networks. Neural Networks 5(6) pp. 961-970

[5] Crow L H, Singpurwalla N D (1984) An empirically developed Fourier series model for describing software failures. IEEE Transactions on Reliability 33 pp. 176-183

[6] Dai YS, Xie M, Poh KL, Yang B (2003) Optimal testing-resource allocation with genetic algorithm for modular software systems. Journal of Systems and Software 66(1) pp. 47-55

[7] Dohi T, Nishio Y, Osaki S (1999) Optimal software release scheduling based on artificial neural networks. Annals of Software Engineering 8 pp. 167-185

[8] Goel AL, Okumoto K (1979) Time dependent error detection rate model for software reliability and other performance measures. IEEE Transactions on Reliability 28(3) pp. 206-211

[9] Gokhale SS, Lyu MR, Trivedi KS (2004) Analysis of software fault removal policies using a non-homogeneous continuous time Markov chain. Software Quality Journal 12(3) pp. 211-230

[10]Guo P, Lyu MR (2004) A pseudoinverse learning algorithm for feedforward neural networks with stacked generalization applications to software reliability growth data. Neurocomputing 56 pp. 101-121

[11]Han CC, Shin KG, Wu J (2003) A fault-tolerant scheduling algorithm for real-time periodic tasks with possible software faults. IEEE Transactions on Computers. 52(3) pp. 362-372

[12]Haykin S (1999) Neural networks: A comprehensive foundation. New York: Macmillan College Publishing Company

[13]Ho SL, Xie M, Goh TN (2003) A study of the connectionist models for software reliability prediction. Computers & Mathematics with Applications 46(7) pp. 1037-1045

[14]Jelinski Z, Moranda PB (1972) Software reliability research. In: Freiberger, W. (eds) Statistical Computer Performance Evaluation, Academic Press, New York

[15]Karunanithi N, Whitley D, Malaiya YK (1992) Prediction of software reliability using connectionist models. IEEE Transactions on Software Engineering 18(7) pp. 563-574

[16]Khoshgoftaar T M, Szabo RM (1996) Using neural networks to predict software faults during testing, IEEE Transactions on Reliability 45(3) pp. 456-462

[17]Kimura M, Toyota T, Yamada S (1999) Economic analysis of software release problems with warranty cost and reliability requirement. Reliability Engineering and System Safety 66(1) pp. 49-55.

[18]Kuo SY, Huang CY, Lyu MR (2001) Framework for modeling software reliability, using various testing-efforts and fault-detection rates. IEEE Transactions on Reliability. 50(3) pp. 310-320.

[19]Levitin G (2005) Reliability and performance analysis of hardware-software systems with fault-tolerant software components. Reliability Engineering & System Safety In Press, Corrected Proof, Available online 27 June 2005.

[20]Littlewood B, Verral J L (1973) A Bayesian reliabiltiy growth model for computer software. Applied Statistics 22 pp. 332-346.

[21]Littlewood B, Wright D, (1997) Some conservative stopping rules for the operational testing of safety critical software. IEEE Transactions on Software Engineering 23(11) pp. 673-683.

[22]Lyu MR (1996) Handbook of software reliability engineering. IEEE Computer Society Press

[23]Musa JD, Iannino A, Okumoto K (1987) Software reliability: measurement, prediction, application. McGraw-Hill, New York.

[24]Pham H (2000) Software reliability. Springer, Singapore; New York

[25]Sitte R (1999) Comparison of software-reliability-growth predictions: neural networks vs parametric-recalibration. IEEE Transactions on Reliability 48(3) pp. 285-291

[26]Schneidewind NF (1975) Analysis of error processes in computer software. Proceedings of International Conference on Reliable Software, 1975, IEEE Computer Society, pp. 337-346

[27]Schneidewind NF (2001) Modelling the fault correction process. Proceedings of the 12th International Symposium on Software Reliability Engineering, 2001, pp.185-190

[28]Takada Y, Matsumoto K, Torii K (1994) A Software-Reliability Prediction Model Using a Neural-Network. Systems and Computers in Japan 25(14) pp. 22-31

[29]Tian L, Noore A (2005) Evolutionary neural network modeling for software cumulative failure time prediction. Reliability Engineering & System Safety 87(1) pp. 45-51

[30]Tian L, Noore A (2005) Modeling distributed software defect removal effectiveness in the presence of code churn. Mathematical and Computer Modelling. 41(4-5) pp. 379-389.

[31]Xie M (1991) Software reliability modelling, World Scientific, Singapore.

[32]Xie M, Hong GY (1999) Software release time determination based on unbounded NHPP model. Computers & Industrial Engineering. 37(1-2) pp. 165-168

[33]Xie M, Hong GY, Wohlin C (1999) Software reliability prediction incorporating information from a similar project. Journal of Systems and Software. 49(1) pp. 43-48.

[34]Xie M, Yang B (2003) A study of the effect of imperfect debugging on software development cost. IEEE Transactions on Software Engineering 29(5) pp. 471-473

[35]Xie M, Zhao M (1992) The Schneidewind software reliability model revisited. Proceedings of 3rd International Symposium on Software Reliability Engineering, 1992, pp. 184-192

[36]Yamada S, Ohba M, Osaki S (1983) S-shaped reliaibility growth modeling for software error detection. IEEE Transactions on Reliability. 32(4) pp. 289-292

[37]Yamada S, Ichimori T, Nishiwaki M (1995) Optimal allocation policies for testing-resource based on a software reliability growth model. Mathematical and Computer Modelling. 22(10-12) pp. 295-301

[38]Zhang GQ, Patuwo BE, Hu MY (1998) Forecasting with artificial neural networks: The state of the art. International Journal of Forecasting. 4(1) pp. 35-62

[39]Zhang XM, Pham H (2005) Software field failure rate prediction before software deployment. Journal of Systems and Software. In Press, Corrected Proof, Available online 12 July 2005.

Computation Intelligence in Online Reliability Monitoring

Ratna Babu Chinnam

Department of Industrial and Manufacturing Engineering, Wayne State University, Detroit

Bharatendra Rai

Ford Motor Company, Dearborn

9.1 Introduction

9.1.1 Individual Component versus Population Characteristics

The lifecycle of a product is generally associated with two key players viz., a producer and an end-user. Although producers and end-users depend on each other, they have their own priorities. For a producer, the characteristics of a population of units are more important than the individual units, whereas for an end-user the opposite is true. The government makes laws for the population of a country, but a parent may be more concerned about its impact on the future of their individual children. A car manufacturer targets consistency in fuel efficiency for a population of cars, whereas a car owner has concerns about the fuel efficiency of his/her car. Similar differences in concerns also apply to tennis racquet manufacturer versus a tennis player or a cutting tool manufacturer versus cutting tool user.

The producer commonly uses time-to-failure data to assess and predict reliability of a population of products. For highly reliable products, when time-to-failure data are difficult to obtain, degradation data are also used for such an analysis. On the other hand, for an end-user assessing and predicting reliability of an individual part or component often assumes more importance. For example, a producer involved in voluminous production of drill-bits needs to assess and monitor reliability on a regular basis to ensure consistent population characterists for end users. However, the end-

R.B. Chinnam and B. Rai: *Computation Intelligence in Online Reliability Monitoring*, Computational Intelligence in Reliability Engineering (SCI) **40**, 223–260 (2007)
www.springerlink.com

user of such drill-bits generally has more interest in the reliability of the individual drill-bits. Voluminous amount of work has been published for reliability modeling and analysis related to population characteristics (Kapur and Lamberson 1977; Lawless 1982; Nelson 1982; Lewis 1987; Elsayed 1996; Meeker and Escobar 1998). A good review of literature on degradation signals can be obtained from Tomsky (1982), Lu and Pantula (1989), Nelson (1990), Lu and Meeker (1993), Tseng et al. (1995), Tang et al. (1995), Chinnam et al. (1996), Lu et al. (1997), Meeker et al. (1998), Wu and Shao (1999), Wu and Tsai (2000), and Gebraeel et al. (2004).

This chapter focuses on monitoring of reliability from end-users viewpoint. For monitoring reliability of individual unit, time-to-failure data are not of much use. For example, Fig. 1 gives a plot of the life of 16 M-1 grade quarter-inch high speed twist drills, measured in number of holes successfully drilled in quarter-inch steel plates, when operated with no coolant at a speed 2000 rpm and a feed 20 inches/min (Chinnam 1999). Even though the drill-bits came from the same manufacturer in the same box, it is obvious from the figure that the dispersion in life (ranging from 17 holes to 58 holes) is far too large with respect to the mean time-to-failure (around 28 holes), and hence, information about the population would be of little value to the end user. In contrast, the end user would greatly benefit from an on-line estimate of the reliability of the drill-bit, to make effective decisions regarding optimal drill bit replacement strategies, essentially lowering production costs by fully utilizing the drill-bit.

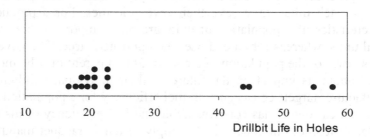

Fig. 1. A plot of life of 16 drill-bits, measured in number of holes successfully drilled

For individual units, condition at different points of time or degradation levels are more useful to arrive at optimal component replacement or maintenance strategies leading to improved system utilization, while reducing the risk and maintenance costs. While gathering data based on direct measurements of the condition or degradation level of a unit is not impossible, such methods are not practical due to the intrusive nature for applications requiring online monitoring of reliability. For example, measuring the amount of wear on a milling insert or a drill bit after every operation would slow down the production drastically. Thus, applications requiring online reliability monitoring of individual units more often use indirect and non-intrusive measurements. The nature and type of such indirect measurements may vary from application to application, however, their selection is critical for an effective decision making process. A basic criterion they need to fulfill, apart from being non-intrusive, is to have a good correlation with the degradation level of the part/component of interest.

9.1.2 Diagnostics and Prognostics for Condition-Based Maintenance

In many physical and electro-mechanical systems, the system or unit under consideration generates degradation signals that contain valuable information about the health/well-being of the system. These degradation signals, such as power consumption of a metal cutting machine tool, error rate of a computer hard disk, temperature of a drill-bit, vibration in machinery, color spectrum from an arc welder, loads acting on a structure, tend to be non-stationary or transitory signals having drifts, trends, abrupt changes, and beginnings and ends of events. Despite considerable advances in intelligent degradation monitoring for the last several decades, on-line condition monitoring and diagnostics are still largely reserved for only the most critical system components and have not found their place in mainstream machinery and equipment health management (Kacprzynski and Roemer, 2000). If one were to talk about predictive maintenance technologies, in particular prognostics and on-line reliability assessment, there exist no robust methods for even the most critical system components. *Diagnostics has traditionally been defined as the ability to detect and classify fault conditions.* Literature is extremely vast in this area. *Prognostics on the contrary is defined here as the capability to provide early detection of the precursor to a failure condition and to manage and predict the progression of this fault condition to component failure.* Recognizing the inability to prevent costly unscheduled equipment breakdowns through Preventive Maintenance (PM) activities and basic diagnostic condition monitoring methods, there seem to be consensus among industry and federal agencies that one of the next great opportunities for product differentiation and successful competition in the world markets lies in true prognostics based condition-based maintenance (CBM).

Condition-Based Maintenance (CBM) is a philosophy of performing maintenance on a machine or system only when there is objective evidence of need or impending failure. CBM typically involves mounting nonintrusive sensors on the component to capture signals of interest and subsequent interpretation of these signals for the purpose of developing a customized maintenance policy. Given recent advances in the areas of nonintrusive sensors, data acquisition hardware, and signal processing algorithms, combined with drastic reductions in computing and networking costs and proliferation of information technology products that integrate factory information systems and industrial networks with web-based visual plant front-ends, it is now possible to realize systems that can deliver cost effective diagnostics, prognostics, and CBM for a variety of industrial systems. The basic elements necessary for successful diagnostics and prognostics for CBM are illustrated in Fig. 2 (Chinnam and Baruah 2004).

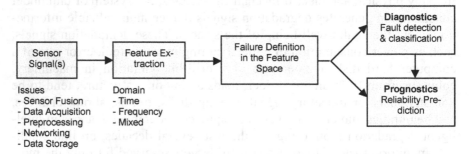

Fig. 2. Basic elements of diagnostics and prognostics for CBM.

Sensor signal(s): Sensing techniques commonly used include touch sensors, temperature, thrust force, vibration, torque, acoustic emission, voltage, noise, vision systems, etc. The data obtained from sensor signals contain useful information about the condition or degradation levels of a unit. As data recording is automated, it is not uncommon to see several thousands of measurements recorded per incremental usage condition. Extraction of appropriate and useful features from such data is critical before models for online monitoring and prediction of individual component reliability can be developed.

Feature extraction: Extraction of useful features typically involves analysis of data in several different domains. Basic time domain signal parameters utilized in conventional diagnostics include amplitude, crest factor, kurtosis, RMS values, and various measures of instantaneous and cumulative energy (Zhou et al. 1995; Quan et al. 1998; Kuo 2000; Dzenis and

Qian 2001; Sung et al. 2002). Frequency domain spectral parameters include the Fourier transform and linear spectral density, while advanced spectral measures include higher-order (or displaced) power spectral density. Extensive research over the last two decades has resulted in a long list of promising features or *Figures-of-Merit* (FOMs) for different applications. For example, Lebold et al. (2000) discussed 14 FOMs for gearbox diagnostics and prognostics, employing a vibration signal alone. The literature offers FOMs for monitoring and diagnosis of mechanical systems such as gearboxes, pumps, motors, engines, and metal cutting tools. In recent years, mixed-domain analysis methods such as Wavelets are gaining popularity for their ability to offer a shorter yet accurate description of a signal by employing scale-based basis functions. Wavelet analysis, while still being researched for machine diagnostics and prognostics (Chinnam and Mohan 2002; Vachtsevanos *et al.* 1999), is well established for such applications as image processing. In the last few years, the *Empirical Mode Decomposition* method has received much attention for its ability to analyze non-stationary and nonlinear time series, something not possible with methods such as Wavelet analysis (Huang et al. 1998).

Failure Definition in the Feature Space: A specified level of degradation in feature space is generally used to define failure. Such a threshold limit is required to assess and predict reliability of a unit. Sometimes the features of interest may consistently show significantly different degradation levels before the physical failure occurs. In such situations, failure definition in the feature space may be easier to determine. In situations where this is not the case, arriving at a failure definition may be more involved. We later discuss a fuzzy inference model to arrive at failure definition in the feature space in Section 4.

Diagnostics: During the diagnostics process, specific FOMs are typically compared to threshold limits (Begg et al. 1999). Additional processing may determine a signature pattern in one, or multiple, fault measure(s). Automated reasoning is often used to identify the faulty type (cracked gear tooth, bearing spall, imbalance etc.), location, and severity. The core problem of diagnostics is essentially a problem of classification (Elverson 1997). Discriminant transformations are often used to map the data characteristic of different failure mode effects into distinct regions in the feature subspace (Byington and Garga 2001). The task is relatively straightforward in the presence of robust FOMs. The literature is vast in this area and commercial technologies are well established. Depending on the application, these systems employ model-based methods, any number of statistical methods, and a variety of computational intelligence methods.

Dimla et al. (1997) provide a critical review of the neural network methods used for tool condition monitoring.

Prognostics: Contrary to diagnostics, the literature on prognostics, called the Achilles' heel of the CBM architecture (Vachtsevanos et al. 1999), is extremely sparse. Unfortunately, the large majority of the literature that employs the term 'prognostics' in the title ends up discussing diagnostics. To achieve prognostics, there need to be features that are suitable for tracking and prediction (Begg et al. 1999). It is for this reason that prognostics is receiving the most attention for systems consisting of mechanical and structural components, for unlike electronic or electrical systems, mechanical systems typically fail slowly as structural faults progress to a critical level (Mathur et al., 2001).

9.2 Performance Reliability Theory

Let $\{y(s)\}$ represent a scalar time-series generated by sampling the performance degradation signal (or a transformation thereof). Suppose that $\{y(s)\}$ can be described by a *nonlinear regressive model* of order p as follows:

$$y(s) = f(y(s-1), y(s-2), \cdots, y(s-p)) + \varepsilon(s) \tag{1}$$

where f is a nonlinear function and $\varepsilon(s)$ is a residual drawn from a *white Gaussian noise process*. In general, the nonlinear function f is unknown, and the only information we have available to us is a set of observables: $y(1), y(2), \cdots, y(S)$, where S is the total length of the time-series. Given the data set, the requirement is to construct a physical model of the time-series. To do so, we can use any number of statistical (for example, Box and Jenkin's auto-regressive integrated moving average (ARIMA) models) or computational intelligence based (for example, feed-forward neural networks such as multi-layer perceptron (MLP)) forecasting techniques as a one-step predictor of order p. Specifically, the model is estimated to make a prediction of the sample $y(s)$, given the immediate past p samples $y(s-1), y(s-2), \cdots, y(s-p)$, as shown by

$$\hat{y}(s) = \tilde{f}(y(s-1), y(s-2), \cdots, y(s-p)) + e(s). \tag{2}$$

The nonlinear function $\tilde{f}$ is the approximation of the unknown function f, built in general to minimize some cost function (J) of the prediction error

$$e(s) = y(s) - \hat{y}(s), \quad p+1 \le s \le S \tag{3}$$

Note that a single model can be potentially built to simultaneously work with many degradation signals and also allow different prediction orders for different signals. Of course, the structure and complexity of the model will increase with an increase in the number of degradation signals jointly modeled.

Now, let $F(t)$ denote the probability that failure of a component takes place at a time or usage less than or equal to t (i.e., $F(t) = P(T \le t)$), where the random variable T denotes the time to failure. From the definition of conditional probability, the conditional reliability that the component will fail at some time or usage $T > t + \Delta t$, given that it has not yet failed at time $T = t$ will be:

$$R\big((t + \Delta t) \,|\, t\big) = \big[1 - P(T \le t + \Delta t)\big] \big/ P(T > t). \tag{4}$$

Let $\mathbf{y} = [y_1, y_2, ..., y_m]$ denote the vector of m degradation signals (or a transformation thereof) being monitored from the system under evaluation. Let $\mathbf{y}^{pcl} = [y_1^{pcl}, y_2^{pcl}, ..., y_m^{pcl}]$ denote the vector of deterministic *performance critical limits* (PCLs), which represent an appropriate definition of failure in terms of the amplitude of the m degradation signals. For any given operating/environmental conditions, *performance reliability* can be defined as "the conditional probability that $\mathbf{y}$ does not exceed $\mathbf{y}^{pcl}$, for a specified period of time or usage." Obviously, the above definition directly applies to the case where the amplitudes of the degradation signals are preferred to be low (lower-the-better signals with higher critical limits), and can be easily extended to deal with higher-the-better signals (with lower critical limits) and nominal-value-is-best signals (with two-sided critical limits), and any combinations between. Without loss of generality, for illustrative purposes, let us make the assumption here that all the m degradation signals are of lower-is-better type signals.

Since a model estimated using past degradation signals collected from other similar components keeps providing us with an estimate of $\mathbf{y}$ into the future, denoted by $\hat{\mathbf{y}}(t_f)$, under the assumption that the change in $\mathbf{y}$

from the current time point (t_c) to the predicted time point (t_f) is either monotonically decreasing or increasing, the reliability that the component or system will operate without failure until t_f is given by:

$$R\left(\left(T \ge t_f\right)| t_c\right) = \int\limits_{y_1=-\infty}^{y_1^{pcl}} \int\limits_{y_2=-\infty}^{y_2^{pcl}} .. \int\limits_{y_m=-\infty}^{y_m^{pcl}} g\left(\hat{\mathbf{y}}\left(t_f\right)\right) dy_1 dy_2 .. dy_m \tag{5}$$

where $g\left(\hat{\mathbf{y}}\left(t_f\right)\right)$ denotes the probability density function of $\hat{\mathbf{y}}(t_f)$. The assumption here is that y_i^{pcl} is a constant for any given i and is independent of $\mathbf{y}(t_f)$. Under these conditions, the failure space is bounded by orthogonal hyper planes. If the independence assumption is not justified, one could use a hyper-surface to define the failure boundary (Lu et al. 2001). If need be, one could even relax the assumption of a deterministic boundary and replace it with a stochastic boundary model. However, such an extension is non-trivial.

For the special case where there exists just one lower-the-better degradation signal, this process is illustrated in Fig. 3 (Chinnam and Baruah 2004). The shaded area of $g\left(\hat{\mathbf{y}}\left(t_f\right)\right)$ at any t_f denotes the conditional-unreliability of the unit. That is, given that the unit has survived until t_c, the shaded area denotes the probability that the unit will fail by t_f. To obtain mean residual life (MRL), using r_{MRL}, the least acceptable reliability, one can estimate t_{MRL}, the time instant/usage at which the reliability of the unit reaches r_{MRL}. Thus, one can calculate the MRL to be the time difference between t_c and t_{MRL}.

9.3 Feature Extraction from Degradation Signals

Feature extraction is an important step in developing effective procedures for online reliability monitoring using degradation signals. In this section we discuss time, frequency and mixed-domain analysis techniques for preprocessing degradation signals and feature extraction.

9.3.1 Time, Frequency, and Mixed-Domain Analysis

As is pointed out in the introduction, the most important and fundamental variables in degradation signal processing are time and frequency. In addition, the degradation signals often tend to be stochastically non-stationary, rendering the fast Fourier transform (FFT) spectrum (a transform that is quite popular for frequency analysis) inadequate, for it can only evaluate an average spectrum over a definite time period and loses the non-stationary characteristics of the signals (Yen and Lin 2000). Given this, in many real world applications, it is far more useful to characterize the signal in both the time- and frequency- domains, simultaneously.

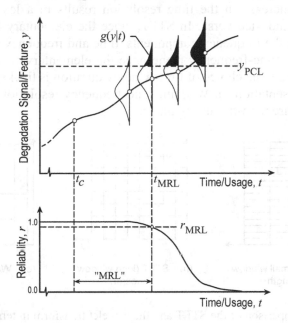

Fig. 3. Degradation signal forecasting model coupled with a failure definition PCL to estimate MRL

Several joint time-frequency (or mixed-domain) alternatives have been proposed in the literature. Some of the alternatives include the short term Fourier transform (STFT) and the Wavelet transform (WT), to name a few. Joint time-frequency methods are conventionally classified into two categories: linear and quadratic (Qian and Chen, 1996). The principle of linear time-frequency representation involves decomposing any signal into a linear expansion of functions that belong to a set of redundant elementary functions. All linear transformations are achieved by comparing the analyzed signal with a set of prudently selected elementary functions. While the functions for STFT are obtained by frequency modulation of sine and cosine waves in STFT, in the WT, the functions are obtained by scaling and shifting the mother wavelet.

The most important relationship in terms of joint time-frequency analysis is the relation between signal's time window duration and frequency bandwidth. Several different definitions are offered in the literature for specifying the time window duration and frequency bandwidth (Qian and Chen, 1996; Akay and Mello, 1998). In general, for mixed-domain methods, there is a tradeoff between time resolution and frequency resolution for there is an upper bound on the product of the two resolutions. In other words, an increase in the time resolution results in a loss of frequency resolution, and vice versa. In STFT, since the elementary function is the same for all the frequency components, time and frequency resolutions are fixed on the time-frequency plane once the elementary function has been chosen. Hence, the choice of time window duration is the key for any good STFT representation. In WT, time and frequency resolutions are not fixed over the entire time-frequency plane.

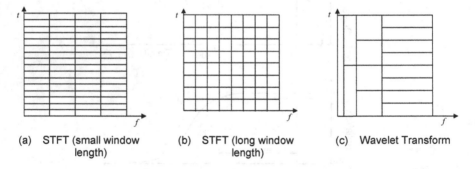

(a) STFT (small window length)

(b) STFT (long window length)

(c) Wavelet Transform

Fig. 4. Comparison of the STFT and the wavelet transform in terms of time and frequency resolution.

The tiling of the windows in the joint time-frequency plane is illustrated for STFT and WT in Fig. 4. While the STFT tilling is linear, the WT tilling is logarithmic. In Fig. 4(a) and 4(b), when the length of window is specified, the time and frequency resolution remains constant throughout the plane. In Fig. 4(c), time and frequency resolution is not fixed over the entire time-frequency domain: time resolution becomes good at higher frequencies whereas frequency resolution becomes good at lower frequency.

9.3.2 Wavelet Preprocessing of Degradation Signals

As discussed in the introduction, degradation signals often tend to be rich in time and frequency components, and hence, lend themselves for better representation in mixed-domain analysis. The treatment on wavelets that follows is borrowed heavily from DeVore and Lucier (1992). The term wavelet denotes a univariate function ψ (multivariate wavelets exist as well), defined on $\mathbf{R}$, which, when subjected to the fundamental operations of shifts (i.e., translation by integers) and dyadic dilation, yields an orthogonal basis of $L_2(\mathbf{R})$. That is, the functions $\psi_{j,k} := 2^{k/2}\psi(2^k \cdot -j)$, $j,k \in \mathbf{Z}$, form a complete orthonormal system for $L_2(\mathbf{R})$. Such functions are generally called orthogonal wavelets, since there are many generalizations of wavelets that drop the requirement of orthogonality.

One can view a wavelet ψ as a "bump" and think of it as having compact support, though it need not. Dilation squeezes or expands the bump and translation shifts it. Thus, $\psi_{j,k}$ is a scaled version of ψ centered at the dyadic integer $j2^{-k}$. If k is large positive, then $\psi_{j,k}$ is a bump with small support; if k is large negative, the support $\psi_{j,k}$ is large. The requirement that the set $\{\psi_{j,k}\}_{j,k \in \mathbf{Z}}$ forms an orthonormal system means that any function $f \in L_2(\mathbf{R})$ can be represented as a series

$$f = \sum_{j,k \in \mathbf{Z}} \langle f, \psi_{j,k} \rangle \, \psi_{j,k} \tag{6}$$

with $\langle f, g \rangle := \int_{\mathbf{R}} f\bar{g}dx$ the usual inner product of two $L_2(\mathbf{R})$ functions. One can view Eq. (6) as building up the function f from the bumps. Bumps corresponding to small values of k contribute to the broad resolution of f; those corresponding to large values of k give finer detail.

The decomposition of Eq. (6) is analogous to the Fourier decomposition of a function $f \in L_2(\mathbf{R})$ in terms of the exponential functions $e_k := e^{ik\cdot}$, but there are important differnces. The exponential functions e_k have global support. Thus, all terms in the Fourier decomposition contribute to the value of f at a point x. On the other hand, wavelets are usually either of compact support or fall off exponentially at infinity. Thus, only the terms in Eq. (6) corresponding to $\psi_{j,k}$ with $j2^{-k}$ near x make a large contribution to x. The representation Eq. (6) is in this sense local.

All this would be of little more than theoretical interest if it were not for the fact that one can efficiently compute wavelet coefficients and reconstruct functions from these coefficients. These algorithms, known as "fast wavelet transforms" are the analogue of the Fast Fourier Transform and follow simply from the refinement of the dilation and shift equation mentioned above.

In summary, the wavelet transform results in many wavelet coefficients, which are a function of scale (or level or frequency) and position. Hence, a wavelet plot is a plot of coefficients on time-scale axis. The higher the scale, the more stretched the wavelet. The more stretched the wavelet, the longer the portion of the signal with which it is compared, and thus the coarser the signal features being measured by the wavelet coefficients. Multiplying each coefficient by the appropriately scaled and shifted wavelet yields the constituent wavelets of the original signal. The coefficients constitute the results of a regression of the original signal performed on the wavelets.

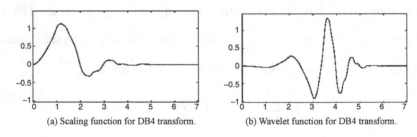

(a) Scaling function for DB4 transform.   (b) Wavelet function for DB4 transform.

Fig. 5. Daubechies DB4 wavelet tranform.

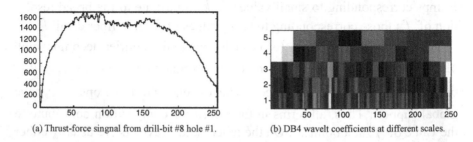

(a) Thrust-force singnal from drill-bit #8 hole #1.   (b) DB4 wavelt coefficients at different scales.

Fig. 6. Thrust-force degradation signal from drill-bit #8 hole #1 and their transformed DB4 wavelet coefficients.

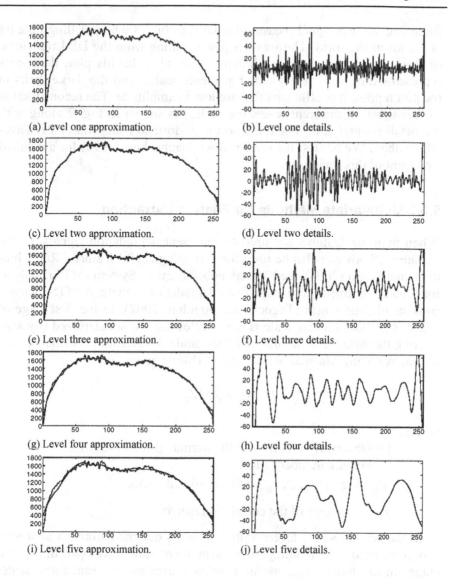

(a) Level one approximation.   (b) Level one details.

(c) Level two approximation.   (d) Level two details.

(e) Level three approximation.   (f) Level three details.

(g) Level four approximation.   (h) Level four details.

(i) Level five approximation.   (j) Level five details.

Fig. 7. Discrete wavelet analysis of thrust-force signal from drill-bit #8 hole #1

The particular wavelet transform considered in this paper is the compactly supported Daubechies' wavelet transform. The transform is compactly supported with extreme phase and highest number of vanishing moments for a given support width. One particular Daubechies' wavelet transform, the DB4 discrete wavelet transform function and its associated scaling function, is shown in Fig. 5. For illustrative purposes, the thrust-force degradation signal measured from drill-bit #8 while drilling hole #4 is shown in Fig. 6(a). The wavelet plot resulting from the DB4 transform of this degradation signal is shown in Fig. 6(b). In this plot, the x-axis represents time while the y-axis represents scales, and the darker cells in the plot represent coefficients that are low in amplitude. The reconstruction of this signal at different levels (or scales) is shown in Fig. 7 along with the detail counterparts. The transform is performed using MatLab's Wavelet Toolbox. We request the reader to see Daubechies (1990) for a detailed treatment of this transform.

9.3.3 Multivariate Methods for Feature Extraction

When multiple features are used to represent degradation signals, multivariate methods can also be used for extracting useful features. Rai, Chinnam, and Singh (2004) used Mahalanobis-Taguchi System (MTS) analysis for predicting drill-bit breakage from degradation signals. A MTS analysis consists of four stages (Taguchi and Jugulum 2002). In the first stage of analysis a measurement scale is constructed from a standardized (by subtracting the mean and dividing by the standard deviation) 'normal' group of features using Mahalanobis distances (MDs) given by,

$$MD_j = D_j^2 = \frac{1}{k} Z_{ij}' C^{-1} Z_{ij} \tag{7}$$

where,

j = Observation number in the normal group (1 to m)
i = Feature number (1 to k)
$Z_{ij} = (z_{1j}, z_{2j}, ..., z_{kj})$ = Standardized vector

C^{-1} = Inverse of the correlation matrix.

In the second stage, larger values of MDs obtained from an abnormal group are used for validating the measurement scale developed in the first stage. In the third stage, useful features are extracted from those under study using signal-to-noise ratio values. A S/N ratio for say qth trial with 't' features present in the combination can be obtained as,

$$\eta_q = -10 \log \left[\frac{1}{t} \sum_{j=1}^{t} \frac{1}{MD_j} \right] \tag{8}$$

For a given feature, an average value of the S/N ratio is determined separately at level-1 indicating presence and at level-2 indicating absence of a feature. Subsequently, gain in S/N ratio values is obtained by taking difference of the two average values as,

$$\text{Gain} = (\text{Avg. S/N Ratio})_{\text{Level-1}} - (\text{Avg. S/N Ratio})_{\text{Level-2}} \tag{9}$$

A positive gain for a feature indicates its usefulness and vice-versa. And finally, in the fourth stage of analysis a threshold value for the MDs is developed from the normal group to enable degradation level prediction using the useful features.

9.4 Fuzzy Inference Models for Failure Definition

Most prognostics methods in the literature for on-line estimation of MRL utilize trending or forecasting models in combination with mechanistic or empirical failure definition models. However, in spite of significant advances made throughout the last century, our understanding of the physics of failure is not quite complete for many electro-mechanical systems. In the absence of sound knowledge for the mechanics of degradation and/or adequate failure data, it is not possible to establish practical failure definition models in the degradation signal space. Under these circumstances, the sort of procedures illustrated in Section 2 is not feasible. However, if there exist domain experts with strong experiential knowledge, one can potentially establish fuzzy inference models for failure definition. In this section, we suggest the incorporation of fuzzy inference models to introduce the definition of failure in the degradation signal space using domain experts with strong experiential knowledge. While the trending or forecasting subcomponent will predict the future states of the system in the degradation signal space, it is now the task of the fuzzy inference model to estimate the reliability associated with that forecast state. If one were to compare this procedure with that discussed in Section 2, it is equivalent to replacing the right hand side of Eq. (5) with a fuzzy inference model.

One might argue that probabilistic models could be potentially used for modeling experiential knowledge of domain experts. However, it is widely accepted that classical probability theory has some fundamental shortcomings when it comes to modeling the nature of human concepts and thoughts, which tend to be abstract and imprecise. While probability theory is developed to model and explain randomness, fuzzy arithmetic and logic is developed to model and explain the uncertain and imprecise nature of abstract thoughts and concepts. Over the last three decades, since Lofti Zadeh authored his seminal paper in 1965 on fuzzy set theory (Zadeh 1965), the scientific community had made major strides in extending the set theory to address applications in areas such as automatic control, data classification, decision analysis, and time series prediction (Jang et al. 1997).

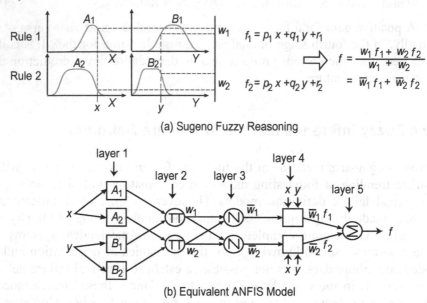

(a) Sugeno Fuzzy Reasoning

(b) Equivalent ANFIS Model

Fig. 8. Sugeno FIM with two inputs (X, Y) and one output (F).

In the context of prognostics and failure definition, Sugeno Fuzzy Inference Model (FIM), illustrated in Fig. 8(a), is particularly attractive for failure definition for three reasons:

1. It makes a provision for incorporating subjective knowledge of domain experts and experienced operators,
2. Model can be viewed as a feed-forward neural network (labeled Adaptive-Network based Fuzzy Inference Systems or ANFIS), and hence, can be adapted using empirical/historical data coupled with gradient search methods (Jang 1993), and
3. Computationally efficient for the absence of a de-fuzzification operator prevalent in other fuzzy inference models.

The illustrated two-input (X and Y) one-output (F) Sugeno FIM carries two membership functions for each of the two input variables, namely A_1, A_2 and B_1, B_2. The model is made of two rules. For example, *Rule-1* states that if X is A_1 and Y is B_1, then the output is given by $f_1 = p_1 x + q_1 y + r_1$. Here, A_1 and B_1 denote linguistic variables (such as "thrust is low" or "vibration is high"). Even though the consequent of each rule constitutes a first-order model, the overall relationship is often highly nonlinear. The equivalent ANFIS model is shown in the illustration as well.

For the application considered here, typically, the number of input variables for the Sugeno FIM will be equal to the number of degradation signals under investigation, and there is one output variable predicting the reliability of the unit (i.e., $r(t)$). The number of membership functions and the number of rules needed to fullyscribe the failure definition will be dictated by the specific application and input from domain experts. In the absence of first-principles models, rules can be initially formulated with the help of domain experts and experienced operators. All the parameters of the Sugeno FIM can be adapted to best describe any historical dataset using the ANFIS framework. For more details regarding Sugeno fuzzy inference models or their ANFIS equivalents (Jang et al. 1997).

9.5 Online Reliability Monitoring with Neural Networks

In general, artificial neural networks are composed of many non-linear computational elements, called nodes, operating in parallel and arranged in patterns reminiscent of biological neural nets (Lippmann 1987). These processing elements are connected by weight values, responsible for modifying signals propagating along connections (also called synapses) and used for the training process. The number of nodes plus the connectivity define the topology/structure of the network, and is intimately connected with the learning algorithm used to train the network Haykin (1999). The higher the number of nodes per layer and/or the number of layers, the higher the ability of the network to extract higher-order statistics (Churchland and Sejnowski 1992) and approximate more complex relationships between inputs and outputs.

One of the most significant properties of a neural network is its ability to learn from its environment that normally involves an iterative process of adjustments applied to the synaptic weights. There is no unique learning algorithm for the design of neural networks and they differ from each other in the way in which the adjustment of synaptic weights takes place. Two popular learning algorithms are the error-correction learning algorithm (in essence a stochastic gradient-descent search technique) used normally for training FFNs such as FIR MLPs discussed in Section 5.2 and the competitive learning algorithm used for training networks such as SOMs discussed in Section 5.3. It is beyond the scope of this chapter to discuss the nature of these learning algorithms in detail, and the reader is referred to Haykin (1999).

Two popular learning paradigms for neural networks involve *supervised* learning and *unsupervised* (self-organized) learning. FFNs, such as the FIR MLP, are trained in a supervised learning mode. In essence, there is a teacher present (metaphorically speaking) to guide the network toward making accurate predictions. Unsupervised learning is performed in a self-organized manner in that no external teacher or critic is required to instruct synaptic changes in the network, and is the case with SOMs. For a more thorough treatment of the general topic of neural networks, the reader is referred to Haykin (1999).

9.5.1 Motivation for Using FFNs for Degradation Signal Modeling

FFNs have proven to be very effective in function approximation and time series forecasting (Wan 1994; Sharda and Patil 1990; Tang et al. 1991; Harnik et al. 1989; Haykin 1999; Cheng and Titterington 1994; Balazinski et al. 2002). They are flexible models that are widely used to model high dimensional, nonlinear data (De Veaux et al. 1998). In fact, FFNs with nonlinear sigmoidal nodal functions are universal approximators (proved by Hornik et al. (1989), using the Stone-Weierstrass theorem), meaning that a network with finite number of hidden layers and finite number of nodes per hidden layer can approximate any continuous function (R^N, R^M) over a compact subset of R^N to arbitrary precision. However, since the FFN model parameters are generally not interpretable, they are not recommended for process understanding. However, if the emphasis is simply on accurate prediction, they tend to be extremely good and tend to outperform most traditional methods. This is not to say that traditional methods cannot be effectively used for modeling degradation signals. The method for on-line estimation of individual component reliability introduced in this chapter is compatible with traditional methods of modeling degradation signals as well. However, there are other motivations for using FFNs for degradation signal modeling. These include their nonparametric properties and superior ability to adapt to changes in surrounding environment (neural network trained to operate in a specific environment can be easily retrained to deal with minor changes in the environmental conditions).

9.5.2 Finite-Duration Impulse Response Multi-layer Perceptron Networks

A FIR MLP is an extension to the popular MLP network in which each scalar synaptic weight in an MLP network is replaced by a FIR synaptic filter. The additional memory allows the network *dynamic* properties necessary to make the network responsive to time-varying signals. A standard MLP network trained using an algorithm such as back-propagation is only capable of learning an input-output mapping that is static, and hence, is only capable of performing nonlinear prediction on a stationary time series (Haykin 1999). However, most degradation signals measured from physical systems, as they degrade with time, tend to be non-stationary. A typical FIR MLP network with an input layer, an output layer, and two hidden layers is shown in Fig. 9(a), whose synaptic FIR filter structure is defined by the signal-flow graph of Fig. 9(b). Here $\mathbf{x}^T = [x_1, x_2, ..., x_n]$ denotes the input vector while $\mathbf{y}^T = [y_1, y_2, ..., y_m]$ is the output vector. $\mathbf{v}^T = [v_1, v_2, ..., v_p]$ and $\mathbf{z}^T = [z_1, z_2, ..., z_q]$ are the outputs at the first and second hidden layers, respectively. $\{\mathbf{w}_{ij}^1\}_{pxn}$, $\{\mathbf{w}_{ki}^2\}_{qxp}$, $\{\mathbf{w}_{lk}^3\}_{mxq}$ are matrices of FIR weight vectors associated with the three layers. For example, $\mathbf{w}_{ij}^1 = [w_{ij}(0), w_{ij}(1), ..., w_{ij}(M)]$, where $w_{ij}(r)$ denotes the weight connected to the rth memory tap of the FIR filter modeling the synapse that connects the input neuron j to neuron i in the first hidden layer. As shown in Fig. 9(b), the index r ranges from 0 to M, where M is the total number of delay units (element z^{-1} represents unit time delay in Fig. 9(b) and s denotes a discrete-time variable) built into the design of the FIR filter. The vectors $\overline{\mathbf{v}} \in R^p$, $\overline{\mathbf{z}} \in R^q$, and $\overline{\mathbf{y}} \in R^m$ are as shown in Fig. 9(a) with $\gamma(\overline{v}_i) = v_i$, $\gamma(\overline{z}_k) = z_k$, and $\gamma(\overline{y}_l) = y_l$ where $\overline{v}_i$, $\overline{z}_k$, and $\overline{y}_l$ are elements of $\overline{\mathbf{v}}$, $\overline{\mathbf{z}}$, and $\overline{\mathbf{y}}$ respectively. Here γ is a sigmoidal nonlinear operator[1], $\overline{v}_i = \sum_{j=1}^n \overline{v}_{ij}$, $\overline{z}_k = \sum_{i=1}^p \overline{z}_{ki}$, and $\overline{y}_l = \sum_{k=1}^q \overline{y}_{lk}$.

During network training, the weights of the network are adjusted using an adaptive algorithm based on a given set of input-output pairs. An error-correction learning algorithm will be briefly discussed here, and readers can see Haykin (1999) for further details and information regarding other training algorithms. If the weights of the networks are considered as ele-

---

[1] The most popular non-linear nodal function for FIR MLP networks is the sigmoid [*unipolar* $\rightarrow f(x) = 1/(1 + e^{-x})$ and *bipolar* $\rightarrow f(x) = (1 - e^{-x})/(1 + e^{-x})$].

ments of a parameter vector θ, the error-correction learning process involves the determination of the vector θ^* which optimizes a performance function J based on the output error.[2] In error-correction learning, the weights are adjusted along the negative gradient of the performance function as follows:

$$\theta^{(s+1)} = \theta^{(s)} - \eta \frac{\partial J^{(s)}}{\partial \theta^{(s)}} \tag{10}$$

where η is a positive constant that determines the rate of learning and the superscript refers to the iteration step. In the literature, a method for determining this gradient for FIR MLP networks is the temporal back-propagation learning, which is not repeated here due its complexity. For further information on the algorithm, see Haykin (1999) or Wan (1990).

9.5.3 Self-Organizing Maps

The principal goal of the SOM developed by Kohonen (1982) is to transform an incoming signal pattern of arbitrary dimension into a one- or two-dimensional discrete map, and to perform this transformation adaptively in a topological ordered fashion. The presentation of an input pattern causes a corresponding "localized group of neurons" in the output layer of the network to be active (Haykin 1999), introducing the concept of a neighborhood.

Let Φ denote a non-linear SOM transformation which maps the spatially *continuous input space X* onto a *spatially discrete output space* (made up of a set of N computation nodes of a lattice) A. Given an input vector **x**, the SOM identifies a best-matching neuron $i(\mathbf{x})$ in the output space A, in accordance with the Map Φ. For information on the unsupervised competitive learning algorithm typically used for training SOMs (Haykin 1999). For a typical SOM, trained in such a fashion, the map Φ has the following properties (Haykin 1999):

Property 1: Approximation of the Input Space–The SOM Φ, represented by the set of synaptic weight vectors $\{\mathbf{w}_j |\, j = 1, 2,..., N\}$, in the output space A, provides a good approximation of the input space X.

Property 2: Topological Ordering–The Map Φ computed by the SOM algorithm is topologically ordered in the sense that the spatial location of a

---

[2] A popular performance function in the literature is the sum of the squared values of the prediction error for all training patterns.

neuron in the lattice corresponds to a particular domain or feature of input patterns.

Property 3: Density Matching–The Map Φ reflects variations in the statistics of the input distribution.

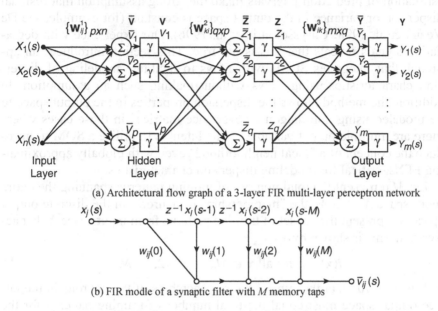

(a) Architectural flow graph of a 3-layer FIR multi-layer perceptron network

(b) FIR modle of a synaptic filter with M memory taps

Fig. 9. A typical network and its structure

9.5.4 Modeling Dispersion Characteristics of Degradation Signals

The proposed approach for on-line performance reliability estimation of physical systems calls for modeling degradation signals as well as the dispersion characteristics of the signals around the degradation models. Globally generalizing neural networks such as FFNs do not easily lend themselves for modeling dispersion characteristics. In contrast, locally generalizing networks such as radial-basis function (RBF) networks and cerebellar model arithmetic computer (CMAC) networks have a naturally well-defined concept of local neighborhoods and lend themselves for modeling dispersion. Such networks have been extended in the literature to include dispersion attributes such as prediction limits (PLs). For example, the validity index (VI) network derived from an RBF network, fits functions (Park and Sandberg 1991) and calculates PLs/error bounds for its predictions (Leonard et al. 1992).

Since globally generalizing neural networks concentrate on global approximation and tend to produce highly compact and effective models, we need a method for estimating prediction limits for these networks. Most of the very few extensions to FFNs discussed in the literature that facilitate estimation of prediction intervals make the strong assumption that residual dispersion or variance in the output space is constant (for example, see De Veaux et al. 1998; Chryssolouris et al. 1996). Our experience with degradation signals reveals that this is not true. This section introduces an approach that integrates SOMs with FFNs to facilitate modeling of dispersion characteristics using FFNs without making such an assumption. In addition, the method allows the dispersion properties in the output space to be modeled using non-diagonal covariance matrices in those cases where there are multiple output variables. The intent is to utilize a SOM to introduce the concept of a "local neighborhood" even with globally approximating FFNs, critical for modeling dispersion characteristics.

Let M represent the total number of training patterns spanning the entire input space X. Let M_j the "membership" of neuron j in the discrete output space A represent the subset of training patterns from input space X that activate it. This is shown by:

$$i(\mathbf{x}) = j \quad \text{for all } \mathbf{x} \in M_j, \quad j = 1, 2, ..., N. \tag{11}$$

It is also true that the sum of the memberships of the neurons in the lattice output space must equal the total number of training patterns for the SOM, as shown by:

$$\sum_{j=1}^{N} M_j = M. \tag{12}$$

The three properties exhibited by SOMs (discussed earlier) provide the motivation to utilize the SOM to break the input space X into N distinct regions (denoted by X_j) that are mutually exclusive, and hence satisfy the following relationship:

$$\sum_{j=1}^{N} X_j = X. \tag{13}$$

All the signal patterns from any given distinct region X_j, when provided as input to the Map Φ, will activate the same output neuron j. This is shown by:

$$i(\mathbf{x}) = j \quad \text{for all } \mathbf{x} \in X_j, \quad j = 1, 2, ..., N. \tag{14}$$

Thus, using SOMs, one can introduce the concept of a "local neighborhood," the resolution depending on the number of neurons (N) in the discrete output space.

From the above definition of local neighborhood, input signal patterns can be associated unambiguously with one of the distinct regions X_j. Assuming that a FFN is being used for function approximation or time series forecasting, an estimate of the covariance matrix for the FFN model residuals within the domain of region X_j is given by:

$$Cov_j = \begin{bmatrix} S_{11} & S_{12} & \cdots & S_{1O} \\ S_{21} & S_{22} & \cdots & S_{2O} \\ \vdots & \vdots & \ddots & \vdots \\ S_{O1} & S_{O2} & \cdots & S_{OO} \end{bmatrix} \tag{15}$$

where:

$S_{pq} = \dfrac{1}{(M_j - 1)} \sum\limits_{k=1}^{M_j} (E_{kp})(E_{kq})$, and denotes the covariance between output variables p and q,

E_{kp} denotes the FFN model residual for output variable p for pattern k,

O denotes the number of output variables predicted by the FFN.

Assuming that the residuals are independent and Gaussian distributed with a constant covariance matrix over the domain of any region but varying from domain to domain, one can even estimate the PLs. In fact, the 1-α quantile is given by the point $\mathbf{x}$ satisfying the following condition:

$$(\mathbf{x} - \boldsymbol{\mu}_j)^T Cov_j^{-1}(\mathbf{x} - \boldsymbol{\mu}_j) \leq \chi_O^2(\alpha) \tag{16}$$

where:

$\chi_O^2(\alpha)$ denotes the (1-α) quantile of the Chi-Square distribution with O degrees of freedom.

$\boldsymbol{\mu}_j$ denotes the mean residual vector for domain X_j, and

Cov_j^{-1} is the inverse of the matrix Cov_j.

Investigation through simulation studies and statistical analysis have revealed that the residuals do tend to exhibit Gaussian distribution in different neighborhoods as long as the overall noise in the data is Gaussian. If the FFN has adequate representational capacity, the fit should not be significantly biased, and the mean residual vector can be a null vector. In a similar fashion, one could also determine the limits of the dispersion of the mean, i.e., the range of possible values for the mean predicted value, rather than the value for a single sample.

9.6 Drilling Process Case Study

A drilling operation was chosen as the physical test-bed for the reason that it is a commonly used machining process. El-Wardany et al. (1996) note that of all the cutting operations performed in the mechanical industries, drilling operations contribute approximately 40%. Broad steps involved and methodology used is briefly described in Fig. 10 as a guide to the case study followed-up with detailed description of the last two steps.

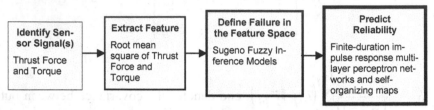

Fig. 10. Broad steps and methodology for online reliability prediction for drilling process

The development of reliable tool-wear sensors has been an active area of metal cutting research (Andrews and Tlusty 1983). Machining literature has shown that there is a strong correlation between thrust-force (and torque) acting on a drill-bit and the bit's future life expectancy (Kim and Kolarik 1992, Dimla 2000, Dimla and Lister 2000). Dimla and Lister (2000) applied multi-layer perceptron neural network for tool-state classification using online data on the cutting forces and vibration, and reported achieving approximately 90% accuracy in tool-state classification. Jantunen (2002) summarizes monitoring methods that have been studied for tool condition monitoring in drilling with thrust force and torque being most popular. Hence, thrust force and torque signals are appropriate degradation signals for estimating on-line drill-bit reliability.

As thousands of data points are recorded for thrust force and torque for each hole drilled, the data are condensed using root mean square value after systematically grouping the data points for each drilled hole. Sun et al (2006) discuss a systematic procedure for sampling the training data that helps to reduce the size of the training data without trading off the generalization performance. For defining failure in feature space, Sugeno fuzzy inference model as explained in Section 4 is used and demonstrated for the drilling process. For reliability prediction, a methodology using finite-duration impulse response multi-layer perceptron neural networks along with self-organizing maps as detailed in Section 5 is demonstrated.

9.6.1 Experimental Setup

A dynamometer was available in-house for measuring on-line the thrust-force and torque acting on the drill-bit. The experimental setup consists of a HAAS VF-1 CNC milling machine, a workstation with LabVIEW software for signal processing, a Kistler 9257B piezo-dynamometer for measuring thrust-force and torque, and a National Instruments PCI-MIO-16XE-10 card for data acquisition. The experimental setup is depicted in Fig. 11.

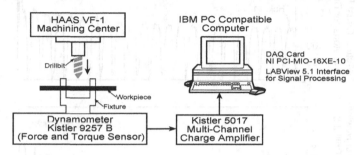

Fig. 11. Experimental setup for capturing thrust-force and torque degradation signals from a ¼" HSS drill-bit.

9.6.2. Actual Experimentation

A series of drilling tests were conducted using quarter-inch drill-bits on a HAAS VF-1 Machining Center. Stainless steel bars with quarter-inch thickness are used as specimens for the tests. The drill-bits were high-speed twist drill-bits with two flutes, and were operated under the following conditions without any coolant: feed-rate of 4.5 inches-per-minute (ipm) and spindle-speed of 800 revolutions-per-minute (rpm).

Twelve drill-bits were used in the experiment. Each drill-bit was used until it reached a state of physical failure, either due to macro chipping or gross plastic deformation of the tool tip due to excessive temperature. Collectively, the drill-bits demonstrated significant variation in life (varying between eight and twenty five successfully drilled holes) even though they came from the same manufacturer in the same box. This further validates the need to develop good on-line reliability estimation methods to help end users arrive at optimal tool or component replacement strategies.

The thrust-force and torque data were collected for each hole from the time instant the drill penetrated the work piece through the time instant the drill tip protruded out from the other side of the work piece. The data was initially collected at 250 Hz and later condensed using RMS techniques to

24 data points per hole, considered normally adequate for the task at hand. Throughout the rest of this paper, in all illustrations, one time unit is equivalent to the time it takes to drill 1/24th of a hole. For illustrative purposes, data collected from drill-bit #8 is depicted in Fig. 12.

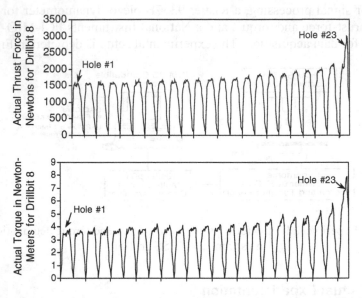

Fig. 12. Plots of thrust-force and torque signals collected from drill-bit #8.

9.6.3 Sugeno FIS for Failure Definition

Experimental data has revealed a lot of variation between drill-bits, in the amplitudes of thrust-force and torque observed during the final hole. This invalidates the concept of a deterministic critical limit for establishing failure definition in the thrust-force and torque signal space. While one could potentially introduce a probabilistic critical plane, here, we utilize fuzzy logic to introduce an FIS failure definition model in the degradation signal space.

It was decided initially to use two membership functions for representing the "low" and "high" linguistic levels for each of the degradation signals. Sigmoid membership functions were considered appropriate for three reasons:

1. They are open-ended on one side,
2. They are monotonous functions (always increase or decrease but not both), and

3. They are compatible with most ANFIS training algorithms.

Past experience suggested that thrust-force typically varies between 0 to 3000 Newtons for the drilling operation at hand. Similarly, it was common to see torque vary between 0 to 6 Newton-meters. Initially, the membership functions were set up to equally divide the ranges of the variables, as illustrated in Fig. 13(a) and 13(c) for thrust-force and torque, respectively. It was expected that ANFIS training would address any misrepresentations in these membership functions.

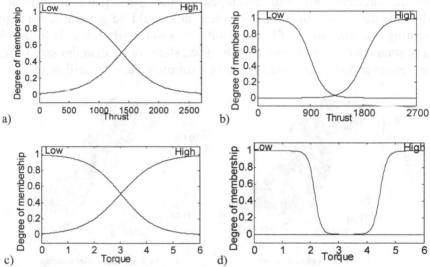

Fig. 13. Plots of membership functions before and after ANFIS training (a) Thrust-force MFs – before training. (b) Thrust-force MFs – after training. (c)Torque MFs – before training. (d) Torque MFs – after training.

Two rules were initially formulated with the understanding that more rules can be added to address any serious violations by the FIS model. The rules are as follows:

IF thrust-force is *low* AND torque is *low*, THEN, drill-bit reliability = 1.0.

IF thrust-force is *high* AND torque is *high*, THEN, drill-bit reliability = 0.0.

Thus, the consequent of each rule constitutes a zero-order model. The resulting FIS model relationship is illustrated in Fig. 14(a). It is clear that the overall relationship is highly non-linear and certainly seems plausible. At this stage, it was decided to extract training data to further refine the FIS model using the ANFIS framework. The training, validation, and testing datasets used for developing the forecasting model were once again exploited to refine the FIS model. The reasoning behind the generation of training data is as follows. Given any drill-bit and the provided operating conditions, it is totally reasonable to assume that the drill-bit will survive the very first hole. This implies that the FIS model should estimate the drill-bit reliability to be 1.0 when exposed to the sort of thrust-force and torque conditions witnessed during the machining of the first hole for all the eight training set drill-bits. Similarly, it is only reasonable to expect the FIS model to estimate the drill-bit reliability to be 0.0 when exposed to the sort of thrust -force and torque conditions witnessed during the machining of the last hole for all the eight training set drill-bits. Thus, in total, 16 data points were developed from the eight training set drill-bits. Validation and testing datasets were also developed similarly, using the corresponding drill-bit data. Note that while labeled data could be generated for representing extreme states of the drill-bits, it is not easily possible to develop any such data for intermediate states (i.e., states other than those representing either an extremely sharp/good or extremely dull/bad drill-bits).

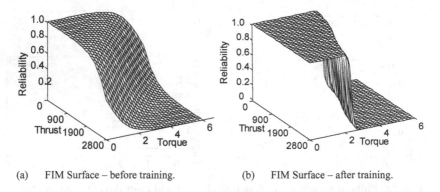

(a)   FIM Surface – before training.   (b)   FIM Surface – after training.

Fig. 14. Failure definition model surface.

Training the ANFIS formulation of our FIS model using these datasets resulted in the final relationship illustrated in Fig. 14(b). The corresponding changes to the membership functions by the ANFIS training algorithms are also illustrated in Fig. 13. Close observation of Fig. 13 and 14 reveals that the FIS model is predominantly utilizing the torque degradation signal in comparison with the thrust-force signal for estimating the on-line reliability of the drill-bit. This is partially attributed to the fact that torque exerted on a drill-bit is more sensitive to most of the failure modes that dominate drilling operations (i.e., it offers better signal-to-noise ratio in comparison with the thrust-force signal).

9.6.4 Online Reliability Estimation using Neural Networks

Structured experiments revealed that a FIR MLP network with the following configuration appeared to be good at maintaining generalization with respect to predicting the thrust force and torque levels into the future:

1. Input layer:
   - Number of neurons: 2
     (One for each of the two (thrust force & torque) degradation signals.)
   - Number of taps per synaptic filter: 15
2. Hidden layer
   - Number of neurons: 25
   - Number of taps per synaptic filter: 5
3. Output layer
   - Number of neurons: 6
     (Three neurons for each of the two degradation signals. First neuron is used for predicting one-step into the future, second neuron for predicting three-steps into the future, and the third neuron is used for predicting six-steps into the future. Networks with simultaneous multi-step predictions into the future outperformed networks with just one-step ahead prediction, in terms of generalization.)

Of the information collected from the 16 drill-bits, information from 12 randomly picked drill-bits was used for training purposes (labeled #1 to #12), and the information from the remaining 4 drill-bits was used for testing purposes (labeled #13 to #16). The network was designed to reduce the mean-square-error associated with testing patterns.

A SOM with a two-dimensional lattice of neurons (8×8) was used in dividing the 42 dimensional continuous input space into 64 distinct regions in an adaptive, topologically ordered fashion. The 42 dimensions are made up of (20+1) dimensional thrust force input vector and (20+1) dimensional torque input vector. The adaptive training scheme and parameter selection process discussed by Haykin (1999) was utilized in training the network. The covariance matrix, for each of the 64 distinct regions, for the FIR MLP model residuals in the output space, has been computed as per the procedure discussed in Section 5.4. Statistical analysis using Chi-Square tests and normal probability plots revealed that the residuals in distinct SOM neighborhoods tend to follow a Gaussian distribution. The reliability integral shown in equation (13) is calculated for these experiments using the Romberg method (Press et al. 1988).

The conditional performance reliability predictions for drill-bit #16 used for testing is shown in Fig. 15. All the conditional performance reliabilities are based on the assumption that the critical plane for the drill-bit with respect to thrust is 700 pounds and critical plane for the drill-bit with respect to torque is 42.5 inch-pounds, levels set from data available from the drill-bit manufacturer and laboratory experiments. Here again, the conditional performance reliability is equivalent to mission reliability where the mission constitutes the probability of successfully drilling the hole for the next $\Delta T = T_f - T_c$, given that it has survived thus far (T_c).

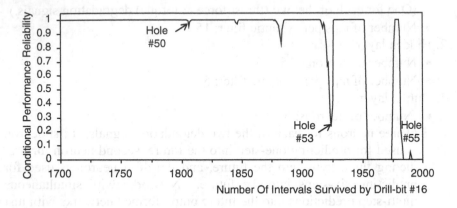

Fig. 15. Conditional performance reliability exhibited by drill-bit #16.

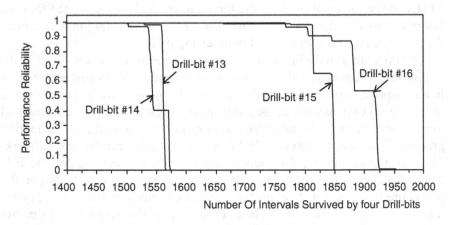

Fig. 16. Performance reliability exhibited by drill-bits #13, #14, #15, and #16.

Fig. 16 depicts the changes in the performance reliabilities (unconditional) for the 4 drill-bits used for testing. These unconditional performance reliabilities are calculated from their respective conditional reliabilities developed above. For example, the performance reliability that a particular drill-bit would survive Z intervals (interval width or sampling period in time units $= 1/\text{sampling rate} = 1/150 = 0.0067$ seconds) is the product of the conditional performance reliabilities of surviving each of these Z intervals (basically the product of Z conditional performance reliabilities). Since the FIR MLP is of prediction order 21, these calculations are based on the assumption that each drill-bit will survive the first 21 intervals (i.e., 0.14 seconds) with certainty.[3]

9.7 Summary, Conclusions and Future Research

Traditional approaches to reliability analysis are based on life tests that record only time-to-failure. With very few exceptions, all such analyses are aimed at estimating a population characteristic or characteristics of a system, subsystem, or component. For some components, it is possible to obtain degradation measurements over time, and these measurements contain useful information regarding component reliability. Then, one can define component failure in terms of a specified level of degradation, and estimate the reliability of that "particular" component based on its unique degradation measures. This chapter demonstrates that fuzzy inference models can be used to introduce failure definition in the degradation signal space using expert opinion and/or empirical data. This is particularly valuable for carrying out prognostics activities in the absence of sound knowledge for the mechanics of degradation and/or lack of adequate failure data. The specific application considered is in-process monitoring of the condition of the drill-bit in a drilling process utilizing the torque and thrust signals. The drilling process case study demonstrates the feasibility of on-line reliability estimation for individual components using the neuro-fuzzy approach. Further, this chapter provides an approach that allows the determination of a component's reliability as it degrades with time by monitoring its degradation measures. The concepts have been implemented using finite-duration impulse response multi-layer perceptron neural networks for modeling degradation measures and self-organizing maps for modeling degradation variation. An approach to compute prediction limits for any feed-forward neural network, critical for on-line performance reliability monitoring of systems using neural networks, is also introduced by combining the net-

---

[3] The prediction order for a FIR MLP is equal to the sum of the memory taps for the input and hidden layers plus one for the current state. For this particular network, $p = 15 + 5 + 1 = 21$.

work with a self-organizing map. Experimental results reveal that neural networks are effective in modeling the degradation characteristics of the monitored drill-bits, and predicting conditional and unconditional performance reliabilities as they degrade with time or usage. In contrast to traditional approaches, this approach to on-line performance reliability monitoring opens new avenues for better understanding and monitoring systems that exhibit failures through degradation. Essentially, implementation of the proposed performance reliability monitoring approach reduces overall operations costs by facilitating optimal component replacement and maintenance strategies.

However, there are still several unanswered questions. For example, there is no evidence that all types of failure modes prevalent in critical equipment could be adequately captured by the proposed Sugeno FIS model. Secondly, the inability to easily generate labeled training data for the ANFIS model from intermediate states (i.e., when the unit is neither brand new nor completely worn out) might jeopardize the interpolation capability of the FIS model. This issue, however, may not be significant from a practical perspective, for in general, there isn't a lot of interest in the intermediate states, at least from the standpoint of CBM. Typically, there is no provision to estimate MRL using the proposed method for the suggested neural network forecasting models are not capable of making long-term forecasts. This is beginning to change with the introduction of the so-called structural learning neural networks (Zimmerman et al. 2002). Means to develop confidence intervals is of paramount importance as well, without which, there is no provision to gauge the accuracy of the overall prognostics procedure.

References

Akay M, Mello C (1998) Time-Frequency and Time-Scale (Wavelets) Analysis Methods: Design and Algorithms, *International Journal of Smart Engineering System Design*, Vol. 1, No. 2, pp. 77-94.

Aminian F, Aminian M (2001) Fault Diagnosis of Nonlinear Analog Circuits Using Neural Networks with Wavelet and Fourier Transforms as Preprocessors, Journal of Electronic Testing: Theory and Applications, Vol. 17, pp. 471-481.

Aminian M, Aminian F (2000) Neural-Network Based Analog-Circuit Fault Diagnosis Using Wavelet Transform as Preprocessor, IEEE Transactions on Circuits and Systems II: Analog and Digital Signal Processing, Vol. 47, pp. 151-156.

Andrews G, Tlusty J (1983) A Critical Review of Sensors for Unmanned Machining", *Annals of the CIRP*, vol 32, pp. 563-572.

Aussem A, Murtagh F (1997) Combining Neural Network Forecasts on Wavelet-Transformed Time Series, *Connection Science*, Vol. 9, pp. 113-121.

Balazinski M, Czogala E, Jemielnia K, Leski J (2002) Tool condition monitoring using artificial intelligence methods, *Engineering Applications of Artificial Intelligence*, 15, 73-80.

Bashir Z, El-Hawary ME (2000) Short Term Load Forecasting by Using Wavelet Neural Networks, *Canadian Conference on Electrical and Computer Engineering*, Vol. 1, pp. 163-166.

Begg C, Merdes T, Byington CS, Maynard KP (1999) Mechanical System Modeling for Failure Diagnostics and Prognosis, *Maintainability and Reliability Conference* (MARCON 99), Gatlinburg, Tennessee, May 10-12.

Byington CS, Garga AK (2001) Data Fusion for Developing Predictive Diagnostics for Electromechanical Systems," in *Handbook of Multisensor Data Fusion*, D.L. Hall and J. Llinas eds., CRC Press, FL: Boca Raton.

Churchland PS, Sejnowski TJ (1992) *The Computational Brain*, Massachusetts: MIT Press.

Cheng B, Titterington DM (1994) Neural Networks: A Review from a Statistical Perspective, *Statistical Sciences*, vol. 9, pp. 2-54.

Chinnam RB (1999) On-line Reliability Estimation of Individual Components Using Degradation Signals, *IEEE Transactions on Reliability*, Vol. 48, No. 4, pp. 403-412.

Chinnam RB, Baruah P (2004) A Neuro-Fuzzy Approach for Estimating Mean Residual Life in Condition-Based Maintenance Systems, Vol. 20, Nos. 1-3, pp. 166-179.

Chinnam RB, Kolarik WJ, Manne VC (1996) Performance Reliability of Tools in Metal Cutting Using the Validity Index Neural Network, *International Journal of Modeling and Simulation*, Vol. 16, No. 4, pp. 210-217.

Chinnam RB, Mohan P (2002) Online Reliability Estimation of Physical Systems Using Neural Networks and Wavelets, Smart Engineering System Design, 4, 253-264.

Chryssolouris G, Lee M, Ramsey A (1996) Confidence Interval Prediction for Neural Network Models, *IEEE Trans. on Neural Networks*, vol 7, pp. 229-232.

Cristea P, et al. (2000) Time Series Prediction with Wavelet Neural Networks, Proceedings of the 5th Seminar on Neural Network Applications in Electrical Engineering. NEUREL 2000, Piscataway, NJ, USA & Belgrade, Yugoslavia.

Daubechies I (1990) The Wavelet Transform, Time-Frequency Localization and Signal Analysis, IEEE Transactions on Information Theory, Vol. 36, pp. 961-1005.

DeVeaux RD, et al. (1998) Prediction Intervals for Neural Networks via Nonlinear Regression, *Technometrics*, vol. 40, pp. 273-282.

Devore RA, Lucier BJ (1992) Wavelets, Acta Numerica, A. Iserles, ed., Cambridge University Press, Vol. 1, pp. 1-56.

Dimla ED (2000) Sensor signals for tool-wear monitoring in metal cutting operations—a review of methods, *International Journal of Machine Tools & Manufacture*, 40, pp. 1073-1098.

Dimla DE, Lister PM (2000) Online metal cutting tool condition monitoring. I: tool-state classification using multi-layer perceptron neural networks, *International Journal of Machine Tools and Manufacture Design, Research and Applications,* Vol. 40, 739-768.

Dimla DE, Lister PM (2000) Online metal cutting tool condition monitoring. II: tool-state classification using multi-layer perceptron neural networks, *International Journal of Machine Tools and Manufacture Design, Research and Applications,* Vol. 40, 769-781.

Dimla DE, Lister PM, Leighton NJ (1997) Neural network solutions to the tool condition monitoring problem in metal cutting - A critical review of methods, *International Journal of Machine Tools & Manufacture,* 37(9), 1219-1241.

Dzenis YA, Qian J (2001) Analysis of microdamage evolution histories in composites, *Int. J. Solids Struct.,* vol. 38, pp.1831-1854.

Elsayed EA (1996) *Reliability Engineering,* Massachusetts: Addison Wesley.

Elverson B (1997) Machinery fault diagnosis and prognosis, MS Thesis, The Pennsylvania State University.

El-Wardany TI, Gao D, Elbestawi MA (1996) Tool condition monitoring in drilling using vibration signature analysis, *International Journal of Machine Tools Manufacturing,* 36, 687-711.

Fan C, et al. (2001) Detection of Machine Tool Contouring Errors Using Wavelet Transforms and Neural Networks, *Journal of Manufacturing Systems,* Vol. 20, pp. 98-112.

Fang H, et al. (2000) Adaptive Control Using Wavelet Neural Networks, Hsi-An Chiao Tung Ta Hsueh/Journal of Xi'an Jiaotong University, Vol. 34, pp. 75-79.

Flores-Pulido L, et al. (2001) Classification of Segmented Images Combining Neural Networks and Wavelet Matching, Proceedings of the SPIE - The International Society for Optical Engineering, Vol. 4305, pp. 90-96.

Gebraeel N, Lawley M, Liu R, Parmeshwaran V (2004) Residual life predictions from vibration-based degradation signals: a neural network approach, *Industrial Electronics, IEEE Transactions on,* 51(3), 694-700.

Gu D, Hu H (2000) Wavelet Neural Network Based Predictive Control for Mobile Robots, Proceedings of IEEE International Conference on Systems, Man, and Cybernetics, Piscataway, NJ, USA.

Haykin S (1999) Neural Networks-A Comprehensive Foundation, 2nd Edition, Prentice Hall.

Hong GS, et al. (1996) Using Neural Network for Tool Condition Monitoring Based on Wavelet Decomposition, *International Journal of Machine Tools & Manufacture,* Vol. 36, pp. 551-566.

Hornik K, Stinchcombe M, White H (1989) Multi-Layer Feed Forward Networks are Universal Approximators, *Neural Networks,* Vol. 2, pp. 359-366.

Huang NE, et al. (1998) The empirical mode decomposition and the Hilbert spectrum for nonlinear and non-stationary time series analysis, *Proc. of R. Soc. Lond.,* Vol. 454, pp. 903-995.

Jang JR, Sun C, Mizutani E (1997) Neuro-Fuzzy and Soft Computing: A Computational Approach to Learning and Machine Intelligence, Prentice Hall, NJ: Upper Saddle River.

Jang JR (1993) ANFIS: Adaptive-Network based Fuzzy Inference Systems, *IEEE Trans. on Systems, Man, and Cybernetics*, vol. 23, pp. 665-685.

Jantunen E (2002) A summary of methods applied to tool condition monitoring in drilling, *International Journal of Machine Tools & Manufacture*, 42, pp. 997-1010.

Jun Z, Fei-Hu Q (1998) A Wavelet Transformation-Based Multichannel Neural Network Method for Texture Segmentation, *Journal of Infrared and Millimeter Waves*, Vol. 17, pp. 54-60.

Kacprzynski GJ, Roemer MJ (2000) Health Management Strategies for 21st Century Condition-Based Maintenance Systems, 13th International Congress on COMADEM, Houston, TX, December 3-8.

Kapur KC, Lamberson LR (1977) *Reliability in Engineering Design*, New York: John Wiley.

Karam M, Zohdy MA (1997) Augmentation of Optimal Control with Recurrent Neural Network and Wavelet Signals, Proceedings of the American Control Conference, Vol. 3, pp. 1551-1555.

Karunaratne PV, Jouny II (1997) Neural Network Face Recognition Using Wavelets, Proceedings of the SPIE - The International Society for Optical Engineering, Vol. 3077, pp. 202-213.

Katic D, Vukobratovic M (1997) Wavelet Neural Network Approach for Control of Non-Contact and Contact Robotic Tasks, Proceedings of 12th IEEE International Symposium on Intelligent Control, New York, NY, USA.

Ke Y, et al. (1998) The Application of Wavelet Neural Network in Spectral Data Compression and Classification, Journal of Infrared and Millimeter Waves, Vol. 17, pp. 215-220.

Kim SY, Kolarik WJ (1992) Real-time Conditional Reliability Prediction from On-line Tool Performance Data, *International Journal of Production Research*, Vol. 30, pp. 1831-1844.

Kobayashi K, et al. (1994) A Wavelet Neural Network with Network Optimizing Function, Transactions of the Institute of Electronics, Information and Communication Engineers D-II, Vol. J77D-II, pp. 2121-2129.

Kohonen T (1982) Self-Organized Formation of Topologically Correct Maps, *Biological Cybernetics*, vol 43, 1982, pp. 59-69.

Kuo RJ (2000) Multi-sensor integration for on-line tool wear estimation through artificial neural networks and fuzzy neural network, *Engineering Applications of Artificial Intelligence*, 13, 249-261.

Lawless JF (1982) Statistical Models and Methods for Lifetime Data, New York: John Wiley.

Lebold M, et al. (2000) Review of Vibration Analysis Methods for Gearbox Diagnostics and Prognostics, *Proceedings of the 54th Meeting of Society for Machinery Failure Prevention Technology*, Virginia Beach, VA, May 1-4, pp. 623-634.

Leonard JA, Kramer MA, Ungar LH (1992) Using Radial Basis Functions to Approximate a Function and Its Error Bounds, *IEEE Trans. on Neural Networks*, vol 3, pp. 624-627.

Lewis EE (1987) *Introduction to Reliability Engineering*, New York: John Wiley.

Lippmann RP (1987) An Introduction to Computing with Neural Nets", *IEEE ASSP Magazine*, vol 4, pp. 4-22.

Lu JC, Meeker WQ (1993) Using Degradation Measures to Estimate a Time-to-Failure Distribution, *Technometrics*, Vol. 35, pp. 161-172.

Lu S, Lu H, Kolarik WJ (2001) Multivariate Performance Reliability Prediction in Real-time, *Reliability Engineering and System Safety*, Vol. 72, No. 1, pp. 39-45.

Lu JC, Pantula SG (1989) A Repeated-Measurements Model for Over-Stressed Degradation Data, Technical Report, North Carolina State University, Dept. of Statistics.

Lu CJ, Park J, Yang Q (1997) Statistical Inference of a Time-to-Failure Distribution Derived From Linear Degradation Data, *Technometrics*, vol 39, pp. 391-400.

Mathur A, et al. (2001) Reasoning and Modeling Systems in Diagnosis and Prognosis, *Proceedings of the SPIE AeroSense Conference*, Orlando, FL, April 16-20.

Matsumoto K, et al. (2001) Machine Fault Diagnosis Using a Neural Network Based on Autocorrelation Coefficients of Wavelet Transformed Signals, Transactions of the Institute of Electrical Engineers of Japan, Part C, Vol. 121-C, pp. 167-176.

McAulay AD, Li J (1992) Wavelet Data Compression for Neural Network Preprocessing, Proceedings of the SPIE - The International Society for Optical Engineering, Vol. 1699, pp. 356-365.

Meeker WQ, Escobar LA, Lu JC (1998) Accelerated Degradation Test: Modeling and Analysis, *Technometrics*, Vol. 40, pp. 89-99.

Meeker WQ, Escobar LA (1998) Statistical Methods for Reliability Data. John Wiley, New York.

Nelson W (1981) Analysis of Performance Degradation Data from Accelerated Life Tests, *IEEE Transactions on Reliability*, Vol. 30, pp.149-155.

Nelson W (1982) Applied Life Data Analysis, New York: John Wiley.

Nelson W (1990) Accelerated Testing-Statistical Models, Test Plans, and Data Analysis, John Wiley & Sons.

Nur TI, et al. (1995) Detection of Tool Failure in End Milling with Wavelet Transformations and Neural Networks (Wt-Nn), International Journal of Machine Tools & Manufacture, Vol. 35, pp. 1137-1147.

Park J, Sandberg IW (1991) Universal Approximation Using Radial-Basis-Function Networks, *Neural Computation*, vol 3, pp. 246-257.

Press WH, Flannery BP, Teukolsky SA, Vetterling WT (1988) Numerical Recipes in C: The Art of Scientific Computing.

Pittner S, et al. (1998) Wavelet Networks for Sensor Signal Classification in Flank Wear Assessment, *Journal of Intelligent Manufacturing*, Vol. 9, pp. 315-322.

Qian S, Chen D (1996) Joint Time-Frequency Analysis, Prentice Hall.

Qing W, et al. (1996) The Application of Wavelet Neural Networks in Optimizing Electromagnetic Devices, Proceedings of the CSEE, Vol. 16, pp. 83-86.

Quan Y, Zhou M, Luo Z (1998) On-line robust identification of tool-wear via multi-sensor neural-network fusion, Engineering Applications of Artificial Intelligence, 11(6), 717-722.

Rai BK, Chinnam RB, Singh N, (2004) Tool-condition monitoring from degradation signals using MTS analysis, Conference proceedings of 21st Robust Engineering Symposium at Novi (Michigan), 343-351.

Schumacher P, Zhang J (1994) Texture Classification Using Neural Networks and Discrete Wavelet Transform, Proceedings of 1st International Conference on Image Processing, Los Alamitos, CA, USA.

Sharda R, Patil RB (1990) Neural Networks as Forecasting Experts: An Empirical Test, *Proceedings of IJCNN Meeting*, pp. 491-494.

Shashidhara HL, et al. (2000) Function Learning Using Wavelet Neural Networks, Proceedings of IEEE International Conference on Industrial Technology 2000, Mumbai, India.

Sun J, Hong GS, Wong YS, Rahman M, Wang ZG (2006) Effective training data selection in tool condition monitoring system, *International Journal of Machine Tools & Manufacture*, 46, pp. 218-224.

Sung DA, Kim CG, Hong CS (2002) Monitoring of impact damages in composite laminates using wavelet transform, *Composites: Part B*, vol. 33, pp.35-43.

Szu HH, et al. (1992) Neural Network Adaptive Wavelets for Signal Representation and Classification, Optical Engineering, Vol. 31, pp. 1907-1916.

Taguchi G, Jugulum R (2002) The Mahalanobis-Taguchi Strategy – A pattern technology system. John Wiley & Sons, New York.

Tan Y, et al. (2000) Dynamic Wavelet Neural Network for Nonlinear Dynamic System Identification, IEEE Conference on Control Applications - Proceedings, Vol. 1, pp. 214-219.

Tang Z, Almeida CD, Fishwick PA (1991) Time Series Forecasting using Neural Networks vs. Box-Jenkins Methodology, *Simulation*, vol 57, pp. 303-310.

Tang LC, Chang DS (1995) Reliability prediction using nondestructive accelerated-degradation data: case study on power supplies. *IEEE Transactions on Reliability*; 44:562-566.

Ting W, Sugai Y (2000) A Wavelet Neural Network for the Approximation of Nonlinear Multivariable Functions, Transactions of the Institute of Electrical Engineers of Japan, Part C, Vol. 120-C, pp. 185-193, 2000.

Tomsky J (1982) Regression Models for Detecting Reliability Degradation, *Proceedings of the Annual Reliability and Maintainability Conference*, pp. 238-244.

Tseng ST, Hamada MS, Chiao CH (1995) Using Degradation Data to Improve Fluorescent Lamp Reliability, *Journal of Quality Technology*, Vol. 27, No. 4, pp. 363-369.

Vachtsevanos G, Wang P, Khiripet N (1999) Prognostication: Algorithms and Performance Assessment Methodologies, *ATP Fall National Meeting Condition-Based Maintenance Workshop*, San Jose, California, November 15-17.

Wan EA (1990) Temporal Backpropagation for FIR Neural Networks, *Proc. of the IEEE Int'l Joint Conf. on Neural Networks*, vol 1, San Diego, CA, pp. 575-580.

Wan EA (1994) Time Series Prediction by Using a Connectionist Network with Internal Delay Lines, *Time Series Prediction: Forecasting the Future and Understanding the Past* (A.S. Weigend, N.A. Gershenfeld, *Ed*), pp. 195-217.

Wu S-J, Shao J (1999) Reliability analysis using the least-squares method in nonlinear mixed-effect degradation models, *Statistica Sinica*; **9**:855-877.

Wu S-J, Tsai T-R (2000) Estimation of time-to-failure distribution from a degradation model using fuzzy clustering, *Quality and Reliability Engineering International*; **16**:261-267.

Xiaoli L, et al. (1997) On-Line Tool Condition Monitoring System with Wavelet Fuzzy Neural Network, *Journal of Intelligent Manufacturing*, Vol. 8, pp. 271-276.

Xu J, Ho DWC (1999) Adaptive Wavelet Networks for Nonlinear System Identification, Proceedings of the 1999 American Control Conference, Piscataway, NJ, USA.

Yang MY, et al. (1997) Automatic ECG Classification Using Wavelet and Neural Networks, Chinese Journal of Medical and Biological Engineering, Vol. 17, pp. 265-275.

Yen GG, Lin K (2000) Wavelet Packet Feature Extraction for Vibration Monitoring, *Industrial Electronics, IEEE Transactions on*, 47(3), 650-667.

Ying J, Licheng J (1995) Wavelet Neural Networks for Functional Optimization Problems, Proceedings of International Conference on Neural Information Processing - ICONIP `95, Beijing, China.

Yong F, Chow TWS (1997) Neural Network Adaptive Wavelets for Function Approximation, Proceedings of European Symposium on Artificial Neural Networks. ESANN '97, Brussels, Belgium.

Yu IK, et al. (2000) Industrial Load Forecasting Using Kohonen Neural Network and the Wavelet Transform, Proceedings of 2000 Universities Power Engineering Conference, Belfast, Ireland.

Zadeh LA (1965) Fuzzy sets, *Information and Control*, Vol. 8, pp. 338-353.

Zhang J, et al. (1995) Wavelet Neural Networks for Function Learning, IEEE Transactions on Signal Processing, Vol. 43, pp. 1485-1497.

Zhang Q, Benveniste A (1991) Approximation by Nonlinear Wavelet Networks, ICASSP 91: 1991 International Conference on Acoustics, Speech and Signal Processing (Cat. No.91CH2977-7), New York, NY, USA.

Zhou Q, Hong GS, Rahman M (1995) A New Tool Life Criterion For Tool Condition Monitoring Using a Neural Network, *Engineering Applications of Artificial Intelligence*, 8, 578-588.

Zimmerman HG, Neuneir R, Gothmann R (2002) Modeling of Dynamic Systems by Error Correction Neural Networks, in Modeling and Forecasting Financial Data: Techniques of Nonlinear Dynamics, A. S. Soofi, L. Cao eds., Kluwer Publishing, MA: Boston, March 2002.

Imprecise Reliability: An Introductory Overview

Lev V. Utkin

Department of Computer Science, St.Petersburg State Forest Technical
Academy, Russia

Frank P.A. Coolen

Department of Mathematical Sciences, Durham University, UK

10.1 Introduction

A lot of methods and models in classical reliability theory assume that all
probabilities are precise, that is, that every probability involved is perfectly
determinable. Moreover, it is usually assumed that there exists some
complete probabilistic information about the system and component
reliability behavior. The completeness of the probabilistic information
means that two conditions must be fulfilled:

1) all probabilities or probability distributions are known or perfectly
   determinable;
2) the system components are independent, i.e., all random variables,
   describing the component reliability behavior, are independent, or,
   alternatively, their dependence is precisely known.

The precise system reliability measures can always (at least
theoretically) be computed if both these conditions are satisfied (it is
assumed here that the system structure is precisely defined and that there is
a known function linking the system *time to failure* (TTF) and TTFs of
components or some logical system function [8]). If at least one of these
conditions is violated, then only interval reliability measures can be
obtained. In reality, it is difficult to expect that the first condition is
fulfilled. If the information we have about the functioning of components
and systems is based on a statistical analysis, then a probabilistic
uncertainty model should be used in order to mathematically represent and
manipulate that information. However, the reliability assessments that are
combined to describe systems and components may come from various

L.V. Utkin and F.P.A. Coolen: *Imprecise Reliability: An Introductory Overview*, Computational
Intelligence in Reliability Engineering (SCI) **40**, 261–306 (2007)
www.springerlink.com

sources. Some of them may be objective measures based on relative frequencies or on well-established statistical models. A part of the reliability assessments may be supplied by experts. If a system is new or exists only as a project, then there are often not sufficient statistical data on which to base precise probability distributions. Even if such data exist, we do not always observe their stability from the statistical point of view. Moreover, failure times may not be accurately observed or may even be missed. Sometimes, failures do not occur at all or occur partially, leading to censored observations of failure times, and the censoring mechanisms themselves may be complex and not precisely known. As a result, only partial information about reliability of system components may be available, for example, the *mean time to failure* (MTTF) or bounds for the *probability of failure* at a time. Of course, one can always assume that the TTF has a certain probability distribution, where, for example, exponential, Weibull and lognormal are popular choices. However, how should we trust the obtained results of reliability analyses if our assumptions are only based on our experiences or on those of experts. One can reply that if an expert provides an interval for the MTTF on the basis of his experience, why should we reject his assumptions concerning the probability distribution of TTFs? The fact is that judgements elicited from experts are usually imprecise and unreliable due to the limited precision of human assessments. Therefore, any assumption concerning a certain probability distribution in combination with imprecision of expert judgements may lead to incorrect results which often cannot be validated due to lack of (experimental) data.

In many situations, it is unrealistic to assume that components of systems are independent. Let us consider two programs functioning in parallel (two-version programming). If these programs were developed by means of the same programming language, then possible errors in a language library of typical functions produce dependent faults in both programs. Several experimental studies show that the assumption of independence of failures between independently developed programs does not hold. However, the main difficulty here is that the degree of dependence between components is unknown, and one typically does not get sufficient data from which to learn about such dependence in detail. Similar examples can be presented for various applications. This implies that the second condition for complete information is also violated in most practical applications, and it is difficult to obtain precise reliability measures for a system, indeed such measures are mostly based on strong assumptions.

Dependence modelling is particularly important for large systems, for example to support high reliability software testing under practical constraints [86]. Wooff *et al* [150] present an approach based on Bayesian graphical modelling to support software testers, and thus enhance the reliability of software systems, in which dependencies are quantified precisely via elicitation of expert judgements. Due to the enormous elicitation task this is difficult to achieve completely in practice, hence imprecise probability assessments may be needed to enable wide-scale implementation of such methods, where imprecision at varying levels of model structuring and belief quantification can be used to guide efficient elicitation. This is an important research topic both from the perspective of statistical theory based on imprecise probability and reliability theory. Another possible way to model and quantify dependence structures is via Bayes linear methods [35], where expectation (`prevision') rather than probability is the central concept. In principle, due to linearity of expectation, it promises to be easier to generalize this statistical framework to allow imprecision than it is for probability theory, but this is still an open topic for research.

One of the tools to cope with imprecision of available information in reliability analysis is *fuzzy reliability theory* [17,18,48,127,129], which is based on using fuzzy and possibilistic models [51], models of fuzzy statistics [143]. However, the framework of this theory does not cover a large variety of possible judgements in reliability. Moreover, it requires to assume a certain type of possibility distributions of TTF or time to repair, and may be unreasonable in a wide scope of cases. Existing models of fuzzy reliability theory meet some difficulties from the practical point of view. Let us consider one of the most powerful models proposed by Cai [17], according to which the TTF of the i-th component is considered to be a fuzzy variable governed by a possibility distribution function $\mu_i(t)$ [51]. Then the reliability measure (possibility of failure before time t) of a series system consisting of n components is defined as $\max_{i=1,\ldots,n} \sup_{u \leq t} \mu_i(u)$. If all components are identical, then the possibility of failure does not depend on n. This controversial result is due to the operations min and max used in calculations, which practitioners cannot accept because it is well known that system reliability decreases with n. In this approach, a similarly problematic property holds for parallel systems. Other problems with this theory are the lack of clear interpretation of the possibility function, and lack of consistent and well founded theory for relating the possibility distribution function to statistical data. Cai [16] proposed a method based on computing the possibilistic likelihood function. However, this method has a shortcoming. By increasing the number of observations, the imprecision of the obtained possibility distribution function does not decrease and may even increase, which is not acceptable for practitioners in reliability analysis. It should be noted that the first point can be explained [107] by interpreting the possibility distribution by means of lower and upper probability distributions [52] and considering conditions of independence of random variables. Some models use fuzzy probabilities to describe the system reliability behavior. This representation can be regarded as a special type of second-order uncertainty models. However, most existing models using fuzzy probabilities also have shortcomings due to unreasonable usage of fuzzy operations and comparison indices. Moreover, the fuzzy sets and possibility theory are often used in reliability analysis as an alternative to the classical probability theory that cannot be accepted by many practitioners. In spite of these shortcomings, fuzzy reliability models can be viewed as an interesting class of models for taking incompleteness of information into account, with a variety of challenging open research problems.

Another approach to reliability analysis under incomplete information, based on the use of *random set and evidence theories* [89], has been proposed in the literature [6,65,93]. Random set theory offers an appropriate mathematical model of uncertainty when the information is not complete or when the result of each observation is not point-valued but set-valued, so that it is not possible to assume the existence of a unique probability measure. However, this approach also does not cover all possible judgements in reliability.

To overcome every difficulty of the methods considered above, the theory of imprecise probabilities [144] and its analogues (the theory of interval statistical models [77], the theory of interval probability [148,149]) can be used, which can be a general and promising tool for reliability analysis.

Coolen [28] provided an insight into imprecise reliability, discussing a variety of issues and reviewing suggested applications of imprecise probabilities in reliability. The idea of using some aspects of imprecise probability theory in reliability had already been considered in the literature. For example, Barlow and Proschan [8] studied a case of the lack of information about independence of components (Frechet bounds [55]) and nonparametric interval reliability analysis of ageing classes of TTF distributions. Barzilovich and Kashtanov [10] considered interval methods for optimal preventive maintenance under incomplete information. It has also been shown [26,27,37] how several commonly used concepts in reliability theory can be generalized, and combined with prior knowledge, through the use of imprecise probabilities in a generalized Bayesian statistical framework. Recently, nonparametric predictive inference has been developed, see Coolen *et al* [34] for an introductory overview, as a coherent statistical framework offering exciting application opportunities in reliability in situations where sufficient data are available. In this approach, only few mathematical assumptions are made, leading to imprecision, and further sources of uncertainty such as censored observations also lead to imprecision. Applications of this approach to maintenance and replacement problems have also been presented. We discuss this approach in more detail in Sec. Imprecise probability models for inference.

Further examples of applications of imprecise probabilities to reliability analysis have been presented by Utkin and Gurov [63,133], we briefly consider some of these examples. Suppose that the following information is available about components of a two-component series system. The MTTF of the first component is 10 hours, and the probability that the second component fails before 2 hours is 0.01. Without additional assumptions, the reliability of the system cannot be determined by means conventional reliability theory because the probability distribution of TTF is unknown. Any assumption about a certain probability distribution of TTF may lead to incorrect results. The reliability can also not be determined by means of methods of fuzzy reliability theory without further assumptions. However, this problem can be solved by using imprecise probabilities, with the restricted information leading to imprecise reliability quantifications.

A main objective of imprecise reliability is the analysis of system reliability using only available information without additional assumptions or, with a minimal number of assumptions. This theory also allows clear insights into the effects of any such further assumptions, as reflected via their effect on the imprecision in the system reliability measures. The following virtues of imprecise probability theory can be pointed out:

1) It is not necessary to make assumptions about probability distributions of random variables characterizing the component reliability behavior (TTFs, numbers of failures in a unit of time, etc.).
2) Imprecise probability theory is completely based on classical probability theory and can be regarded as its generalization. Therefore, imprecise reliability models can be interpreted in terms of classical probability theory. Conventional reliability models can be regarded as a special case of imprecise models.
3) Imprecise probability theory provides a unified tool (natural extension) for computing system reliability measures under partial information about the component reliability behavior.
4) Imprecise probability theory provides a generalization of possibility theory and evidence theory, and allows us to explain and understand some results of these approaches in reliability analysis.

5) Reliability measures that are different in kind can be combined and involved into the natural extension in a straightforward way. This implies that quite different reliability measures and estimates can be combined for computing the system reliability measures.

6) Imprecise probability theory allows us to obtain the best possible bounds for the system reliability given any information about component reliability and dependence structures.

7) The possible large imprecision of resulting system reliability measures reflects the available incompleteness of initial information and can direct the search for effective additional information sources.

At the same time, we can not assert that imprecise probability theory is the best and unique tool for reliability analysis under incomplete information. Ben-Haim [12,13] developed info-gap decision theory which has been successfully applied to solving some reliability problems. Info-gap models differ from the models of possibility, random set, and probability theories using real-valued measures functions defined on the space of events, which express either a probability or a possibility for each event in the space. An info-gap model of uncertainty is a family of nested sets. Each set corresponds to a particular degree of knowledge-deficiency, according to its level of nesting. There are no measure functions in an info-gap model of uncertainty.

This introductory overview of imprecise reliability is not intended as an exhaustive and comprehensive review of the literature. Instead, its aim is to show that imprecise reliability theory offers exciting opportunities and has been developed, yet this process is still at a relatively early stage, in particular with regard to (large-scale) practical applications.

10.2 System Reliability Analysis

Consider a system consisting of n components. Suppose that partial information about reliability of components is represented as a set of lower and upper expectations $\underline{\mathbf{E}}f_{ij}$ and $\overline{\mathbf{E}}f_{ij}$, $i = 1,...,n$, $j = 1,...,m_i$, of functions f_{ij}. Here m_i is a number of judgements that are related to the i-th component; $f_{ij}(X_i)$ is a function of the random TTF X_i of the i-th component or some different random variable, describing the i-th component reliability and corresponding to the j-th judgement about this component. For example, the interval-valued probability that a failure is in the interval $[a,b]$ can be represented as expectations of the indicator function $I_{[a,b]}(X_i)$ such that $I_{[a,b]}(X_i) = 1$ if $X_i \in [a,b]$ and $I_{[a,b]}(X_i) = 0$ if $X_i \notin [a,b]$. The lower and upper MTTFs are expectations of the function $f(X_i) = X_i$.

Denote $\mathbf{X} = (x_1,...,x_n)$ and $X = (X_1,...,X_n)$. Here $x_1,...,x_n$ are values of random variables $X_1,...,X_n$, respectively. It is assumed that the random variable X_i is defined on a sample space Ω and the random vector X is defined on a sample space $\Omega^n = \Omega \times ... \times \Omega$. If X_i is the TTF, then $\Omega = \mathbf{R}_+$. If X_i is a random state of a multi-state system [9], then $\Omega = \{1,...,L\}$, where L is a number of states of the multi-state system. In the case of a discrete TTF, $\Omega = \{1,2,...\}$, i.e. $\Omega = \mathbf{Z}_+$. According to Barlow and Proschan [8], the system TTF can be uniquely determined by the component TTFs. Then there exists a function $g(X)$ of the component lifetimes characterizing the system reliability behavior. The same holds for a multi-state system. If X_i is a random state, then a state of the multi-state system is determined by states of its components, i.e., there exists a function $g(X)$ called a structure function.

In terms of imprecise probability theory the lower and upper expectations can be regarded as *lower and upper previsions*. The functions f_{ij} and g can be regarded as *gambles* (the case of unbounded gambles is studied by Troffaes and de Cooman [96]). The lower and upper previsions $\underline{\mathbf{E}}f_{ij}$ and $\overline{\mathbf{E}}f_{ij}$ can be also viewed as bounds for an unknown precise prevision $\mathbf{E}f_{ij}$ which will be called a *linear prevision*. Since the function g is the system TTF, then, for computing the reliability measures (such as the probability of failure, MTTF, k-th moment of TTF), it is necessary to find lower and upper previsions of a gamble $h(g)$, where the function h is defined by the system reliability measure which has to be found. For example, if this measure is the probability of failure before time t, then $h(g) = I_{[0,t]}(g)$.

If we assume that the vector X is governed by some unknown joint density $\rho(\mathbf{X})$, then $\underline{\mathbf{E}}h(g)$ and $\overline{\mathbf{E}}h(g)$ can be computed by solving the following optimization problems (*natural extension*):

$$\underline{\mathbf{E}}h(g) = \min_{\mathbf{p}} \int_{\Omega^n} h(g(\mathbf{X}))\rho(\mathbf{X})d\mathbf{X},$$

$$\overline{\mathbf{E}}h(g) = \max_{\mathbf{p}} \int_{\Omega^n} h(g(\mathbf{X}))\rho(\mathbf{X})d\mathbf{X},$$

subject to

$$\rho(\mathbf{X}) \geq 0, \ \int_{\Omega^n} \rho(\mathbf{X}) d\mathbf{X} = 1,$$

$$\underline{\mathbf{E}} f_{ij} \leq \int_{\Omega^n} f_{ij}(x_i) \rho(\mathbf{X}) d\mathbf{X} \leq \overline{\mathbf{E}} f_{ij}, \ i \leq n, \ j \leq m_i.$$

Here the minimum and maximum are taken over the set $\mathbf{P}$ of all possible density functions $\{\rho(\mathbf{X})\}$ satisfying the above constraints, i.e., solutions to the problems are defined on the set $\mathbf{P}$ of densities that are consistent with partial information expressed in the form of the constraints. These optimization problems mean that we only have to find the largest and smallest possible values of $\mathbf{E}h(g)$ over all densities from the set $\mathbf{P}$.

If the considered random variables are discrete and the sample space Ω^n is finite, then integrals and densities in the optimization problems are replaced by sums and probability mass functions, respectively.

It should be noted that only joint densities are used in the above optimization problems because, in a general case, we may not be aware whether the variables $X_1, ..., X_n$ are dependent or not. If it is known that components are independent, then $\rho(\mathbf{X}) = \rho_1(x_1) \times \cdots \times \rho_n(x_n)$. In this case, the set $\mathbf{P}$ is reduced and consists only of the densities that can be represented as a product of marginal densities. This results in more precise reliability assessments. The manner in which the condition of independence influences on the precision of assessments is often an interesting topic of study, as it may provide useful insights into the effect of independence assumptions.

If the set $\mathbf{P}$ is empty, this means that the set of available evidence is *conflicting* and the optimization problems become irrelevant, hence this method would not provide any solutions. For example, if two experts provide [10,12] and [14,15] as bounds for the MTTF of a component, this information is clearly conflicting because these bounds produce non-intersecting sets of probability distributions, so the set $\mathbf{P}$ of common distributions is empty. There are several ways to cope with conflicting evidence. One is to localize the conflicting evidence and discard it, another is to somehow correct the conflicting evidence making it non-conflicting [102]. A third possibility is to introduce some beliefs to every judgement and to deal with second-order hierarchical models [109,110] which will be considered below.

The dual optimization problems for computing the lower $\underline{\mathbf{E}}h(g)$ and upper $\overline{\mathbf{E}}h(g)$ previsions of $h(g)$ are [300,133]:

$$\underline{\mathbf{E}}h(g) = \max\left\{c + \sum_{i=1}^{n}\sum_{j=1}^{m_i}\left(c_{ij}\underline{\mathbf{E}}f_{ij} - d_{ij}\overline{\mathbf{E}}f_{ij}\right)\right\},$$

subject to $c_{ij}, d_{ij} \in \mathbf{R}_+$, $i = 1, ..., n$, $j = 1, ..., m_i$, $c \in \mathbf{R}$, and $\forall \mathbf{X} \in \Omega^n$,

$$c + \sum_{i=1}^{n} \sum_{j=1}^{m_i} \left(c_{ij} - d_{ij} \right) f_{ij} \leq h(g(\mathbf{X})).$$

The dual optimization problem for computing the upper prevision $\overline{\mathbf{E}}h(g)$ of the system function $h(g)$ is

$$\overline{\mathbf{E}}h(g) = \min \left\{ c + \sum_{i=1}^{n} \sum_{j=1}^{m_i} \left(c_{ij} \overline{\mathbf{E}} f_{ij} - d_{ij} \underline{\mathbf{E}} f_{ij} \right) \right\},$$

subject to $c_{ij}, d_{ij} \in \mathbf{R}_+$, $i = 1, ..., n$, $j = 1, ..., m_i$, $c \in \mathbf{R}$, and $\forall \mathbf{X} \in \Omega^n$,

$$c + \sum_{i=1}^{n} \sum_{j=1}^{m_i} \left(c_{ij} - d_{ij} \right) f_{ij} \geq h(g(\mathbf{X})).$$

Here c, c_{ij}, d_{ij} are optimization variables such that c corresponds to the constraint $\int_{\Omega^n} \rho(\mathbf{X}) d\mathbf{X} = 1$, c_{ij} corresponds to the constraint $\int_{\Omega^n} f_{ij}(x_i) \rho(\mathbf{X}) d\mathbf{X} \leq \overline{\mathbf{E}} f_{ij}$, and d_{ij} corresponds to the constraint $\underline{\mathbf{E}} f_{ij} \leq \int_{\Omega^n} f_{ij}(x_i) \rho(\mathbf{X}) d\mathbf{X}$. It turns out that dual optimization problems are simpler in comparison with primal ones in many applications, because this representation allows avoidance of situations with infinite numbers of optimization variables.

Most reliability measures (probabilities of failure, MTTFs, failure rates, moments of TTF, etc.) can be represented in the form of lower and upper previsions or expectations. Eachmeasure is defined by a gamble f_{ij}. Precise reliability information is a special case of imprecise information when lower and upper previsions of the gamble f_{ij} coincide, i.e.,

$$\underline{\mathbf{E}} f_{ij} = \overline{\mathbf{E}} f_{ij}.$$

For example, let us consider a series system consisting of two components. Suppose that the following information about reliability of components is available. The probability of the first component failure before 10 hours is 0.01. The MTTF of the second component is between 50 and 60 hours. It can be seen from the example that the available information is heterogeneous and it is impossible to find system reliability measures on the basis of conventional reliability models without using additional assumptions about probability distributions. At the same time, this information can be formalized as follows:

$$\underline{\mathbf{E}} I_{[0,10]}(X_1) = \overline{\mathbf{E}} I_{[0,10]}(X_1) = 0.01, \quad \underline{\mathbf{E}} X_2 = 50, \quad \overline{\mathbf{E}} X_2 = 60,$$

or

$$0.01 \le \int_{\mathbf{R}_+^2} I_{[0,10]}(x_1)\rho(x_1,x_2)\mathrm{d}x_1\mathrm{d}x_2 \le 0.01,$$

$$50 \le \int_{\mathbf{R}_+^2} x_2\rho(x_1,x_2)\mathrm{d}x_1\mathrm{d}x_2 \le 60.$$

If it is known that components are statistically independent, then the constraint $\rho(x_1,x_2) = \rho_1(x_1)\rho_2(x_2)$ is added. The above constraints form a set $\mathbf{P}$ of possible joint densities. Suppose that we want to find the probability of system failure after time 100 hours. This measure can be regarded as previsions of the gamble $I_{[100,\infty)}(\min(X_1,X_2))$, i.e., $g(X) = \min(X_1,X_2)$ and $h(g) = I_{[100,\infty)}(g)$. Then the objective functions are of the form:

$$\underline{E}h(g) = \min_{\mathbf{P}} \int_{\mathbf{R}_+^2} I_{[100,\infty)}(\min(x_1,x_2))\rho(x_1,x_2)\mathrm{d}x_1\mathrm{d}x_2,$$

$$\overline{E}h(g) = \max_{\mathbf{P}} \int_{\mathbf{R}_+^2} I_{[100,\infty)}(\min(x_1,x_2))\rho(x_1,x_2)\mathrm{d}x_1\mathrm{d}x_2.$$

Solutions to the problems are $\underline{E}h(g) = 0$ and $\overline{E}h(g) = 0.59$, which are the sharpest bounds for the probability of system failure after time 100 hours based solely on the given information. If there is no information about independence, then optimization problems for computing $\underline{E}h(g)$ and $\overline{E}h(g)$ can be written as

$$\underline{E}h(g) = \max\{c + 0.01c_{11} - 0.01d_{11} + 50c_{21} - 60d_{21}\},$$

subject to $c_{11}, d_{11}, c_{21}, d_{21} \in \mathbf{R}_+$, $c \in \mathbf{R}$, and $\forall(x_1,x_2) \in \mathbf{R}_+^2$,

$$c + (c_{11} - d_{11})I_{[0,10]}(x_1) + (c_{21} - d_{21})x_2 \le I_{[100,\infty)}(\min(x_1,x_2)),$$

and

$$\overline{E}h(g) = \min\{c + 0.01c_{11} - 0.01d_{11} + 60c_{21} - 50d_{21}\},$$

subject to $c_{11}, d_{11}, c_{21}, d_{21} \in \mathbf{R}_+$, $c \in \mathbf{R}$, and $\forall(x_1,x_2) \in \mathbf{R}_+^2$,

$$c + (c_{11} - d_{11})I_{[0,10]}(x_1) + (c_{21} - d_{21})x_2 \ge I_{[100,\infty)}(\min(x_1,x_2)),$$

The solutions to these problems are $\underline{E}h(g) = 0$ and $\overline{E}h(g) = 0.99$. This example clearly shows the possible influence of independence assumptions.

Another method for computing $\underline{E}h(g)$ and $\overline{E}h(g)$ is based on an assertion that optimal densities in the primal optimization problems are the weighted sums of Dirac functions [138] which have unit area concentrated in the immediate vicinity of some point. In this case, the infinite dimensional optimization problems are reduced to a problem with a finite number of variables equal to the number of constraints (pieces of evidence). The optimization problems, unfortunately, become non-linear, but it turns out that in some special cases [108,114,117,140] their solution is rather simple. If there is no information about independence of components, then

$$\underline{E}h(g) = \min_{c_k, \mathbf{X}_k} \sum_{k=1}^{N+1} c_k h(g(\mathbf{X}_k)),$$

$$\overline{E}h(g) = \max_{c_k, \mathbf{X}_k} \sum_{k=1}^{N+1} c_k h(g(\mathbf{X}_k)),$$

subject to

$$\sum_{k=1}^{N+1} c_k = 1, \ c_k \geq 0, \ k = 1,...,N+1,$$

$$\underline{E}f_{ij} \leq \sum_{k=1}^{N+1} c_k f_{ij}(x_i^{(k)}) \leq \overline{E}f_{ij}, \ j \leq m_i, \ i \leq n,$$

where $\mathbf{X}_k = (x_1^{(k)},...,x_n^{(k)}) \in \mathbf{R}_+^n$, $c_k \in \mathbf{R}_+$, $N = \sum_{i=1}^{n} m_i$.

Here $\mathbf{X}_k$, c_k are optimization variables. If components are independent, then

$$\underline{E}h(g) = \min_{c_j, \mathbf{X}_j} \sum_{l_1=1}^{m_1+1} ... \sum_{l_n=1}^{m_n+1} h(g(x_1^{(l_1)},...,x_n^{(l_n)})) \prod_{v=1}^{n} c_{l_v}^{(v)},$$

$$\overline{E}h(g) = \max_{c_j, \mathbf{X}_j} \sum_{l_1=1}^{m_1+1} ... \sum_{l_n=1}^{m_n+1} h(g(x_1^{(l_1)},...,x_n^{(l_n)})) \prod_{v=1}^{n} c_{l_v}^{(v)},$$

subject to

$$\sum_{k=1}^{m_l+1} c_k^{(l)} = 1, \ c_k^{(l)} \geq 0, \ l = 1,...,n,$$

$$\underline{E}f_{ij} \leq \sum_{l=1}^{m_i+1} f_{ij}(x_i^{(l)}) c_l^{(i)} \leq \overline{E}f_{ij}, \ j \leq m_i, \ i \leq n.$$

Let us introduce the notion of the *imprecise reliability model* of the i-th component as a set of m_i available lower and upper previsions and corresponding gambles

$$\mathbf{M}_i = \langle \underline{E}_{ij}, \overline{E}_{ij}, f_{ij}(X_i), j = 1,...,m_i \rangle = \wedge_{j=1}^{m_i} \mathbf{M}_{ij} = \wedge_{j=1}^{m_i} \langle \underline{E}_{ij}, \overline{E}_{ij}, f_{ij}(X_i) \rangle.$$

Our aim is to get the imprecise reliability model $\mathbf{M} = \langle \underline{E}, \overline{E}, h(g(X)) \rangle$ of the system. This can be done by using the natural extension which will be regarded as a transformation of the component imprecise models to the system model and denoted $\wedge_{i=1}^{n} \mathbf{M}_i \rightarrow \mathbf{M}$. The models in the above considered example are $\mathbf{M}_1 = \langle 0.01, 0.01, I_{[0,10]}(X_1) \rangle$, $\mathbf{M}_2 = \langle 50, 60, X_2 \rangle$,

$$\mathbf{M} = \langle \underline{\mathbf{E}}, \overline{\mathbf{E}}, I_{[100,\infty)}(\min(X_1, X_2)) \rangle .$$

Different forms of optimization problems for computing system reliability measures are studied by Utkin and Kozine [138]. However, if the number of judgements about component reliability behavior, $\sum_{i=1}^{n} m_i$, and the number of components, n, are large, optimization problems for computing $\underline{\mathbf{E}}h(g)$ and $\overline{\mathbf{E}}h(g)$ cannot be practically solved due to their extremely large dimensionality. This fact restricts the application of imprecise calculations to reliability analysis. Therefore, simplified algorithms for approximate solutions to such optimization problems must be developed, together with analytical solutions for some special types of systems and initial information. Some efficient algorithms are proposed by Utkin and Kozine [115,137]. The main idea underlying these algorithms is to decompose the difficult non-linear optimization problems into several linear programming problems which are easy to solve. For example, in terms of the introduced imprecise reliability models, an algorithm given in [115] allows us to replace the complex transformation $\wedge_{i=1}^{n} \mathbf{M}_i \to \mathbf{M}$ by a set of $n+1$ simple transformations

$$\mathbf{M}_i \to \mathbf{M}_i^0 = \langle \underline{\mathbf{E}}, \overline{\mathbf{E}}, h(X_i) \rangle, \ i = 1, ..., n,$$

$$\wedge_{i=1}^{n} \mathbf{M}_i^0 \to \mathbf{M}.$$

10.3 Judgements in Imprecise Reliability

The judgements considered above can be related to *direct* ones, which are a straightforward way to elicit the imprecise reliability characteristics of interest. Moreover, the condition of independence of components can be related to *structural* judgements. However, there is a wide variety of possible judgements [76] that imprecise reliability theory can deal with, and other types of initial information have to be pointed out.

Comparative judgements are based on comparison of reliability measures concerning one or two components [76,99]. An example of a comparative judgement related to one component is the probability of the i-th component failure before time t is less than the probability of the same component failure in time interval $[t_1, t_2]$. This judgement can be formally represented as $\underline{\mathbf{E}}(I_{[t_1, t_2]}(X_i) - I_{[0,t]}(X_i)) \geq 0$. An example of a comparative judgement related to two components is the MTTF of the i-th component is less than the k-th component MTTF, which can be rewritten as $\underline{\mathbf{E}}(X_k - X_i) \geq 0$. By using the property of previsions $\overline{\mathbf{E}}X = -\underline{\mathbf{E}}(-X)$, for instance, the last comparative judgement can be rewritten as $\overline{\mathbf{E}}(X_i - X_k) \le 0$.

Many reliability measures are based on *conditional probabilities* or *conditional previsions*, for example, failure rate, mean residual TTF, probability of residual TTF, etc. Moreover, experts sometimes find it easier to quantify uncertainties using probabilities of outcomes conditionally on the occurrence of other events. The lower and upper residual MTTFs can be formally represented as $\underline{\mathbf{E}}(X - t \mid I_{[t,\infty)}(X))$ and $\overline{\mathbf{E}}(X - t \mid I_{[t,\infty)}(X))$, where $X - t$ is the residual lifetime. The lower and upper probabilities of residual TTF after time z (lower and upper residual survivor functions) are similarly written as $\underline{\mathbf{E}}(I_{[z,\infty)}(X - t) \mid I_{[t,\infty)}(X))$ and $\overline{\mathbf{E}}(I_{[z,\infty)}(X - t) \mid I_{[t,\infty)}(X))$. It should be noted that the imprecise conditional reliability measures may be computed from unconditional ones by using the generalized Bayes rule [144]. For example, if lower $\underline{E}X$ and upper $\overline{E}X$ MTTFs are known, then the lower and upper residual MTTFs produced by the generalized Bayes rule are $\max\{0, \underline{E}X - t\}$ and $\overline{E}X$, respectively. A more detailed description of conditional judgements in reliability analysis can be found in [136].

It is often reasonable to assume that lifetime probability distribution functions are unimodal. Therefore, additional information about *unimodality* of lifetime probability distributions may be taken into account for imprecise reliability calculations [104,106]. Implementing an unimodality condition on discrete probability distributions into imprecise reliability calculations has been studied in [106].

Some qualitative or quantitative judgements about *kurtosis, skewness*, and *variance* can also be taken into account in imprecise reliability calculations [104]. For example, we may know that the component TTF typically has a flat density function, which is rather constant near zero, and very small for larger values of the variable (negative kurtosis). This qualitative judgement can be represented by the lower and upper previsions $\underline{E}X^2 = \overline{E}X^2 = h$ together with $\overline{\mathbf{E}}(X^4 - 3h^2) \le 0$, where $h \in [\inf X^2, \sup X^2]$. If, for instance, we know that data are skewed to the right (positive skewness), then this information can be formalized by the lower and upper previsions $\underline{E}X = \overline{E}X = h$ together with $\overline{\mathbf{E}}(3hX^2 - X^3) \le 2h^3$. If we know that the variance of the component TTF is less than the expectation squared, then additional constraints to optimization problems for computing lower and upper previsions are of the form: $\underline{E}X = \overline{E}X = h$ and $\overline{E}X^2 \leq 2h^2$. In such cases, the natural extension can be conveniently formulated as a parametric linear optimization problem with the parameter h.

Experts are often asked about k%-*quantiles* of the TTF X, i.e., they supply points x_i such that $\Pr\{X \leq x_i\} = k/100$. As pointed out by Dubois and Kalfsbeek [50], experts are often more confident at supplying intervals rather than point-values, becauseeith knowledge is often restricted. So experts may provide intervals for quantiles in the form $[\underline{x}_i, \overline{x}_i]$. This information can be written as

$$\Pr\{X \leq [\underline{x}_i, \overline{x}_i]\} = q_i,$$

and it can be interpreted as I do not know the true value of the quantile exactly, but I belief one of the values in the interval $[\underline{x}_i, \overline{x}_i]$ to be its true value. It is worth noting that the considered model of uncertainty differs from standard uncertainty models used in the imprecise probability theory, where there exists an interval of previsions of a certain gamble. In the models of quantiles, the gamble is viewed as a set of gambles for which the same previsions are defined. Thodel is represented as the union of imprecise models

$$\vee_{t \in [\underline{x}_i, \overline{x}_i]} \langle q_i, q_i, I_{[0,t]}(X) \rangle.$$

The symbol $\vee_{t \in [\underline{x}_i, \overline{x}_i]}$ means that at least one of the models $\langle q_i, q_i, I_{[0,t]}(X) \rangle$ is true. Then arbitrary reliability measures may be computed by using the natural extension. For example, if there are n judgements about imprecise quantiles ($q_1 \leq ... \leq q_n$) and a sample space of TTF of a component is bounded by values x_0 and x_N, then the lower and upper MTTFs of the component are

$$\underline{E}X = q_1 x_0 + \sum_{i=1}^{n} (q_{i+1} - q_i) \max_{k=1,...,i} \underline{x}_k, \quad q_{n+1} = 1,$$

$$\overline{E}X = (1 - q_n) x_N + \sum_{i=1}^{n} (q_i - q_{i-1}) \min_{k=i,...,n} \overline{x}_k, \quad q_0 = 0.$$

10.4 Imprecise Probability Models for Inference

Standard models for inference usually require a large number of observations of events, e.g. failures, or assume that an appropriate precise prior probability distribution is available (for Bayesian models). A possible way to avoiding these assumptions is by use of imprecise probability models or models with imprecise prior distribution for statistical inference [37]. As an alternative to the kind of models also used in robust Bayesian analysis [14], which provide useful models in the imprecise probability context although requiring a different interpretation of the lower and upper bounds for inferences, Coolen [24] presents a generalization by including a further parameter which explicitly controls the level of imprecision in case of updating with newly available data. For all these models, computation of lower and upper previsions, as required for many reliability applications, may seem to involve complex nonlinear optimization problems, in particular if multi-dimensional parameters are involved. Coolen [25] shows how these optimization problems can be replaced by relatively straightforward one-dimensional search problems, independent of the dimensionality of the original parameter space, which makes such methods far more readily available for use in imprecise reliability.

The *imprecise Dirichlet model* (IDM) was introduced by Walley [145] as a model for objective statistical inference from multinomial data. In the IDM, prior or posterior uncertainty about the multinomial distribution parameter θ are described by sets of Dirichlet distributions, and inferences about events are summarized by lower and upper probabilities which depend on the choice of a hyperparameter s. The hyperparameter determines how quickly upper and lower probabilities of events converge as statistical data accumulate. There are several arguments [145] in favour of $1 \le s \le 2$. The IDM avoids some shortcomings of alternative objective models, either frequentist or Bayesian. Coolen [27] presented a generalization of the IDM, suitable for lifetime data including right-censored observations, which are common in reliability theory and survival analysis. The resulting imprecise inferences typically encompass frequentist results for the same setting. For statistical inference on interval-valued data, Utkin [116,119] considered a set of IDMs produced by these data. The set of IDMs in this case does not require to divide the time-axis into a number of intervals for constructing the multinomial model. These intervals are produced by bounds of interval-valued data. The following example illustrates the above. Suppose that we observe $N = 5$ intervals of TTF $A_1 = [10,14]$, $A_2 = [12,16]$, $A_3 = [9,11]$, $A_4 = [12,14]$, $A_5 = [13,\infty)$. Then the lower and upper probabilities of an arbitrary interval A are of the form:

$$\underline{P}(A \mid s) = \frac{\sum_{i \,:\, A_i \subseteq A} 1}{N + s}, \quad \overline{P}(A \mid s) = \frac{\sum_{i \,:\, A_i \cap A \neq \varnothing} 1 + s}{N + s}.$$

Let $A = [0,14]$ and $s = 1$. Then $\underline{P}([0,14] \mid 1) = 3/6$, $\overline{P}([0,14] \mid 1) = 1$. It can be seen from the above expressions that the lower and upper probabilities do not depend on the division of the time-axis into intervals and right-censored observations (A_5) can be analyzed by the set of IDMs.

Quaeghebeur and de Cooman [88] applied the main ideas underlying the IDM to all distributions belonging to the exponential family, and constructed similar imprecise probability models for sampling from these distributions. Although the IDM has been established as an attractive model for statistical inference using imprecise probabilities, in reliability and other application areas, it has several serious shortcomings that were raised both by Walley himself in the paper introducing the IDM and by several discussion contributors to this paper [145]. These shortcomings were mostly apparent in situations where one has few observations, as is regularly the case in reliability problems. Recently, Coolen and Augustin [31] presented an alternative imprecise probability model for inference in case of multinomial data, which overcomes the reported shortcomings of the IDM. Applications of this model to reliability problems have not yet been presented.

Walley [145] proposed a *bounded derivative model* for statistical inference about a real-valued parameter in problems where there is little or no prior information. Prior ignorance about the parameter is modelled by a set of all continuous probability density functions for which the derivative of the log-density is bounded by positive constant. This is also a promising model, which as far as we are aware has not yet been applied to reliability problems.

For restricting to a set of possible distribution functions of TTF, and for formalizing judgements about the *ageing* aspects of lifetime distributions, various nonparametric or semi-parametric classes of probability distributions can be used. In particular the classes of all IFRA (increasing failure rate average) and DFRA (decreasing failure rate average) distributions have been studied by Barlow and Proschan [8]. In order to formalize judgements about the ageing aspects of lifetime distributions, new flexible classes of distributions, denoted as $\mathbf{H}(r,s)$ *classes* [64,131,132,134], have been proposed and investigated. The probability distribution of the component (or system) lifetime X can be written as $H(t) = \Pr(X \geq t) = \exp(-\Lambda(t))$, where $\Lambda(t) = \int_0^t \lambda(x)\mathrm{d}x$ and $\lambda(t)$ is the time-dependent failure rate, also known as the hazard rate. Let r and s be numbers such that $0 \leq r \leq s \leq +\infty$. A probability distribution belongs to a class $\mathbf{H}(r,s)$ with parameters r and s if $\Lambda(t)/t^r$ increases and $\Lambda(t)/t^s$ decreases as t increases. In particular, $\mathbf{H}(1,+\infty)$ is the class of all IFRA distributions; $\mathbf{H}(r,s)$ with $1 \leq r < s$ is the class of all IFRA distributions whose failure rate increases with rate bounded by r and s; $\mathbf{H}(0,1)$ is the class of all DFRA distributions; $\mathbf{H}(r,s)$ with $r < s \leq 1$ is the class of all DFRA distributions whose failure rate decreases with rate bounded by r and s; and $\mathbf{H}(r,s)$, $r < 1 < s$ is a class containing distributions whose failure rate is non-monotone. Inferences for such classes, and solutions to corresponding computational problems, were presented by Utkin and others in the papers referred to above. To make these promising distributional classes available for imprecise reliability analysis in practice, a number of interesting research problems are still open, including the important question of how to fit such classes to available data.

From statistical perspective, imprecise probability enables inferential methods based on relatively few mathematical assumptions, in particular in situations where data are available. During the last decade, Coolen, with a number of co-authors, has developed *nonparametric predictive inference* (NPI), where inferences are directly on future observable random quantities, e.g. the random time to failure of the next system. In this approach, imprecision depends in an intuitively logical way on the available data, as it decreases if information is added, yet aspects as censoring or grouping of data result in an increase of imprecision. Foundations of NPI, including proofs of its consistency in theory of interval probability, are presented by Augustin and Coolen [4]. An introduction to NPI in reliability is presented in [34], and theory for dealing with right-censored observations in NPI in [41], with applications to some specific reliability problems presented in [39,40]. This framework is also suitable for guidance on high reliability demonstration, answering the important question of how many failure-free observations are required in order to accept a system in a critical operation [32]. The fact that, in such situations, imprecise reliability theory allows decisions to be based on the more pessimistic one of the lower and upper probabilities, e.g. lower probability of failure-free operation over a period of specified length, is an intuitively attractive manner for dealing with indeterminacy. Recently, Coolen also considered probability safety assessment from similar perspective [30].

In early work, Coolen and Newby [38] showed how NPI can also be applied for support of replacement decisions for technical systems, which is often a core reliability activity. Along such lines, Coolen-Schrijner and Coolen [32,42,43,44,45] investigated NPI-based alternatives to established replacement strategies based on the length of time a system has been in operation. These methods are fully adaptive to available failure data, and imprecision is reflected in bounds of cost functions. In addition, their results provide clear insights into the influence of a variety of assumptions which are often used for the more established methods, and which may frequently be rather unrealistic if considered in detail. Hence, the fact that their NPI-based method can do without most of such assumptions and still be useful under quite a reasonable data requirement is interesting, and suggests that further development of NPI-based methods for imprecise reliability is an interesting topic of research.

10.5 Second-order Reliability Models

Natural extension is a powerful tool for analyzing the system reliability on the basis of available partial information about the component reliability. However, it has a disadvantage. Let us imagine that two experts provide the following judgements about the MTTF of a component: (1) MTTF is not greater than 10 hours; (2) MTTF is not less than 10 hours. The natural extension produces the resulting MTTF $[0,10] \cap [10,\infty) = 10$. In other words, the absolutely precise MTTF is obtained from extremely imprecise initial data. This is unrealistic in the practice of reliability analysis. The reason of such results is that probabilities of judgements are assumed to be 1. If we assign some different probabilities to judgements, then we obtain more realistic assessments. For example, if the belief to each judgement is 0.5, then, according to [73], the resulting MTTF is greater than 5 hours. Let us consider another example. Suppose that many experts, say 1000, provide the same interval for some probability of failure, say [0.9, 0.99] and one expert provides the interval [0, 0.89]. Clearly, these judgements are conflicting and the set of probability distributions produced by these intervals is empty. As a result, we can not use the natural extension. Of course, we can use the so called *unanimity rule* defined as the envelope of the expert previsions [97], which is guaranteed to exist, but leads to extremely imprecise results (in the considered example, the resulting interval is [0, 0.99]). On the other hand, it is intuitively obvious that our belief to the judgement supplied by the last expert is rather low in comparison with our belief to the judgement provided by 1000 experts, and the unreliable judgement could be removed from consideration. One might say that this example is highly artificial. Of course, the example is given here only for illustration purposes. However, what to do if only 2 experts instead of 1000 provide the interval [0.9, 0.99] and one expert provides the interval [0, 0.99]. In this case, it is difficult to remove the contradictory interval. Of course, the inconsistency of the assessments in this artificial example were trivial, but in practice, with a variety of assessments on possibly different random variables, it may actually be difficult to discover whether or not the assessments are inconsistent, providing a further difficulty. However, in case of precise judgements, it is extremely unlikely that different assessments, even when made by a single expert, are consistent, so the generalization to interval reliability offers powerful methods for checking and dealing with realistic uncertainty judgements.

The above examples imply that in order to obtain accurate and realistic system reliability assessments it is necessary to take into account some vagueness of information about the component reliability measures, i.e., to assume that expert judgements and statistical information about reliability of a system or its components may be unreliable. one possible solution is the use of *second-order uncertainty models,* also known as *hierarchical uncertainty models,* on which much attention has been focused in recent years, particularly in the statistics literature. These models describe the uncertainty of a random quantity by means of two levels. For example, suppose that an expert provides a judgement about the mean level of component performance [131]. If this expert sometimes provides incorrect judgements, we have to take into account some degree of belief to this judgement. In this case, the information about the mean level of component performance can be considered on the first level of the hierarchical model (first-order information) and the degree of belief to the expert judgements is considered on the second level (second-order information). Many papers are devoted to the theoretical [62,81,147] and practical [58,82] aspects of second-order uncertainty models. Lindqvist and Langseth [79] investigated monotone multi-state systems under the assumption that probabilities of the component states (first-order probabilities) can be regarded as random variables governed by the Dirichlet probability distribution (second-order probabilities). A comprehensive review of hierarchical models is given in [49], where it is argued that Bayesian hierarchical models are most common [61]. However, the use of Bayesian hierarchical models may be unrealistic in problems where only partial information is available about the system behavior.

Troffaes and de Cooman [97] specify and discuss two general ways for approaching the problem of aggregating expert opinions: *axiomatic* and *ad hoc.* Axiomatic approaches aim at deriving a preferably unique rule of aggregation from axioms or properties that this rule should satisfy. Ad hoc approaches are not as much concerned with axioms: one simply proposes or derives a mathematical formula, together with some form of justification. Both approaches have shortcomings and virtues, but axiomatic ones can be justified for various applications and initial data, whereas ad hoc approaches depend on specific applications and data.

Various methods of the pooling of assessments, taking into account the quality of experts, are available in the literature [23,57,80]. These methods use the concept of precise probabilities for modelling uncertainty, and the quality of experts is modelled by means of *weights* assigned to each expert in accordance with some rules. It should be noted that most of these rules use some available information about correctness of previous expert opinions. This might meet several difficulties. First, the behavior of experts is unstable, i.e., exact judgements related to a system elicited from an expert do not mean that this expert will provide results of the same quality for new systems. Second, when experts provide imprecise values of an evaluated quantity, the weighted rules can lead to controversial results. For instance, if an expert with a small weight, say 0.1, provides a very large interval, say [0, 10], for a quantity (covering its sample space), it is obvious that this expert is too cautious and the interval he supplies is non-informative, although this interval covers a true value of the quantity. On the other hand, if an expert with a large weight, say 0.9, supplies a very narrow interval, say [5, 5.01], the probability that true value of the quantity lies in this interval is rather small. We can see that the values of weights contradict with the probabilities of provided intervals. It should be noted that sometimes we do not know anything about the quality of experts, and assignment of weights might meet some psychological difficulties. This implies that weights for experts as measures of the quality of their expertise should not normally be interpreted as measures of the quality of provided opinions [113,116,119].

Most axiomatic second-order uncertainty models assume that there is a precise second-order probability distribution (or possibility distribution). Moreover, most models use precise probabilities for the first-level uncertainty quantification. Unfortunately, such information is often absent in many applications and additional assumptions may lead to some inaccuracy in results. A study of some tasks related to the homogeneous second-order models without any assumptions about probability distributions has been presented by Kozine and Utkin [73,75]. However, these models are of limited use due to homogeneity of gambles considered on the first-order level. A hierarchical uncertainty model for combining different types of evidence was proposed by Utkin [103,109], where the second-order probabilities can be regarded as confidence weights and the first-order uncertainty is modelled by lower and upper previsions of different gambles. However, the proposed model [103,109] supposes that the second-order initial information is analyzed only for one random variable. At the same time, the reliability applications suppose that there is a set of random variables (component TTFs) described by a second-order uncertainty model, and it is necessary to find a model for some function of these variables (system TTF). Suppose that we have a set of weighted expert judgements related to some measures $\mathbf{E}f_{ij}(X_i)$ of the component reliability behavior, $i = 1,...,n$, $j = 1,...,m_i$, i.e., there are lower and upper previsions $\underline{\mathbf{E}}f_{ij}$ and $\overline{\mathbf{E}}f_{ij}$. Here n is the number of components, m_i is the number of judgements that are related to the i-th component. Suppose that each expert is characterized by an interval of probabilities $[\underline{\gamma}_{ij}, \overline{\gamma}_{ij}]$. Then the judgements can be represented as

$$\Pr\left\{\underline{\mathbf{E}}f_{ij} \leq \mathbf{E}f_{ij} \leq \overline{\mathbf{E}}f_{ij}\right\} \in [\underline{\gamma}_{ij}, \overline{\gamma}_{ij}], \ i \leq n, \ j \leq m_i.$$

Here the set $\{\underline{\mathbf{E}}f_{ij}, \overline{\mathbf{E}}f_{ij}\}$ contains the *first-order previsions*, the set $\{\underline{\gamma}_{ij}, \overline{\gamma}_{ij}\}$ contains the *second-order probabilities*. Our aim is to produce new judgements which can be regarded as combinations of available ones. In other words, the following tasks can be solved:

1) Computing the probability bounds $[\underline{\gamma}, \overline{\gamma}]$ for some new interval $[\underline{\mathbf{E}}g, \overline{\mathbf{E}}g]$ of the system linear prevision $\mathbf{E}g$.

2) Computing an average interval $[\underline{\mathbf{E}\mathbf{E}}g, \overline{\mathbf{E}\mathbf{E}}g]$ for the system linear prevision $\mathbf{E}g$ (reduction of the second-order model to first-order one).

An imprecise hierarchical reliability model of systems has been studied by Utkin [111]. This model supposes that there is no information about independence of components. A model taking into account the possible independence of components leads to complex non-linear optimization problems. However, this difficulty can be overcome by means of approaches proposed in [112,121]. Some hierarchical models of reliability taking into account the imprecision of parameters of known lifetime distributions are investigated in [118,120].

10.6 Reliability of Monotone Systems

A system is called *monotone* if it does not become better by a failure of one or more components. Various results have been obtained for computing imprecise reliability measures of typical monotone systems based on some particular types of initial information.

Some results concerning the reliability of typical systems are given in [70,71]. If initial information about reliability of components is restricted by lower and upper MTTFs, then the lower and upper system MTTFs have been obtained in explicit form for *series* and *parallel* systems [98,128].

The MTTFs of *cold standby* systems have been obtained by Utkin and Gurov [63,133]. The cold standby systems do not belong to a class of monotone systems. Nevertheless, we consider these systems as typical ones. It is worth noticing that expressions in the explicit form have been derived for the cases of independent components and complete lack of information about independence.

Suppose that the probability distribution functions of the component TTFs X_i are known only at some points t_{ij}, i.e., the available initial information is represented in the form of lower $\underline{E}I_{[0,t_{ij}]}(X_i)$ and upper $\overline{E}I_{[0,t_{ij}]}(X_i)$ previsions, $i=1,...,n$, $j=1,...,m_i$. Here t_{ij} is the j-th point of the i-th component TTF. Explicit expressions for lower and upper probabilities of the system failures before some time t have been obtained for series, parallel [114], *m-out-of-n* [117], cold standby [108] systems. For example, the lower and upper probabilities of the n-component parallel system failure before time t, for independent components, are

$$\underline{E}I_{[0,t]}\left(\max_{i=1,...,n} X_i\right) = \prod_{i=1}^{n} \underline{E}I_{[0,t_{iw_i}]}(X_i),$$

$$\overline{E}I_{[0,t]}\left(\max_{i=1,...,n} X_i\right) = \prod_{i=1}^{n} \overline{E}I_{[0,t_{iv_i}]}(X_i),$$

and, in case of complete lack of knowledge about independence,

$$\underline{E}I_{[0,t]}\left(\max_{i=1,...,n} X_i\right) = \max_{i=1,...,n} \underline{E}I_{[0,t_{iw_i}]}(X_i),$$

$$\overline{E}I_{[0,t]}\left(\max_{i=1,...,n} X_i\right) = \min\left(1, \sum_{i=1}^{n} \overline{E}I_{[0,t_{iv_i}]}(X_i)\right),$$

where $v_i = \min\{j : t_{ij} \geq t\}$ and $w_i = \max\{j : t_{ij} \leq t\}$.

General expressions for the reliability of arbitrary monotone systems under the same conditions are given by Utkin [110]. Moreover, it is proved that the lower (upper) bound for the system reliability of arbitrary monotone systems by given lower and upper points of probability distributions of the component TTFs depends only on these upper (lower) points. This result allows us to simplify the system reliability analysis.

It is interesting to study a case when the initial information about reliability of components is given in the form:

$$\underline{p}_{ij} \leq \Pr\{\underline{\alpha}_{ij} \leq X_i \leq \overline{\alpha}_{ij}\} \leq \overline{p}_{ij}, \quad i=1,...,n, \quad j=1,...,m_i,$$

where

$$[\underline{\alpha}_{i1},\overline{\alpha}_{i1}] \subset [\underline{\alpha}_{i2},\overline{\alpha}_{i2}] \subset ... \subset [\underline{\alpha}_{im_i},\overline{\alpha}_{im_i}], \quad i=1,...,n.$$

So there are nested intervals $[\underline{\alpha}_{ij}, \overline{\alpha}_{ij}]$, with interval probabilities $[\underline{p}_{ij}, \overline{p}_{ij}]$ for the event that the failure of the i-th component is inside these intervals. If we denote $v_i = \max\{j : \underline{\alpha}_{ij} \geq t\}$ and $w_i = \max\{j : \overline{\alpha}_{ij} \leq t\}$, then the lower and upper probabilities, for instance, of the n-component series system failure before time t, under the assumption of independent components, are

$$\underline{EI}_{[0,t]}\left(\min_{i=1,\dots,n} X_i\right) = 1 - \prod_{i=1}^{n}(1 - \underline{p}_{iw_i}),$$

$$\overline{EI}_{[0,t]}\left(\min_{i=1,\dots,n} X_i\right) = 1 - \prod_{i=1}^{n}\underline{p}_{iv_i}.$$

If there is no information about independence, then

$$\underline{EI}_{[0,t]}\left(\min_{i=1,\dots,n} X_i\right) = \max_{i=1,\dots,n} \underline{p}_{iw_i},$$

$$\overline{EI}_{[0,t]}\left(\min_{i=1,\dots,n} X_i\right) = 1 - \max\left(0, \sum_{i=1}^{n}\underline{p}_{iv_i} - (n-1)\right).$$

It can be seen that the lower and upper bounds for the system unreliability depend only on the lower probabilities of the nested intervals. This implies that knowledge of upper probabilities does not give any useful information in this case. The same is valid for arbitrary monotone systems. Moreover, the initial information can be regarded as the possibility and necessity measures [51]. It is proved that the system reliability measures $\underline{EI}_{[0,t]}(\cdot)$ and $\overline{EI}_{[0,t]}(\cdot)$ also can be regarded as the possibility and necessity measures. This result allows us to obtain and to explain the reliability measures of systems by fuzzy initial data.

10.7 Multi-state and Continuum-state Systems

The reliability behavior of many system can be formalized by means of multi-state and continuum-state models which can be viewed as an extension of binarystate models [78]. Let L be a set representing levels of component performance ranging from perfect functioning, $\sup L$, to complete failure, $\inf L$. A *general* model of the structure function of a system consisting of n multi-state components can be written as $S : L^n \to L$. If $L = \{0,1\}$, we have a classical *binary* system; if $L = \{0,1,\dots,m\}$, we have a *multi-state* system; if $L = [0,T]$, $T \in \mathbf{R}^+$, we have a *continuum* system. The i-th component may be in a state $x_i(t)$ at arbitrary time t. This implies that the component is described by the random process $\{X_i(t), t \geq 0\}$, $X_i(t) \in L$. Then the probability distribution function of the i-th component states at time t is defined as the mapping $F_i : L \to [0,1]$ such that $F_i(r,t) = \Pr\{X_i(t) \geq r\}$, $\forall r \in L$. The state of the system at time t is determined by states of its n components, i.e., $S(\mathbf{X}) = S(X_1, \ldots, X_n) \in L$.

The *mean level of component performance* is defined as $\mathbf{E}\{X_i(t)\}$. For a system, we write the *mean level of system performance* $\mathbf{E}\{S(\mathbf{X})\}$. Suppose that probability distributions of the component states are unknown and we have only partial information in the form of lower $\underline{\mathbf{E}}\{X_i(t)\}$ and upper $\overline{\mathbf{E}}\{X_i(t)\}$ mean levels of component performance. It is proved by Utkin and Gurov [131], that the number of states in this case does not influence on the mean level of system performance which is defined only by boundary states $\inf L$ and $\sup L$. This implies that reliability analysis of multi-state and continuum-state systems by such initial data is reduced to analysis of a binary system. A number of expressions for these systems have been obtained in explicit form [131].

At the same time, incomplete information about reliability of the multi-state and continuum-state components can be represented as a set of reliability measures (precise or imprecise) defined for different time moments. For example, interval probabilities of some states of a multi-state unit at time t_1 may be known. How to compute the probabilities of states at time t_2 without any information about the probability distribution of time to transitions between states? This problem has been solved by using the imprecise probabilities models [139].

10.8 Fault Tree Analysis

Fault tree analysis (FTA) is a logical and diagrammatic method to evaluate the probability of an accident resulting from sequences and combinations of faults and failure events. Fault tree analysis can be regarded as a special case of event tree analysis. A comprehensive study of event trees by representing initial information in the framework of convex sets of probabilities has been proposed by Cano and Moral [22]. Therefore, this work may be a basis for investigating fault trees. One of the advantages of imprecise fault tree analysis is a possibility to consider dependent events in a straightforward way, although complete lack of knowledge about the level of dependence is likely to lead to too much imprecision for practical use of such methods. However, the influence of any additional assumptions about dependence will then easily show in the final results, which in itself may provide valuable information, as well as guidance on the information requirement for practically useful conclusions.

Other substantial topics include the influence of events in a fault tree on a top event, and the influence of uncertainty of the event description on uncertainty of the top event description. This may be done by introducing and computing importance measures of events and uncertainty importance measures of their description. However, we are not aware of any reported study on this topic within the framework of interval reliability, which clearly suggests an important area of research.

10.9 Repairable Systems

Reliability analysis of *repairable* systems often involves difficult computational tasks, even when based on precise initial information. In addition, such analyses tend to require a substantial information input, often not in line with practical experience as many state descriptors are typically not observable or directly measurable without serious effort, if at all. A simple repairable process with instantaneous repair (*the time to repair* (TTR) is equal to 0), and under complete lack of information about dependence of random TTFs X_i, has been studied in [133]. According to this work, if the lower and upper MTTFs of a system are known, then the time-dependent lower $\underline{B}(t)$ and upper $\overline{B}(t)$ *mean time between failures* (MTBF) before time t are $\underline{B}(t) = 0$,

$$\overline{B}(t) = \min_{1 \le k < +\infty} \left(\overline{\mathrm{E}X} \sum_{i=1}^{k} \frac{1}{i} + \min\left(\frac{t - k\overline{\mathrm{E}X}}{k+1}, \frac{t - k\overline{\mathrm{E}X}}{k} \right) \right).$$

These bounds are of limited interest because $\underline{B}(t) = 0$ and $\overline{B}(t)$ becomes very large for large values of t (with $\overline{B}(t) \to \infty$ for $t \to \infty$), due to the lack of information about dependence.

Another basic and interesting model for repairable systems, based on interval-valued Markov chains, has been considered by Kozine and Utkin [72,74]. Some results on optimal preventive maintenance under incomplete information are presented in [10]. Useful preventive replacement guidelines for situations where failure data are available are presented within the NPI framework, as discussed in Sec. Imprecise probability models for inference. A quite general approach for reliability analysis of repairable systems, proposed by Gurov and Utkin, is to substitute the optimal density functions of TTF and time to repair, which are weighted sums of Dirac functions [138], into integral equations modelling arbitrary repairable systems, and to solve obtained optimization problems. Let us illustrate this approach for computing the lower and upper probabilities of the working state at time t (the time-dependent availability) under condition that the distributions of TTF and TTR are unknown and only the precise MTTF, denoted a, and the precise *mean time to repair*, denoted b, are specified. For the component, the following system of integral equations holds:

$$\begin{cases} y_0(s,t) = \int_0^t f(x+s)y_1(0,t-x)\mathbf{d}x + f(t+s) \\ \quad y_1(\tau,t) = \int_0^t g(x+\tau)y_0(0,t-x)\mathbf{d}x \end{cases}.$$

Here $f(x)$ and $g(x)$ are unknown densities of the TTF and TTR such that $\int_0^\infty xf(x)\mathbf{d}x = a$ and $\int_0^\infty xg(x)\mathbf{d}x = b$. The probability of the working state $p_0(t)$ at time t is computed as

$$p_0(t) = \int_0^\infty y_0(s,t)\mathbf{d}s.$$

The optimal densities $f_o(x)$ and $g_o(x)$ are in the classes of densities of the form (the weighted sums of Dirac functions $\delta(x-c)$, see Sec. System reliability analysis):

$$f_o(x) = \frac{x_2 - a}{x_2 - x_1}\delta(x - x_1) + \frac{a - x_1}{x_2 - x_1}\delta(x - x_2),$$

$$g_o(x) = \frac{z_2 - b}{z_2 - z_1}\delta(x - z_1) + \frac{b - z_1}{z_2 - z_1}\delta(x - z_2).$$

Here $x_1, x_2, z_1, z_2 \in \mathbf{R}_+$ are optimization variables. Then the lower (upper) bound for $p_0(t)$ is computed by minimizing (maximizing) $p_0(t)$ over all possible values of x_1, x_2, z_1, z_2 after substituting the densities $f_o(x)$ and $g_o(x)$ into integral equations.

Although it is possible, in principle, to analyze arbitrary systems in this manner, this approach requires extremely complex non-linear optimization problems. An efficient and practical approach for imprecise reliability analysis of repairable systems remains an open problem.

10.10 Structural Reliability

A probabilistic model of *structural reliability* and safety has been introduced by Freudenthal [56]. Following his work, a number of studies have been carried out to compute probability of failure under different assumptions about initial information. The problem of structural reliability can be stated as follows. Let Y represent a random variable describing the strength of a system and let X represent a random variable describing the stress or load placed on the system. By assuming that X and Y are defined on $\mathbf{X}$ and $\mathbf{Y}$, respectively, system failure occurs when the stress on the system exceeds the strength of the system: $\Phi = \{(x \in \mathbf{X}, y \in \mathbf{Y}) : x \geq y\}$. Here Φ is a region where the combination of system parameters leads to an unacceptable or unsafe system response. Then the reliability of the system is determined as $R = \Pr\{X \leq Y\}$, and the unreliability is determined as $Q = \Pr\{X > Y\} = 1 - R$.

Uncertainty of parameters in engineering design was successfully modelled by means of interval analysis [84]. Several authors [7,85] used fuzzy set and possibility theories to cope with a lack of complete statistical information about stress and strength. The main idea of their approaches is to consider the stress and strength as fuzzy variables or fuzzy random variables. Another approach to structural reliability analysis based on using random set and evidence theories has been proposed by several authors [6,65,92]. Several solutions to structural problems by means of random set theory have been presented in [93,94,95].

A more general approach to structural reliability analysis using imprecise probabilities was proposed by Utkin and Kozine [140,141]. This approach allows us to utilize a wider class of partial information about structural parameters, which includes possible data about probabilities of arbitrary events, expectations of the random stress and strength and their functions. At the same time, this approach allows us to avoid additional assumptions about probability distributions of the random parameters because the identification of precise probability distributions requires more information than what experts or limit statistical data are able to supply.

For example, if interval-valued probabilities

$$\underline{p}_i \leq \Pr\{X \leq \alpha_i\} \leq \overline{p}_i, \ \underline{q}_j \leq \Pr\{Y \leq \beta_j\} \leq \overline{q}_j,$$

of the stress X and strength Y are known at points α_i, $i = 1,...,n$, and β_j, $j = 1,...,m$, then the interval-valued stress-strength reliability, based on complete lack of information about dependence of X and Y, is

$$\underline{R} = \max_{i=1,\dots,n} \max\left(0, \underline{p}_i - \overline{q}_{j(i)}\right), \ j(i) = \min\{j \ : \ \alpha_i \le \beta_j\},$$

$$\overline{R} = 1 - \max_{k=1,\dots,m} \max\left(0, \underline{q}_k - \overline{p}_{l(k)}\right), \ l(k) = \min\{l \ : \ \beta_k \le \alpha_l\}.$$

If X and Y are independent, then

$$\underline{R} = \sum_{i=1}^{n} (\underline{p}_i - \underline{p}_{i-1})(1 - \overline{q}_{j(i)}), \ j(i) = \min\{j \ : \ \alpha_i \le \beta_j\},$$

$$\overline{R} = 1 - \sum_{k=1}^{m} (\underline{q}_k - \underline{q}_{k-1})(1 - \overline{p}_{l(k)}), \ l(k) = \min\{l \ : \ \beta_k \le \alpha_l\}.$$

Utkin [112] investigated stress-strength reliability analysis based on unreliable information about statistical parameters of stress and strength in the form of a second-order hierarchical uncertainty model. However, there are cases when properties of probability distributions of the stress and strength are known, for example, from their physical nature, but some parameters of the distributions must be assigned by experts. If experts provide intervals of possible parameter values, and these experts are considered to be absolutely reliable, then the problem of structural reliability analysis is solved by standard interval arithmetic. Often, however, it will be necessary to take into account the available information about the quality of experts, to obtain more credible assessments of the stress-strength reliability. An approach for computing the stress-strength reliability under these conditions is considered in [118].

10.11 Software Reliability

Software reliability has been studied extensively in the literature with the objective of improving software performance [19,90,151]. In the last decades, various *software reliability growth models* have been developed based on testing or debugging processes, but no model can be accurate for all situations. This fact is due to the unrealistic assumptions in each model. A comprehensive critical review on probabilistic software reliability models (PSRMs) was proposed by Cai *et al* [20]. Authors argued that fuzzy software reliability models (FSRMs) should be developed in place of PSRMs because the software reliability behavior is fuzzy in nature as a result of the uniqueness of software. This point is explained in three ways. First, any two copies of software exhibit no differences. Second, software never experiences performance deterioration without external intervention. Third, a software debugging process is never replicated. Due to the uniqueness of software and the environment of its use, frequentist statistical methods are rarely suitable for software reliability inferences. In addition, a large variety of factors contribute to the lack of success of existing PSRMs. To predict software reliability from debugging data, it is necessary to simultaneously take into account the test cases, characteristics of software, human intervention, and debugging data. It is impossible to model all four aspects precisely because of the extremely high complexity behind them [20].

To take into account the problems described above, Cai et al [21] proposed a simple FSRM and validated it. Central in this FSRM are the random time intervals between software failures, which are considered to be fuzzy variables governed by membership functions. Extensions of Cai's FSRMs taking into account the programmer's behavior (possibility of error removal and introduction) and combined fuzzy-probabilistic models have been investigated by Utkin et al [135].

Available PSRMs and FSRMs can be incorporated into more general imprecise software reliability models (ISRMs) [105], by application of the theory of imprecise probabilities. A family of non-countably many probability distributions constrained by some lower and upper distributions is constructed and analyzed in the ISRM. Let X_i be the random time interval between the $(i-1)$-th and i-th software failures. It is supposed that there exist lower and upper $\overline{P_i}(x\,|\,\theta_i)$ probability distributions of the random variable X_i with parameters θ_i and these distributions produce a set $\mathbf{R}_i$ of distributions such that $\underline{P_i}(x\,|\,\theta_i) = \min_{\mathbf{R}_i} P_i(x)$, $\overline{P_i}(x\,|\,\theta_i) = \max_{\mathbf{R}_i} P_i(x)$. Let $\{x_1,...,x_n\}$ be the successive intervals between failures. It is assumed that $\theta_i = f(i,\theta)$, where f is some function characterizing the software reliability growth. The main aim is to find the function $f(i)$ and its parameters θ. It is proved that the maximum of the likelihood function by the lack of information about independence of random times between software failures is determined as follows:

$$\max_{\theta} \max_{\mathbf{R}_1 \cup \mathbf{R}_2 \cup ... \cup \mathbf{R}_n} L(x_1,...,x_n\,|\,\theta) = \max_{\theta} \min_{i=1,...,n} \left\{ \overline{P_i}(x_i\,|\,\theta) - \underline{P_i}(x_i\,|\,\theta) \right\}.$$

If random variables are independent, then

$$\max_{\theta} \max_{\mathbf{R}_1 \cup \mathbf{R}_2 \cup ... \cup \mathbf{R}_n} L(x_1,...,x_n\,|\,\theta) = \max_{\theta} \prod_{i=1}^{n} \left\{ \overline{P_i}(x_i\,|\,\theta) - \underline{P_i}(x_i\,|\,\theta) \right\}.$$

It is also proved that in the case of right-censoring times for software failures the upper probabilities in the above expressions are replaced by 1.

ISRMs can be regarded as a generalization of the well known

probabilistic and possibilistic models. Moreover, they allow us to explain some peculiarities of known models, for example, taking into account the condition of independence of times to software failures, which are often hidden or can be explained intuitively. For example, the ISRM explains why FSRMs, as stated in [20], allow us to take into account a lot of factors influencing the software reliability. At the same time, PSRMs and FSRMs can be regarded as some boundary cases. Indeed, too rigid and often unrealistic assumptions are introduced in PSRMs, namely, times to software failure are independent and governed by a certain distribution. In FSRMs, it is assumed that the widest class of possible distributions of times to software failure is considered and there is no information about independence. It is obvious that the functions $\overline{P}_i(x)$, $\underline{P}_i(x)$ in the ISRM cannot be chosen arbitrarily because maximization of $L(\cdot|\theta)$ over parameters θ would give $\overline{P}_i(x)=1$, $\underline{P}_i(x)=0$. This implies that the functions $\overline{P}_i(x)$, $\underline{P}_i(x)$ must be constrained. For example, they may be connected by means of common parameters. Another possibility is to restrict the degree of imprecision $\max_x\left\{\overline{P}_i(x)-\underline{P}_i(x)\right\}\leq\varepsilon$. It should be noted that such constraints can also be used to arrive again at the PSRM and the FSRM. The PSRM assumes $\overline{P}_i(x)=\underline{P}_i(x)$ and $\varepsilon=0$. In the FSRM, we have identical parameters θ for lower and upper distributions.

Although quantification of software reliability metrics can provide useful insights into both the likely software performance and the quality of its development, uptake of such mathematical models has remained limited. This is mostly due to the crucial practical circumstances under which software developers and testers operate, typically with short turn-around times and huge time pressures. In addition, many models which have been suggested for supporting the activities, are based on unrealistic assumptions, e.g. independence assumptions underlying partition testing. Rees *et al.* [86] and Coolen *et al.* [36] have described such practical circumstances in detail, and report on a method employing Bayesian graphical models to support software testing of large-scale systems that require high reliability, with complex tasks and huge time pressures, technical details of the statistical aspects are described in [150]. This approach is fully subjective, with the testers' activities central to the model. As such, building the models requires substantial subjective inputs, which provides a bottle-neck to wide-scale practical application due to the enormous time pressures. So far, case studies have used a variety of methods to limit the elicitation effort, and the effect of assumptions has been studied by sensitivity analyses. It is recognized that imprecise probabilistic methods can offer much benefit to this approach in future, putting less pressure on experts to provide reasonably coherent information on very many variables. In addition, the effects of differing levels of imprecision, at different input places of the models, on the overall test strategies that result from such exercises, can be studied in order to decide where best to focus detailed elicitation effort. This is an exciting area of future research, requiring algorithms for manipulating Bayesian graphical models with imprecise probabilities. Although research on this latter issue has been ongoing in the statistical and computer science literatures for several years, it is not yet at the stage that it can be implemented to realistic large-scale software reliability models, due to the often complex dependence structures in these models. A possible way around this problem might be the use of Bayes linear methods [59], where useful in combination with full Bayesian models, to model complex dependence structures. This would have the benefit of the fact that previsions, the core concept in Bayes linear methods, are linear functionals, which would make inclusion of imprecision more straightforward, both in principle and from computational perspective. Coolen *et al* [35] present a first approach for such Bayes linear modelling for software reliability, Goldstein and Shaw [60] have shown how Bayes linear and Bayesian methods can be combined. Generalizing these approaches to include imprecision, hence further reducing elicitation effort and more clearly reporting levels of indeterminacy, is also an exciting topic for future research.

10.12 Human Reliability

Human reliability [66,67] is defined as the probability for a human operator to perform correctly required tasks in required conditions and not to assume tasks which may degrade the controlled system. Human reliability analysis aims at assessing this probability. Fuzzy or possibilistic descriptions of human reliability behavior are presented in [83]. Human behaviour has been described also by means of evidence theory [91]. Cai [18] noted the following factors of human reliability behavior contributing to the fuzziness:
1) inability to acquire and process an adequate amount of information about systems;
2) vagueness of the relationship between people and working environments;
3) vagueness of human thought process;
4) human reliability behavior is unstable and vague in nature because it

depends on human competence, activities, and experience.

These factors can also be addressed via imprecision, so imprecise probability theory might be successfully applied to human reliability analysis. Moreover, the behavioural interpretation of lower and upper previsions may well be suitable for describing human behavior. However, we are not aware of any research reported on imprecise human reliability, suggesting another stream of interesting research topics.

10.13 Risk Analysis

Risk of an unwanted event happening is often defined as the product of the probability of the occurrence of this event multiplied by its consequences, assuming that these consequences can be combined into a simple metric. The consequences may include financial cost, elapsed time, etc. One of the main objectives of performing risk analyses is to support decision-making processes. Risk analysis provides a basis for comparing alternative concepts, actions or system configurations under uncertainty [5,11]. A variety of methods has been developed for estimating losses and risks. When events occur frequently and when they are not very severe, it is relatively simple to estimate the risk exposure of an organization, as well as a reasonable premium when, for instance, an insurance transaction is made [53]. Commonly used methods rely on variations of the principle of maximizing expected utility, tacitly assuming that all underlying uncertainty can adequately be described by a precise and completely known probability measure. However, when the uncertainty is complex and the quality of the estimates is poor, the customary use of such rules together with over-precise data could be harmful as well as misleading. Therefore, it is necessary to extend the principle of maximizing expected utility to deal with complex uncertainty. Imprecise probability theory provides an efficient way for realizing such an extension.

The imprecision of information about unwanted events leads to consideration of minimal and maximal values of risk, which can be regarded as lower and upper previsions of consequences whose computation by complex events is studied in [101]. Some methods of handling partial information in risk analysis have been investigated by several authors [53,54]. Risk analysis under hierarchical imprecise uncertainty models has been studied by Utkin and Augustin [122], where two types of the second-order uncertainty models of states of nature are considered. The first type assumes that first-order uncertainty is modelled by lower and upper previsions of different gambles and the second-order probabilities can be regarded as confidence weights of judgements on the first-order level. The second type assumes that some aspects of the probability distribution of the states of nature is known, for example, from their physical nature, but (some) parameters of the probability distribution must be defined by experts, and there is some degree of our belief to each expert's judgement whose value is determined by experience and competence of the expert. New procedures for risk analysis under different conditions of partial information about states of nature in the framework of imprecise probabilities have been studied by Utkin and Augustin [123,124,125,126].

In situations where risk can be assessed via experiments, with the emphasis on low risk situations where systems are only released for practical operation following a number of tests without failures, the NPI framework (see Sec. Imprecise probability models for inference) provides useful guidelines on required test effort, in particular via the use of lower probabilities of corresponding future successful operation, to take indeterminacy into account ('to err on the side of safety', so to say). Some initial results in this area have been presented [30,33]. A further interesting topic, which has remained largely neglected as far as we are aware, is the fact that consequences, and their impact on life, are often not known in great detail. In particular where random features are studied with information occurring at different moments in time, it is natural to also take learning about such consequences and impacts into account. This also typically suggests that, at least at early stages (e.g. when designing a new chemical process), indeterminacy about such risks may well be modelled via imprecision, and it should be possible to take adaptive metrics for such risks into account. It is possible that the Bayesian adaptive utility framework [47], which was developed in the seventies within economics contexts, may provide an attractive solution to this problem. However, adaptive utility has not yet been generalized to allow imprecision, even more its uptake has been almost nonexistent, quite possibly due to both the computational complexities involved and the foundational aspects. Work in this direction has recently been initiated, and we hope to report on progress in the near future, where we will also particularly focus on applications in risk and reliability.

10.14 Security Engineering

Security engineering is concerned with whether a system can survive accidental or intentional attacks on it from outside (e.g. from users or virus intruders). In particular, computer security deals with the social regulations, managerial procedures and technological safeguards applied to computer hardware, software and data, to assure against accidental or deliberate unauthorized access to, and dissemination of, computer system resources (hardware, software, data) while they are in storage, processing or communication [68]. An important problem in security engineering is the quantitative evaluation of security efficiency. An interesting and valuable approach to measuring and predicting the operational security of a system was proposed by Brocklehurst *et al.* [15]. According to this approach, the behavior of a system should be considered from owner's and attacker's points of view. From the attacker's point of view, it is necessary to consider the *effort* (E) expended by the attacking agent and the *reward* (R) an attacker would get from breaking into the system. Effort includes financial cost, elapsed time, experience, ability of attacker, and could be expressed in such terms as mean effort to next security breach, probability of successfully resisting an attack, etc. Examples of rewards are personal satisfaction, gain of money, etc. From the owner's point of view, it is necessary to consider the system *owner's loss* (L) which can be interpreted as an *infimum selling price for a successful attack*, and the *owner's expenses* (Z) on the security means which include, for instance, anti-virus programs, new passwords, encoding, etc. The expenses come out in terms of time used for system verification, for maintenance of anti-virus software, as well as in terms of money spent on the protection. The expenses can be interpreted as a *supremum buying price for a successful attack*. Brocklehurst *et al.* [15] proposed to consider also the viewpoint of an all-knowing, all-seeing *oracle*, as well as the owner and attacker. This viewpoint could be regarded as being in a sense the true security of the system in the testing environment.

From the above, we can say that four variables are the base for obtaining security measures: effort, rewards, system owner's loss, owner's expenses. Moreover, their interpretation coincides with the behavioural interpretation of lower (expenses) and upper (system owner's loss) previsions and linear previsions (the all-knowing oracle). Therefore, imprecise probability theory provides an interesting and logical framework for quantifying such security measures [100,142]. Because of the increasing importance of security engineering, this also provides exciting opportunities for (research into) theory and application of imprecise methods.

10.15 Concluding Remarks and Open Problems

In recent years, many results have been presented which enable application of imprecise probability theory to reliability analyses of various systems, many of such results have been discussed here. Imprecise reliability theory is being developed step-by-step, mostly addressing problems from the existing reliability literature. However, the state-of-the-art is only a visible top of the iceberg called the imprecise reliability theory and there are many open theoretical and practical problems, which should be solved in future. Several exciting areas for future research have been indicated in the earlier sections, let us now say a bit more on this, and mention some further related topics of research.

It is obvious that modern systems and equipment are characterized by complexity of structures and variety of initial information. This implies that, on the one hand, it is impossible to adjust all features of a real system to the considered framework. On the other hand, introduction of some additional assumptions for constructing a reasonable model of a system may cancel all advantages of imprecise probabilities. Where are limits for introducing additional assumptions (simplification) in construction of a model? How do possible changes of initial information and assumptions influence the results of system reliability calculations? It is obvious that such questions relate to the informational aspect of imprecise reliability. The same can be said about necessity of studying the effects of possible estimation errors of initial data on resulting reliability measures. This leads to introducing and determining uncertainty importance measures.

Another important point is how to solve the optimization problems if the function $h(g(\mathbf{X}))$ is not expressed analytically in explicit form and can be computed only numerically. For example, this function may be a system of integral equations (repairable system). One of the ways to solve the corresponding optimization problems is the well-known simulation technique. However, the development of effective simulation procedures for solving the considered optimization problems is an open problem.

Many results of imprecise reliability are based either on the assumption of independence of components, or complete lack of information about independence. However, the imprecise probability theory allows us to take into account more subtle types of dependence [46,77] and, thereby, to make reliability analysis more flexible and adequate. Therefore, a clear interpretation and development of dependence concepts imprecise reliability theory is also an open problem, which has to be solved in future.

In spite of the fact that many algorithms and methods for reliability analysis of various systems have been developed, they are rather theoretical and cover some typical systems, typical initial evidence, and typical situations. At the same time, real systems are more complex. Therefore, practical approaches to analyze real systems by imprecise reliability methods have to be developed, which is likely to require development of appropriate approximate computational methods.

In order to achieve a required level of system reliability by minimal costs, it is possible to include redundant components in systems. To optimize cost and reliability metrics, the number of redundant components in a system can be determined, together with optimal system structures. Various algorithms for determining the optimal number of redundant components are available in the literature. However, most results assume that there exists complete information about reliability. Therefore, the development of efficient algorithms of optimization by partial information is also an open problem.

A similar problem is the *product quality control* which needs a trade-off between a better product quality and lower production costs by system constraints related to operating feasibility, product specifications, safety and environmental issues. Here results obtained by Augustin [2,3], concerning decision making under partial information about probabilities of states of nature, and results by Quaeghebeur and de Cooman [87], extending some aspects of game theory, might be a basis for investigating this problem. Quality control, in particular the use of control charts, has also been considered within the nonparametric predictive inferential framework [1]. This is also an exciting research area with many open problems, and with imprecision appearing naturally related to limited information. Clearly, ensuring high quality output in production processes can greatly enhance reliability. Even earlier than that, reliability often depends on the actual design of components and systems. At such an early stage, modelling uncertainties via precise probabilities is often extremely restricted, in particular when the designs involve revolutionary products. This is another area where imprecise reliability theory may offer exciting opportunities.

It should be noted that the list of open problems can be extended. However, most problems include at least some optimization problems (natural extension), which are often very complex. This may well be the reason why imprecise probability and reliability was not greatly developed earlier in the twentieth century. Nowadays, with the ever increasing computer power, complex optimization problems do not need to stop further development of appropriate methods for dealing with uncertainty, even though such problems still may need detailed consideration, and the need to develop approximate methods will remain. We believe that these are exciting times for imprecise reliability theory, as so much more can be achieved now than before. Therefore, the time is also right to take on challenges of actual applications, with all careful modelling and complex computational aspects involved. We look forward to these challenges, and hope that many fellow researchers also take up some of these challenges.

References

1. Arts GRJ, Coolen FPA, van der Laan P (2004) Nonparametric predictive inference in statistical process control. Quality Technology and Quantitative Management 1, 201-216.
2. Augustin T (2001) On decision making under ambiguous prior and sampling information. In: de Cooman G, Fine T, Seidenfeld T (eds.): Imprecise Probabilities and Their Applications. Proc. of the 2nd Int. Symposium ISIPTA'01. Ithaca, USA, pp. 9-16.
3. Augustin T (2002) Expected utility within a generalized concept of probability- a comprehensive framework for decision making under ambiguity. Statistical Papers 43, 5-22.
4. Augustin T, Coolen FPA (2004). Nonparametric predictive inference and interval probability. Journal of Statistical Planning and Inference 124, 251-272.
5. Aven T, Korte J (2003) On the use of risk and decision analysis to support decision-making. Reliability Engineering and System Safety 79, 289-299.
6. Bae H-R, R. Grandhi, Canfield R (2004) Epistemic uncertainty quantification techniques including evidence theory. Computers and Structures 82(13-14), 1101-1112.
7. Bardossy A, Bogardi I (1989) Fuzzy fatigue life prediction. Structural Safety 6, 25-38.
8. Barlow R, Proschan F (1975), Statistical Theory of Reliability and Life Testing: Probability Models. New York: Holt, Rinehart and Winston.
9. Barlow R, Wu A (1978) Coherent systems with multistate components. Math. Ops. Res. 3, 275-281.
10. Barzilovich E, Kashtanov V (1971) Some Mathematical Problems of the Complex System Maintenance Theory. Moscow: Sovetskoe Radio. (in Russian).
11. Bedford T, Cooke RM (2001) Probabilistic Risk Analysis: Foundations and Methods. Cambridge University Press.
12. Ben-Haim Y (1996) Robust Reliability in the Mechanical Sciences. Berlin: Springer-Verlag.
13. Ben-Haim Y (2004) Uncertainty, probability and information-gaps. Reliability Engineering and System Safety 85, 249-266.
14. Berger JO (1990) Robust Bayesian analysis: sensitivity to the prior. J of Statistical Planning and Inference 25, 303-328.

15. Brocklehurst S, Littlewood B, Olovsson T, Jonsson E (1994) On measurement of operational security. Technical Report PDCS TR 160, City University, London and Chalmers University of Technology, Goteborg.
16. Cai K-Y (1993) Parameter estimations of normal fuzzy variables. Fuzzy Sets and Systems 55, 179-185.
17. Cai K-Y (1996) Introduction to Fuzzy Reliability. Boston: Kluwer Academic Publishers.
18. Cai K-Y (1996) System failure engineering and fuzzy methodology. An introductory overview. Fuzzy Sets and Systems 83, 113-133.
19. Cai K-Y (1998), Software Defect and Operational Profile Modeling. Dordrecht/Boston: Kluwer Academic Publishers.
20. Cai K-Y, Wen C, Zhang M (1991) A critical review on software reliability modeling. Reliability Engineering and System Safety 32, 357-371.
21. Cai K-Y, Wen C, Zhang M (1993) A novel approach to software reliability modeling. Microelectronics and Reliability 33, 2265-2267.
22. Cano, A, S. Moral (2002) Using probability trees to compute marginals with imprecise probability. Int J of Approximate Reasoning 29, 1-46.
23. Cooke RM (1991) Experts in Uncertainty. Opinion and Subjective Probability in Science. New York: Oxford University Press.
24. Coolen FPA (1994) On Bernoulli experiments with imprecise prior probabilities. The Statistician 43, 155-167.
25. Coolen FPA (1994). Bounds for expected loss in Bayesian decision theory with imprecise prior probabilities. The Statistician 43, 371-379.
26. Coolen FPA (1996) On Bayesian reliability analysis with informative priors and censoring. Reliability Engineering and System Safety 53, 91-98.
27. Coolen FPA (1997) An imprecise Dirichlet model for Bayesian analysis of failure data including right-censored observations. Reliability Engineering and System Safety 56, 61-68.
28. Coolen FPA (2004) On the use of imprecise probabilities in reliability. Quality and Reliability Engineering International 20(3), 193- 202.
29. Coolen FPA (2006). On nonparametric predictive inference and objective Bayesianism. J of Logic, Language and Information, to appear.
30. Coolen FPA (2006) On probabilistic safety assessment in case of zero failures. J of Risk and Reliability, to appear.
31. Coolen FPA, Augustin T (2005) Learning from multinomial data: a nonparametric predictive alternative to the Imprecise Dirichlet Model. In: Cozman FG, Nau R, Seidenfeld T (eds.): Proc. of the 4th Int. Symposium on Imprecise Probabilities and their Applications, ISIPTA'05. Pittsburgh, USA, pp. 125-134.
32. Coolen FPA, Coolen-Schrijner P (2000) Condition monitoring: a new perspective. J of the Operational Research Society 51, 311-319.
33. Coolen FPA, Coolen-Schrijner P (2005) Nonparametric predictive reliability demonstration for failure-free periods. IMA J of Management Mathematics 16, 1-11.

34. Coolen FPA, Coolen-Schrijner P, Yan KJ (2002) Nonparametric predictive inference in reliability. Reliability Engineering and System Safety 78, 185-193.
35. Coolen FPA, Goldstein M, Munro M (2001) Generalized partition testing via Bayes linear methods. Information and Software Technology 43, 783-793.
36. Coolen FPA, Goldstein M, Wooff DA (2005) Using Bayesian statistics to support testing of software systems. In: Proc. 16th Advances in Reliability Technology Symposium, Loughborough, UK, 109-121.
37. Coolen FPA, Newby M (1994) Bayesian reliability analysis with imprecise prior probabilities. Reliability Engineering and System Safety 43, 75-85.
38. Coolen FPA, Newby MJ (1997) Guidelines for corrective replacement based on low stochastic structure assumptions. Quality and Reliability Engineering International 13, 177-182.
39. Coolen FPA, Yan KJ (2003) Nonparametric predictive inference for grouped lifetime data. Reliability Engineering and System Safety 80, 243-252.
40. Coolen FPA, Yan KJ (2003) Nonparametric predictive comparison of two groups of lifetime data. In: de Cooman G, Fine T, Seidenfeld T (eds.): Proc. of the 3rd International Symposium on Imprecise Probabilities and Their Applications, ISIPTA'03. Lugano, Switzerland, 148-161.
41. Coolen FPA, Yan KJ (2004) Nonparametric predictive inference with right-censored data. J of Statistical Planning and Inference 126, 25-54.
42. Coolen-Schrijner P, Coolen FPA (2004) Nonparametric predictive inference for age replacement with a renewal argument. Quality and Reliability Engineering International 20, 203-215.
43. Coolen-Schrijner P, Coolen FPA (2004) Adaptive age replacement based on nonparametric predictive inference. J of the Operational Research Society 55, 1281-1297.
44. Coolen-Schrijner P, Coolen FPA (2006) Nonparametric adaptive age replacement with a one-cycle criterion. Reliability Engineering and System Safety, to appear.
45. Coolen-Schrijner P, Coolen FPA, Shaw SC (2006) Nonparametric adaptive opportunity-based age replacement strategies. J of the Operational Research Society, to appear.
46. Couso I, Moral S, Walley P (1999) Examples of independence for imprecise probabilities. In: de Cooman G, Cozman F, Moral S, Walley P (eds.): ISIPTA '99- Proc. of the First Int. Symposium on Imprecise Probabilities and Their Applications. Zwijnaarde, Belgium, pp. 121-130.
47. Cyert RM, DeGroot MH (1987). Bayesian Analysis and Uncertainty in Economic Theory. Chapter 9: Adaptive Utility. Chapman and Hall, pp. 127-143.
48. de Cooman G (1996) On modeling possibilistic uncertainty in two-state reliability theory. Fuzzy Sets and Systems 83, 215-238.
49. de Cooman G (2002) Precision-imprecision equivalence in a broad class of imprecise hierarchical uncertainty models. J of Statistical Planning and Inference 105(1), 175-198.

50. Dubois D, Kalfsbeek H (1990) Elicitation, assessment and pooling of expert judgement using possibility theory. In: Manikopoulos C (ed.): Proc. of the 8th Int. Congress of Cybernetics and Systems. Newark, NJ, pp. 360-367.

51. Dubois D, Prade H (1988), Possibility Theory: An Approach to Computerized Processing of Uncertainty. Plenum Press, New York.

52. Dubois D, Prade H (1992) When upper probabilities are possibility measures. Fuzzy Sets and Systems 49, 65-74.

53. Ekenberg L, Boman M, Linnerooth-Bayer J (1997) Catastrophic risk evaluation. Interim report IR-97-045, IIASA, Austria.

54. Ferson S, Ginzburg L, Kreinovich V, Nguyen H, Starks S (2002) Uncertainty in risk analysis: Towards a general second-order approach combining interval, probabilistic, and fuzzy techniques. In: Proc. of FUZZ-IEEE'2002, Vol. 2. Honolulu, Hawaii, pp. 1342-1347.

55. Frechet M (1935) Generalizations du theoreme des probabilities totales. Fundamenta Mathematica 25, 379-387.

56. Freudenthal A (1956) Safety and the probability of structural failure. Transactions ASCE 121, 1337-1397.

57. Genest C, Zidek J (1986) Combining probability distributions: A critique and an annotated bibliography. Statistical Science 1, 114-148.

58. Gilbert L, de Cooman G, Kerre E (2000) Practical implementation of possibilistic probability mass functions. In: Proc. of Fifth Workshop on Uncertainty Processing (WUPES 2000). Jindvrichouv Hradec, Czech Republic, pp. 90-101.

59. Goldstein M (1999) Bayes linear analysis. In: Kotz S, Read CB, Banks DL (eds.): Encyclopaedia of Statistical Sciences, update vol. 3. New York: Wiley, pp. 29-34.

60. Goldstein M, Shaw SC (2004) Bayes linear kinematics and Bayes linear graphical models. Biometrika 91, 425-446.

61. Good I (1980) Some history of the hierarchical Bayesian methodology. In: Bernardo J, DeGroot M, Lindley D, Smith A (eds.): Bayesian Statistics. Valencia: Valencia University Press, pp. 489-519.

62. Goodman IR, Nguyen HT (1999) Probability updating using second order probabilities and conditional event algebra. Information Sciences 121(3-4), 295-347.

63. Gurov SV, Utkin LV (1999) Reliability of Systems under Incomplete Information. Saint Petersburg: Lubavich Publ. (in Russian).

64. Gurov SV, Utkin LV, Habarov SP (2000) Interval probability assessments for new lifetime distribution classes. In: Proceedings of the 2nd Int. Conf. on Mathematical Methods in Reliability, Vol. 1. Bordeaux, France, pp. 483-486.

65. Hall J, Lawry J (2001) Imprecise probabilities of engineering system failure from random and fuzzy set reliability analysis. In: de Cooman G, Fine T, Seidenfeld T (eds.): Imprecise Probabilities and Their Applications. Proc. of the 1st Int. Symposium ISIPTA'01. Ithaca, USA, pp. 195-204.

66. Hollnagel E (1993) Human reliability analysis. Context and control. London: Academic Press.

67. Holmberg, J, Hukki K, Norros L, Pulkkinen U, Pyy P (1999) An integrated approach to human reliability analysis- decision analytic dynamic reliability model. Reliability Engineering and System Safety 65, 239-250.
68. Hsiao D, Kerr S, Madnick S (1979) Computer Security. New York: Academic Press.
69. Khintchine A (1938) On unimodal distributions. Izv. Nauchno-Isled. Inst. Mat. Mech. 2, 1-7.
70. Kozine I (1999) Imprecise probabilities relating to prior reliability assessments. In: de Cooman G, Cozman F, Moral S, Walley P (eds.): ISIPTA'99 - Proc of the First International Symposium on Imprecise Probabilities and Their Applications. Zwijnaarde, Belgium, pp. 241-248.
71. Kozine I, Filimonov Y (2000) Imprecise reliabilities: Experiences and advances. Reliability Engineering and System Safety 67, 75-83.
72. Kozine I, Utkin LV (2000) Generalizing Markov chains to imprecise previsions. In: Proc. of the 5th Int Conference on Probabilistic Safety Assessment and Management. Osaka, Japan, pp. 383-388.
73. Kozine I, Utkin LV (2001) Constructing coherent interval statistical models from unreliable judgements. In: Zio E, Demichela M, Piccini N (eds.): Proc. of the European Conference on Safety and Reliability ESREL2001, Vol. 1. Torino, Italy, pp. 173-180.
74. Kozine I, Utkin LV (2002) Interval-valued finite Markov chains. Reliable Computing 8(2), 97-113.
75. Kozine I, Utkin LV (2002) Processing unreliable judgements with an imprecise hierarchical model. Risk Decision and Policy 7(3), 325-339.
76. Kozine I, Utkin LV (2003) Variety of judgements admitted in imprecise statistical reasoning. In: Proc. of the 3-rd Safety and Reliability Int. Conf. Vol. 2. Gdynia, Poland, pp. 139-146.
77. Kuznetsov VP (1991) Interval Statistical Models. Moscow: Radio and Communication. (in Russian).
78. Levitin G, Lisnianski A (2003) Multi-state system reliability. Assessment, optimization and applications. Singapore, New Jersey: World Scientific.
79. Lindqvist B, Langseth H (1998) Uncertainty bounds for a monotone multistate system. Probability in the Engineering and Informational Sciences 12, 239-260.
80. Mosleh A, Apostolakis G (1984) Models for the use of expert opinions. In: Waller R, Covello V (eds.): Low Probability / High Consequence Risk Analysis. New York: Plenum Press.
81. Nau RF (1992) Indeterminate probabilities on finite sets. The Annals of Statistics 20, 1737-1767.
82. Nguyen H, Kreinovich V, Longpre L (2001) Second-order uncertainty as a bridge between probabilistic and fuzzy approaches. In: Proceedings of the 2nd Conf. of the European Society for Fuzzy Logic and Technology EUSFLAT'01. England, pp. 410-413.
83. Onisawa T (1988) An approach to human reliability in man-machine system using error possibility. Fuzzy Sets and Systems 27, 87-103.

84. Penmetsa R, Grandhi R (2002) Efficient estimation of structural reliability for problems with uncertain intervals. Int J of Computers and Structures 80, 1103-1112.
85. Penmetsa R, Grandhi R (2003) Uncertainty propagation using possibility theory and function approximation. Int J of Mechanics of Structures and Machines 31(2), 257-279.
86. Rees K, Coolen FPA, Goldstein M, Wooff DA (2001) Managing the uncertainties of software testing: a Bayesian approach. Quality and Reliability Engineering International 17, 191-203.
87. Quaeghebeur E, de Cooman G (2003) Game-theoretic learning using the imprecise Dirichlet model. In: Bernard J-M, Seidenfeld T, Zaffalon M (eds.): Proc. of the 3rd Int. Symposium on Imprecise Probabilities and Their Applications, ISIPTA'03. Lugano, Switzerland, pp. 450-464.
88. Quaeghebeur E, de Cooman G (2005) Imprecise probability models for inference in exponential families. In: Cozman FG, Nau R, Seidenfeld T (eds.): Proc. of the 4th Int. Symposium on Imprecise Probabilities and Their Applications, ISIPTA'04. Pittsburgh, USA, pp. 287-296.
89. Shafer G (1976) A Mathematical Theory of Evidence. Princeton University Press.
90. Singpurwalla ND, Wilson SP (1999) Statistical Methods in Software Engineering: Reliability and Risk. New York: Springer.
91. Tanaka K, Klir G (1999) A design condition for incorporating human judgement into monitoring systems. Reliability Engineering and System Safety 65, 251-258.
92. Tonon F, Bernardini A (1998) A random set approach to optimisation of uncertain structures. Computers and Structures 68, 583-600.
93. Tonon F, Bernardini A, Elishakoff I (1999) Concept of random sets as applied to the design of structures and analysis of expert opinions for aircraft crash. Chaos, Solutions and Fractals 10(11), 1855-1868.
94. Tonon F, Bernardini A, Mammino A (2000) Determination of parameters range in rock engineering by means of Random Set Theory. Reliability Engineering and System Safety 70, 241-261.
95. Tonon F, Bernardini A, Mammino A (2000) Reliability analysis of rock mass response by means of Random Set Theory. Reliability Engineering and System Safety 70, 263-282.
96. Troffaes M, de Cooman G (2002) Extension of coherent lower previsions to unbounded random variables. In: Proc. of the Ninth International Conference IPMU 2002 (Information Processing and Management). Annecy, France, pp. 735-742.
97. Troffaes M, de Cooman G (2003) Uncertainty and conflict: A behavioural approach to the aggregation of expert opinions. In: Vejnarova J (ed.): Proc. of the 6th Workshop on Uncertainty Processing (WUPES '03). pp. 263-277.
98. Utkin LV (1998) General reliability theory on the basis of upper and lower previsions. In: Ruan D, Abderrahim H, D'hondt P, Kerre E (eds.): Fuzzy Logic and Intelligent Technologies for Nuclear Science and Industry. Proc. of the 3rd Int. FLINS Workshop. Antwerp, Belgium, pp. 36-43.

99. Utkin LV (2000) Imprecise reliability analysis by comparative judgements. In: Proc. of the 2nd Int. Conf. on Mathematical Methods in Reliability, Vol. 2. Bordeaux, France, pp. 1005-1008.

100. Utkin LV (2000) Security analysis on the basis of the imprecise probability theory. In: Cottam M, Harvey DW, Pape R, Tait J (eds.): Foresight and Precaution. Proc. of ESREL 2000, Vol. 2. Rotterdam, pp. 1109-1114.

101. Utkin LV (2001) Assessment of risk under incomplete information. In: Proc. of Int. Scientific School "Modelling and Analysis of Safety, Risk and Quality in Complex Systems". Saint Petersburg, Russia, pp. 319-322.

102. Utkin LV (2002) Avoiding the conflicting risk assessments. In: Proc. of Int. Scientific School "Modelling and Analysis of Safety, Risk and Quality in Complex Systems". Saint Petersburg, Russia, pp. 58-62.

103. Utkin LV (2002) A hierarchical uncertainty model under essentially incomplete information. In: Grzegorzewski P, Hryniewicz O, Gil M (eds.): Soft Methods in Probability, Statistics and Data Analysis. Heidelberg, New York: Phisica-Verlag, pp. 156-163.

104. Utkin LV (2002) Imprecise calculation with the qualitative information about probability distributions. In: Grzegorzewski P, Hryniewicz O, Gil M (eds.): Soft Methods in Probability, Statistics and Data Analysis. Heidelberg, New York: Phisica-Verlag, pp. 164-169.

105. Utkin LV (2002) Interval software reliability models as generalization of probabilistic and fuzzy models. In: German Open Conf. on Probability and Statistics. Magdeburg, Germany, pp. 55-56.

106. Utkin LV (2002) Involving the unimodality condition of discrete probability distributions into imprecise calculations. In: Proc. of the Int. Conf. on Soft Computing and Measurements (SCM'2002), Vol. 1. St. Petersburg, Russia, pp. 53-56.

107. Utkin LV (2002) Some structural properties of fuzzy reliability models. In: Proc. of the Int. Conf. on Soft Computing and Measurements (SCM'2002), Vol. 1. St. Petersburg, Russia, pp. 197-200.

108. Utkin LV (2003) Imprecise reliability of cold standby systems. Int. J of Quality and Reliability Management 20(6), 722-739.

109. Utkin LV (2003) Imprecise second-order hierarchical uncertainty model. Int. J of Uncertainty, Fuzziness and Knowledge-Based Systems 11(3), 301-317.

110. Utkin LV (2003) Reliability of monotone systems by partially defined interval probabilities. In: Proc. of the 3-rd Safety and Reliability Int. Conf., Vol. 2. Gdynia, Poland, pp. 187-194.

111. Utkin LV (2003) A second-order uncertainty model for the calculation of the interval system reliability. Reliability Engineering and System Safety 79(3), 341-351.

112. Utkin LV (2003) A second-order uncertainty model of independent random variables: An example of the stress-strength reliability. In: Bernard J-M, Seidenfeld T, Zaffalon M (eds.): Proc. of the 3rd Int. Symposium on Imprecise Probabilities and Their Applications, ISIPTA'03. Lugano, Switzerland, pp. 530-544.

113. Utkin LV (2004) Belief functions and the imprecise Dirichlet model. In: Batyrshin I, Kacprzyk J, Sheremetov L (eds.): Proc. of the Int. Conf. on Fuzzy Sets and Soft Computing in Economics and Finance, Vol. 1. St.Petersburg, Russia, pp. 178-185.

114. Utkin LV (2004) Interval reliability of typical systems with partially known probabilities. European J of Operational Research 153(3), 790-802.

115. Utkin LV (2004) A new efficient algorithm for computing the imprecise reliability of monotone systems. Reliability Engineering and System Safety 86(3), 179-190.

116. Utkin LV (2004) Probabilities of judgements provided by unknown experts by using the imprecise Dirichlet model. Risk, Decision and Policy 9(4), 391-400.

117. Utkin LV (2004) Reliability models of m-out-of-n systems under incomplete information. Computers and Operations Research 31(10), 1681-1702.

118. Utkin LV (2004) An uncertainty model of the stress-strength reliability with imprecise parameters of probability distributions. Zeitschrift für Angewandte Mathematik und Mechanik (Applied Mathematics and Mechanics) 84(10-11), 688- 699.

119. Utkin LV (2005) Extensions of belief functions and possibility distributions by using the imprecise Dirichlet model. Fuzzy Sets and Systems 154(3), 413-431.

120. Utkin LV (2005) A hierarchical model of reliability by imprecise parameters of lifetime distributions. Int J of Reliability, Quality and Safety Engineering 12(2), 167-187.

121. Utkin LV (2005) Imprecise second-order uncertainty model for a system of independent random variables. Int J of Uncertainty, Fuzziness and Knowledge-Based Systems 13(2), 177-194.

122. Utkin LV, Augustin T (2003) Decision making with imprecise second-order probabilities. In: Bernard J-M, Seidenfeld T, Zaffalon M (eds.): Proc. of the 3rd Int. Symposium on Imprecise Probabilities and Their Applications, ISIPTA'03. Lugano, Switzerland, pp. 545-559.

123. Utkin LV, Augustin T (2003) Risk analysis on the basis of partial information about quantiles. In: Modelling and Analysis of Safety and Risk in Complex Systems. Proc. of the Third Int. Scientific School MA SR-2003. IPME RAS, St. Petersburg, Russia, pp. 172-178.

124. Utkin LV, Augustin T (2004) Fuzzy decision making using the imprecise Dirichlet model. In: Batyrshin I, Kacprzyk J, Sheremetov L (eds.): Proc. of the Int. Conf. on Fuzzy Sets and Soft Computing in Economics and Finance, Vol. 1. St.Petersburg, Russia, pp. 186-193.

125. Utkin LV, Augustin T (2005) Decision making under imperfect measurement using the imprecise Dirichlet model. In: Cozman FG, Nau R, Seidenfeld T (eds.): Proc. of the 4th Int. Symposium on Imprecise Probabilities and Their Applications, ISIPTA'05. Pittsburgh, USA, pp. 359-368

126. Utkin LV, Augustin T (2005) Efficient algorithms for decision making under partial prior information and general ambiguity attitudes. In: Cozman FG,

Nau R, Seidenfeld T (eds.): Proc. of the 4th Int. Symposium on Imprecise Probabilities and Their Applications, ISIPTA'05. Pittsburgh, USA, pp. 349-358.

127. Utkin LV, Gurov SV (1996) A general formal approach for fuzzy reliability analysis in the possibility context. Fuzzy Sets and Systems 83, 203-213.

128. Utkin LV, Gurov SV (1998) New reliability models on the basis of the theory of imprecise probabilities. In: The 5th Int. Conf. on Soft Computing and Information/Intelligent Systems IIZUKA'98, Vol. 2. Iizuka, Japan, pp. 656-659.

129. Utkin LV, Gurov SV (1998) Steady-state reliability of repairable systems by combined probability and possibility assumptions. Fuzzy Sets and Systems 97(2), 193-202.

130. Utkin LV, Gurov SV (1999) Imprecise reliability models for the general lifetime distribution classes. In: de Cooman G, Cozman F, Moral S, Walley P (eds.): Proc. of the First Int. Symposium on Imprecise Probabilities and Their Applications. Zwijnaarde, Belgium, pp. 333-342.

131. Utkin LV, Gurov SV (1999) Imprecise reliability of general structures. Knowledge and Information Systems 1(4), 459-480.

132. Utkin LV, Gurov SV (2000) Generalized ageing lifetime distribution classes. In: Cottam M, Harvey DW, Pape R, Tait J (eds.): Foresight and Precaution. Proc. of ESREL 2000, Vol. 2. Rotterdam, pp. 1539-1545.

133. Utkin LV, Gurov SV (2001) New reliability models based on imprecise probabilities. In: Hsu C (ed.): Advanced Signal Processing Technology. World Scientific, Chapt. 6, pp. 110-139.

134. Utkin LV, Gurov SV (2002) Imprecise reliability for the new lifetime distribution classes. J of Statistical Planning and Inference 105(1), 215-232.

135. Utkin LV, Gurov SV, Shubinsky M (2002) A fuzzy software reliability model with multiple-error introduction and removal. Int J of Reliability, Quality and Safety Engineering 9(3), 215-228.

136. Utkin LV, Kozine I (2000) Conditional previsions in imprecise reliability. In: Ruan D, Abderrahim H, D'Hondt P (eds.): Intelligent Techniques and Soft Computing in Nuclear Science and Engineering. Bruges, Belgium, pp. 72-79.

137. Utkin LV, Kozine I (2001) Computing the reliability of complex systems. In: de Cooman G, Fine T, Seidenfeld T (eds.): Imprecise Probabilities and Their Applications. Proc. of the 2st Int. Symposium ISIPTA'01. Ithaca, USA, pp. 324-331.

138. Utkin LV, Kozine I (2001) Different faces of the natural extension. In: de Cooman G, Fine T, Seidenfeld T (eds.): Imprecise Probabilities and Their Applications. Proc. of the 2nd Int. Symposium ISIPTA'01. Ithaca, USA, pp. 316-323.

139. Utkin LV, Kozine I (2002) A reliability model of multi-state units under partial information. In: (eds.): Proc. of the Third Int. Conf. on Mathematical Methods in Reliability (Methodology and Practice). Trondheim, Norway, pp. 643-646.

140. Utkin LV, Kozine I (2002) Stress-strength reliability models under incomplete information. Int J of General Systems 31(6), 549-568.
141. Utkin LV, Kozine I (2002) Structural reliability modelling under partial source information. In: Langseth H, Lindqvist B (eds.): Proc. of the Third Int. Conf. on Mathematical Methods in Reliability (Methodology and Practice). Trondheim, Norway, pp. 647-650.
142. Utkin LV, Shubinsky IB (2000) Unconventional Methods of the Information System Reliability Assessment. St. Petersburg: Lubavich Publ. (in Russian).
143. Viertl R (1996) Statistical Methods for Non-Precise Data. Boca Raton, Florida: CRC Press.
144. Walley P (1991) Statistical Reasoning with Imprecise Probabilities. London: Chapman and Hall.
145. Walley P (1996) Inferences from multinomial data: Learning about a bag of marbles. J of the Royal Statistical Society, Series B58, 3-57.
146. Walley P (1997) A bounded derivative model for prior ignorance about a real-valued parameter. Scandinavian J of Statistics 24(4), 463-483.
147. Walley P (1997) Statistical inferences based on a second-order possibility distribution. Int J of General Systems 9, 337-383.
148. Weichselberger K (2000) The theory of interval-probability as a unifying concept for uncertainty. Int J of Approximate Reasoning 24, 149-170.
149. Weichselberger K (2001) Elementare Grundbegriffe einer allgemeineren Wahrscheinlichkeitsrechnung, Vol. I Intervallwahrscheinlichkeit als umfassendes Konzept. Heidelberg: Physika.
150. Wooff DA, Goldstein M, Coolen FPA (2002). Bayesian graphical models for software testing. IEEE Transactions on Software Engineering 28, 510-525.
151. Xie M (1991) Software Reliability Modeling. Singapore: World Scientific.

Posbist Reliability Theory for Coherent Systems

Hong-Zhong Huang

School of Mechanical and Electronic Engineering, University of Electronic Science and Technology of China, P.R. China

Xin Tong

Beijing Institute of Radio Metrology Measurement
Beijing, 100854, P. R. of China

Ming J Zuo

Department of Mechanical Engineering, University of Alberta, Canada

11.1 Introduction

The conventional reliability theory is built on the probability assumption and the binary-state assumption [1]. It has been successfully used for solving various reliability problems. However, it is not suitable when the failure probabilities concerned are very small (e.g., 10^{-7}) or when there is a lack of sufficient data. As a result, researchers have been searching for new models and new reliability theories that overcome the shortcomings of the classical probabilistic definition of reliability. Among others, we mention the works by Tanaka et al. [2], Singer [3], Onisawa [4], Cappelle and Kerre [5], Cremona and Gao [6], Utkin and Gurov [7], Cai et al [1, 8, 9], Huang [10-12], and Huang et al [13-18]. All these researchers have attempted to define reliability in terms other than the probabilistic definition. The fuzzy set concept represents a new paradigm of accounting for uncertainty. Two new assumptions in these definitions include the fuzzy-state assumption and the possibility assumption. The fuzzy state assumption indicates that the state of a piece of equipment can be represented by a fuzzy variable. The possibility assumption indicates that the reliability of a piece of equipment needs to be measured subjectively. These two new assump-

H.-Z. Huang et al.: *Posbist Reliability Theory for Coherent Systems*, Computational Intelligence in Reliability Engineering (SCI) **40**, 307–346 (2007)
www.springerlink.com

tions have been used in place of the conventional probability and the binary-state assumption. Considering these developments in the past 10 years, we can divide the fuzzy reliability theory into the following three categories [19]:

(1) The profust reliability theory: It is based on the probability assumption and the fuzzy-state assumption.

(2) The posbist reliability theory: It is based on the possibility assumption and the binary-state assumption.

(3) The posfust reliability theory: it is based on the possibility assumption and the fuzzy-state assumption.

For some systems, equipment, and components, it is very difficult to obtain necessary statistical data for conventional reliability analysis [11]. In addition, the failure occurrence patterns do not follow the statistical behavior of probability. In these situations, subjective evaluation of reliability by experts based on their engineering judgment is more significant than objective statistics. Cai et al. [8, 9, 19] used the mathematical notions of possibility and fuzzy variables and developed the theory of posbist reliability. They then provided a preliminary discussion of the posbist reliability of typical system structures including series, parallel, k-out-of-n, and fault-tolerant systems. They also demonstrated the advantages of the posbist reliability theory over conventional reliability theory. However, these works [8, 9, 19] were confined to nonrepairable systems. Utkin and Gurov [7] proposed a general approach to formalize posbist reliability analysis based on a system of functional equations according to Cai's theory. Cooman [20] introduced the notion of possibilistic structure function based on the concept of the classical, two-valued structure function and studied the possibilistic uncertainty of the states of a system and its components. Cremona and Gao [6] presented a new reliability theory for measuring and analyzing structural possibilistic reliability similar in methodology to the probabilistic reliability theory. Moller et al [21] applied the possibility theory to safety assessment of structures considering non-stochastic uncertainties and subjective estimates of the objective functions by experts. Savoia [22] presented a method for structural reliability analysis using the possibility theory and the fuzzy number approach. Guo et al. [23] developed a new model of structural possibilistic reliability based on the possibility theory and fuzzy interval analysis.

Since Zadeh [24] introduced the mathematical framework of possibility theory in 1978, many important theoretical and practical advances have been achieved in this field. The possibility theory has been applied to the fields of artificial intelligence, knowledge engineering, fuzzy logic, automatic control, and other fields. Many researchers have reported constructive achievements in application of the possibility theory to reliability analysis and safety assessment [5-8, 21-23].

Now that various frameworks of the possibilistic reliability theory have been constructed, how does one apply them to real-life systems or structures? Developing the possibility distributions from practical data is the first fundamental issue associated with applications of the possibilistic reliability theory. On one hand, the concept of possibility distribution plays a role, in the theory of possibilistic reliability, that is analogous, though not completely, to that of a probability distribution in the theory of probability reliability. Developing possibility distributions is an important step in possibilistic reliability theory. However, it might be difficult, if not impossible, to come up with a general method for developing possibility distributions which will work for all applications. We usually combine or use several methods for constructing possibility distributions in order to obtain all the possibility distribution functions of the fuzzy variables concerned.

In this chapter, we provide a detailed analysis of the posbist reliability theory and illustrate its applications in system reliability analysis. The lifetime of the system is treated as a fuzzy variable defined on the possibility space (U, Φ, P_{oss}) and the universe of discourse is expanded to $(-\infty, +\infty)$. As suggested by Dubios and Prade [25], we approximate the possibility distribution function (i.e., the membership function) $\mu_X(x)$ by two functions $L(x)$ and $R(x)$ with a point of intersection at $\max \mu(x) = 1$, i.e., the $L - R$ type possibility distribution function is adopted. The lifetime of the system is assumed to be a Gaussian fuzzy variable, which is a special $L - R$ type fuzzy variable. Under these conditions, the posbist reliability analysis of typical systems including series, parallel, series-parallel, parallel-series, and cold standby systems is provided in details. We will see in Section 3 that the expansion of the universe of discourse from $(0, +\infty)$ to $(-\infty, +\infty)$ does not affect the nature of the problems to be solved. On the contrary, it makes the proofs originally given in [8, 9] much more straightforward and the complexity of calculation is greatly reduced.

In this chapter, based on posbist reliability theory, event failure behavior is characterized in the context of possibility measures. A model of posbist fault tree analysis (posbist FTA) is proposed for predicting and diagnosing failures and evaluating reliability and safety of systems. The model of posbist FTA in posbist reliability theory plays a role that is analogous — though not completely — to that of probist FTA (Conventional FTA) in probist reliability theory (The reliability theory based on the PRObability assumption and the BInary-STate assumption is probist reliability theory, i.e., conventional reliability theory). As will be seen in the sequel, the model of posbist FTA constructed in this chapter, where the failure behavior of the basic events is characterized in the context of possibility measures, is different from various reported models of fuzzy probist FTA, where the basic events are considered as fuzzy numbers. Furthermore, it will be noted that the proposed model corresponds to posbist reliability theory developed by Cai [8].

The estimation of possibility distributions is a crucial step in the application of possibilistic reliability theory (for example, posbist reliability theory). Because the concept of membership function is closely related to the concept of possibility distributions [24], we believe that in principle, all methods developed for generating membership functions can be used to construct relevant possibility distributions. We will further discuss the properties of the methods for constructing possibility distributions. A method used to generate the $L - R$ type possibility distribution is applied to the possibilistic reliability analysis of fatigue of mechanical parts. Finally, an example is given to illustrate the application of this method to generating the possibility distribution of the fatigue lifetime of gears.

11.2 Basic Concepts in the Possibility Context

The assumptions of the posbist reliability theory include (1) the system failure behavior is fully characterized in the context of the possibility theory and (2) at any instant of time the system is in one of two crisp states: perfectly functioning or completely failed [8].

The concept of the posbist reliability theory was introduced in details in [8]. For ease of reference, we list several basic definitions related to this theory.

Definition 1 [8]: A fuzzy variable X is a real valued function defined on a possibility space (U, Φ, P_{oss}) $X : U \to R = (-\infty, +\infty)$.

Its membership function μ_X is a mapping from R to the unit interval $[0,1]$ with $\mu_X(x) = P_{oss}(X = x), x \in R$.

Thus, a fuzzy set X is defined as $X = \{x, \mu_X(x)\}$.

Based on X, the distribution function of X is given by

$\pi_X(x) = \mu_X(x)$.

Definition 2 [8]: The possibility distribution function of a fuzzy variable X, denoted by π_X, is a mapping from R to the unit interval $[0,1]$ such that $\pi_X(x) = \mu_X(x) = P_{oss}(X = x)$, $x \in R$

Definition 3 [8]: Given a possibility space (U, Φ, P_{oss}), the sets $A_1, A_2, \cdots, A_n \subset \Phi$ are said to be mutually unrelated if for any permutation of the set $\{1, 2, \cdots, n\}$, denoted by $\{i_1, i_2, \cdots, i_k\} (1 \leq k \leq n)$, the following equation holds:

$$P_{oss}\left(A_{i_1} \cap A_{i_2} \cdots \cap A_{i_k}\right) = \min\left(P_{oss}\left(A_{i_1}\right), P_{oss}\left(A_{i_2}\right), \cdots, P_{oss}\left(A_{i_k}\right)\right).$$

Definition 4 [8]: Given a possibility space (U, Φ, P_{oss}), the fuzzy variables $X_1, X_2, \cdots, X_n$ are said to be mutually unrelated if for any permutation of the set $\{1, 2, \cdots, n\}$, denoted by $\{i_1, i_2, \cdots, i_k\} (1 \leq k \leq n)$, the sets

$$\left\{X_{i_1} = x_1\right\}, \left\{X_{i_2} = x_2\right\}, \cdots, \left\{X_{i_k} = x_k\right\},$$

where $(x_1, x_2, \cdots, x_k \in R)$, are mutually unrelated.

11.2.1 Lifetime of the System

When the conventional binary-state assumption is adopted, the failure of a system is defined precisely. However, in practice, the instant of time when a system failure occurs may be uncertain and we may be unable to determine it accurately. In this case, it has to be characterized in the context of a possibility measure. According to the existence theorem of the possibility space [26], we can reasonably assume that there exists a single possibility space (U, Φ, P_{oss}) to characterize all the uncertainties of the times of failure of the system and its components. Accordingly, the lifetimes of the system and its components are treated as Nahmias' fuzzy variables defined on the common possibility space.

Definition 5 [8]: Given a possibility space (U, Φ, P_{oss}), the lifetime of a system is a non-negative real-valued fuzzy variable

$$X : U \to R^+ = (0, +\infty)$$

with possibility distribution function $\mu_X(x) = P_{oss}(X = x), x \in R^+$.

The posbist reliability of a system is then defined as the possibility that the system performs its assigned functions satisfactorily during a predefined exposure period under a given environment [8], that is,

$$R(t) = P_{oss}(X > t) = \sup_{u > t} P_{oss}(X = u) = \sup_{u > t} \mu_X(u), t \in R^+. \tag{1}$$

To simplify calculations when dealing with real-life problems, we may expand the universe of discourse of the lifetime of a system from $(0, +\infty)$ to $(-\infty, +\infty)$, i.e., $X : U \rightarrow R = (-\infty, +\infty)$.

In the following sections, we will show that this expansion makes the proofs originally given in [8, 9] much simplified and the complexity of calculation greatly reduced without affecting the nature of the problems to be solved.

11.2.2 State of the System

Formally, we assume that the state of the system is determined completely by the states of the components, so the structure function of a system of n components is denoted by

$$\phi = \phi(X)$$
$$X = (X_1, X_2, \cdots, X_n) \tag{2}$$

where X is the system state vector and X_i represents the state of component i.

Assume that $X_1, X_2, \cdots, X_n$ and ϕ are all binary fuzzy variables defined on possibility space (U, Φ, P_{oss})

$$X_i : U \rightarrow \{0,1\}, i = 1, 2, \cdots, n$$
$$\phi : U \rightarrow \{0,1\}.$$

Then we assume

$$X_i = \begin{cases} 1, & \text{if the component } i \text{ is functioning} \\ 0, & \text{if the component } i \text{ is failed} \end{cases}$$

and $\quad \phi = \begin{cases} 1, & \text{if the system is functioning} \\ 0, & \text{if the system is failed} \end{cases}$

According to above-mentioned analysis, the system posbist reliability, denoted by R, is defined as

$$R = P_{oss}(\phi = 1) \tag{3}$$

and the system posbist unreliability, denoted by F, is defined as

$$F = P_{oss}(\phi = 0). \tag{4}$$

Furthermore, we note that the system reliability defined in terms of system states coincides with the system reliability defined in terms of system lifetimes. Refer to [8] for more details on posbist reliability theory in terms of system states.

11.3 Posbist Reliability Analysis of Typical Systems

Suppose that X is the lifetime of the system. Assume that the lifetimes of the components, denoted by $X_1, X_2, \cdots, X_n$, are mutually unrelated. Furthermore, we assume that X_i is a Guassian fuzzy variable. Its possibility distribution function is given by the following equation and illustrated in Fig. 1.

$$\mu_{X_i}(x) = \begin{cases} \exp\left(-\left(\dfrac{m_i - x}{b_i}\right)^2\right), & x \le m_i \\ \exp\left(-\left(\dfrac{x - m_i}{b_i}\right)^2\right), & x > m_i \end{cases} \tag{5}$$

$$m_i, b_i > 0, \ i = 1, 2, \cdots, n$$

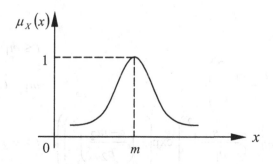

Fig. 1. Possibility distribution function of X_i

11.3.1 Posbist Reliability of a Series System

Consider a series system consisting of n components. The lifetime of the system depends on the lifetimes of the components as follows:

$$X = \min(X_1, X_2, \cdots, X_n)$$
(6)

Theorem 1 [8]: Consider a series system of two components. Let the system lifetime be X with possibility distribution function μ_X and X_1, X_2 be the lifetimes of the two components, which are mutually unrelated, Guassian fuzzy variable with continuous possibility distribution functions $\mu_{X_1}(x)$ and $\mu_{X_2}(x)$, respectively, defined on possibility space (U, Φ, P_{oss}). Then, there exists a unique pair $(a_1, a_2), a_1 \le a_2, a_1, a_2 \in R^+$, such that

$$\mu_X(x) = \begin{cases} \max\left(\mu_{X_1}(x), \mu_{X_2}(x)\right), & x \le a_1 \\ \mu_{X_1}(x), & a_1 < x \le a_2 \\ \min\left(\mu_{X_1}(x), \mu_{X_2}(x)\right), & x > a_2 \end{cases} \tag{7}$$

As a result, the posbist reliability of the series system of two components is

$$R_s(t) = \sup_{u > t} \mu_X(u) \quad = \begin{cases} 1, & t \le a_1 \\ \mu_{X_1}(t), & a_1 < t \le a_2 \\ \min\left(\mu_{X_1}(t), \mu_{X_2}(t)\right), & t > a_2 \end{cases}$$

$$= \begin{cases} 1, & t \le m_1 \\ \exp\left(-\left(\dfrac{t - m_1}{b_1}\right)^2\right), & m_1 < t \le m_2 \\ \min\left(\exp\left(-\left(\dfrac{t - m_1}{b_1}\right)^2\right), \exp\left(-\left(\dfrac{t - m_2}{b_2}\right)^2\right)\right), & t > m_2 \end{cases} \tag{8}$$

For a series system of n components, if $X_1, X_2, \cdots, X_n$ are mutually unrelated and $\mu_{X_1}(x) = \mu_{X_2}(x) = \cdots = \mu_{X_n}(x)$, that is, $X_1, X_2, \cdots, X_n$ are identically and independently distributed components, applying Theorem 1 recursively, we can easily arrive at

$$\mu_X(x) = \mu_{X_1}(x). \tag{9}$$

Thus, the posbist reliability of a series system of n components can be expressed as:

$$R_s(t) = \sup_{u>t} \mu_X(u) = \sup_{u>t} \mu_{X_1}(u) = \begin{cases} 1, & t \le a_1 \\ \mu_{X_1}(t), & t > a_1 \end{cases}$$

$$= \begin{cases} 1, & t \le m_1 \\ \exp\left(-\left(\dfrac{t - m_1}{b_1}\right)^2\right), & t > m_1 \end{cases} \tag{10}$$

11.3.2 Posbist Reliability of a Parallel System

Consider a parallel system consisting of n components. The lifetime of the system can be expressed as a function of the lifetimes of the components as follows:

$$X = \max(X_1, X_2, \cdots, X_n). \tag{11}$$

Theorem 2 [8]: Consider a parallel system of two components. Let the system lifetime be X with possibility distribution function μ_X and X_1, X_2 be the lifetimes of the two components, which are mutually unrelated, Guassian fuzzy variable with continuous possibility distribution functions $\mu_{X_1}(x)$ and $\mu_{X_2}(x)$, respectively, defined on possibility space (U, Φ, P_{oss}) . Then, there exists a unique pair $(a_1, a_2), a_1 \le a_2, a_1, a_2 \in R^+$, such that

$$\mu_X(x) = \begin{cases} \min(\mu_{X_1}(x), \mu_{X_2}(x)), & x \le a_1 \\ \mu_{X_2}(x), & a_1 < x \le a_2 \\ \max(\mu_{X_1}(x), \mu_{X_2}(x)), & x > a_2 \end{cases} \tag{12}$$

As a result, the posbist reliability of the parallel system of two components is

$$R_p(t) = \sup_{u>t} \mu_X(u) = \begin{cases} 1, & t \le a_2 \\ \max(\mu_{X_1}(t), \mu_{X_2}(t)), & t > a_2 \end{cases}$$

$$= \begin{cases} 1, & t \le m_2 \\ \max\left(\exp\left(-\left(\dfrac{t - m_1}{b_1}\right)^2\right), \exp\left(-\left(\dfrac{t - m_2}{b_2}\right)^2\right)\right), & t > m_2 \end{cases} \tag{13}$$

Similarly, for a parallel system of n components, if $X_1, X_2, \cdots, X_n$ are mutually unrelated and

$$\mu_{X_1}(x) = \mu_{X_2}(x) = \cdots = \mu_{X_n}(x),$$

that is, $X_1, X_2, \cdots, X_n$ are identically and independently distributed components, then, we can easily express the posbist reliability of a parallel system of n components as

$$R_p(t) = \sup_{u>t} \mu_X(u) = \sup_{u>t} \mu_{X_1}(u) = \begin{cases} 1, & t \le a_1 \\ \mu_{X_1}(t), & t > a_1 \end{cases}$$

$$= \begin{cases} 1, & t \le m_1 \\ \exp\left(-\left(\dfrac{t-m_1}{b_1}\right)^2\right), & t > m_1 \end{cases}. \qquad (14)$$

11.3.3 Posbist Reliability of a Series-parallel Systems

Consider a series-parallel system which is a series system of m subsystems where each subsystem consists of n parallel components. Assume that $X_1, X_2, \cdots, X_{n \times m}$ are variables with the same continuous possibility distribution function, that is, $\mu_{X_1}(x) = \mu_{X_2}(x) = \cdots = \mu_{X_{n \times m}}(x)$.

Then, for every subsystem, which has a parallel structure, its posbist reliability can be obtained with Eq. (15) as:

$$R_{pi}(t) = \sup_{u>t} \mu_{X_1}(u) = \begin{cases} 1, & t \le a_1 \\ \mu_{X_1}(t), & t > a_1 \end{cases}$$

$$= \begin{cases} 1, & t \le m_1 \\ \exp\left(-\left(\dfrac{t-m_1}{b_1}\right)^2\right), & t > m_1 \end{cases} \quad 1 \le i \le n. \qquad (15)$$

Further, according to Eq. (10), we can express the posbist reliability of the series-parallel system as

$$R_{sp}(t) = R_{pi}(t) = \sup_{u>t} \mu_{X_1}(u) = \begin{cases} 1, & t \le m_1 \\ \exp\left(-\left(\dfrac{t-m_1}{b_1}\right)^2\right), & t > m_1 \end{cases}. \qquad (16)$$

11.3.4 Posbist Reliability of a Parallel-series System

Consider a parallel-series system which is a parallel system of m subsystems where each subsystem consists of n components connected in series. Assume that $X_1, X_2, \cdots, X_{n \times m}$ are variables with the same continuous possibility distribution function, that is, $\mu_{X_1}(x) = \mu_{X_2}(x) = \cdots = \mu_{X_{n \times m}}(x)$.

Then, for every subsystem, which has a series structure, we can express its posbist reliability, according to Eq. (10), as:

$$R_{si}(t) = \sup_{u>t} \mu_{X_1}(u) = \begin{cases} 1, & t \le a_1 \\ \mu_{X_1}(t) & t > a_1 \end{cases}$$

$$= \begin{cases} 1, & t \le m_1 \\ \exp\left(-\left(\dfrac{t-m_1}{b_1}\right)\right), & t > m_1 \end{cases} \qquad 1 \le i \le n. \qquad (17)$$

Further, according to Eq. (14), we can express the posbist reliability of the parallel-series system as

$$R_{ps}(t) = R_{si}(t) = \sup_{u>t} \mu_{X_1}(u) = \begin{cases} 1, & t \le m_1 \\ \exp\left(-\left(\dfrac{t-m_1}{b_1}\right)^2\right), & t > m_1 \end{cases}. \qquad (18)$$

In summary, we can see that the posbist reliability of a series, parallel, series-parallel, or parallel-series system consisting of identically and independently distributed components has the same membership function as each component.

11.3.5 Posbist Reliability of a Cold Standby System

The operating mechanism of a cold standby system with n components is as follows. At any instant of time, only one operative component is required and the other operative components are in standby if they are not failed. Suppose that the components are activated sequentially from component 1 to component n. A component in standby does not fail or deteriorate. A failure of the system occurs only when there is no operative component left. We also assume that the sensing and switching mechanism is absolutely reliable. Such a cold standby system is depicted in Fig. 2.

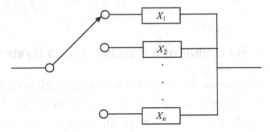

Fig. 2 A cold standby system

The lifetime X of such a cold standby system can be expressed as a sum of the lifetimes of the n individual components, i.e.,

$$X = X_1 + X_2 + \cdots + X_n$$
(19)

Then, the posbist reliability of the system is

$$R(t) = P_{oss}(X > t) = \sup_{u > t} P_{oss}(X = u) = \sup_{u > t} \mu_X(u). \tag{20}$$

For a cold standby system with two components, since X_1, X_2 are $L - R$ type fuzzy numbers, according to the addition operation of $L - R$ type fuzzy numbers [24], we have

$$\mu_X(x) = \mu_{(X_1 + X_2)}(x) = \begin{cases} \exp\left(-\left(\dfrac{m_1 + m_2 - x}{b_1 + b_2}\right)^2\right), & x \le m_1 + m_2 \\[4mm] \exp\left(-\left(\dfrac{x - (m_1 + m_2)}{b_1 + b_2}\right)^2\right), & x > m_1 + m_2 \end{cases} \tag{21}$$

As a result, the posbist reliability of the system can be expressed as

$$R_{cr}(t) = \sup_{u > t} \mu_X(u) = \sup_{u > t} \mu_{(X_1 + X_2)}(u)$$

$$
= \begin{cases} 1, & t \le m_1 + m_2 \\ \exp\left(-\left(\dfrac{t-(m_1+m_2)}{b_1+b_2}\right)^2\right), & t > m_1 + m_2 \end{cases} .
$$

(22)

Similarly, for a cold standby system with n components, we have

$$
\mu_X(x) = \mu_{(X_1+X_2+\cdots+X_n)}(x) = \begin{cases} \exp\left(-\left(\dfrac{\displaystyle\sum_{i=1}^{n} m_i - x}{\displaystyle\sum_{i=1}^{n} b_i}\right)^2\right), & x \le \displaystyle\sum_{i=1}^{n} m_i \\[4mm] \exp\left(-\left(\dfrac{x - \displaystyle\sum_{i=1}^{n} m_i}{\displaystyle\sum_{i=1}^{n} b_i}\right)^2\right), & x > \displaystyle\sum_{i=1}^{n} m_i \end{cases} .
$$

(23)

Then, we can easily express the posbist reliability of the system as

$$
R_{cr}(t) = \sup_{u>t} \mu_X(u) = \sup_{u>t} \mu_{(X_1+X_2+\cdots+X_n)}(u)
$$

$$
= \begin{cases} 1, & t \le \displaystyle\sum_{i=1}^{n} m_i \\[4mm] \exp\left(-\left(\dfrac{t - \displaystyle\sum_{i=1}^{n} m_i}{\displaystyle\sum_{i=1}^{n} b_i}\right)^2\right), & t > \displaystyle\sum_{i=1}^{n} m_i \end{cases} .
$$

(24)

11.4 Posbist Fault Tree Analysis of Coherent Systems

Conventional fault tree analysis was first applied to the analysis of system reliability by Watson in 1961. A fault tree is a logic diagram consisting of a top event and a structure delineating the ways in which the top event may occur. Up to now, the scope of conventional FTA has expanded from the aviation/space industry and nuclear industry to electronics, electric power, and the chemical industry as well as mechanical engineering, traffic, architecture, etc. It is a mature tool for analyzing coherent systems.

The pioneering work on fuzzy fault tree analysis (fuzzy FTA) belongs to Tanaka et al. [2]. They treated probabilities of basic events as trapezoidal fuzzy numbers, and applied the fuzzy extension principle to calculating the probability of the top event. At the same time, they defined an index function analogous to importance measures for evaluating the extent to which a basic event contributes to the top event. Singer [3] analyzed fuzzy reliability by using $L - R$ type fuzzy numbers. He considered the relative frequencies of the basic events as fuzzy numbers and used possibility instead of probability measures. However, these approaches cannot be applied to a fault tree with repeated events. In order to deal with repeated basic events, Soman and Misra [27] provided a simple method for fuzzy fault tree analysis based on the $\alpha -$ cut method, also known as resolution identity. This method was then extended to deal with multistate fault tree analysis [28]. Sawyer and Rao [29] used the $\alpha -$ cut method to calculate the failure probability of the top event in fuzzy fault tree analysis of mechanical systems. Huang et al [14] employed fuzzy fault tree to analyze railway traffic safety. Many other results on fuzzy FTA are reported in [30-34].

There is one common characteristic in the above-mentioned works: the notion of fuzziness is introduced to conventional fault tree analysis and the probabilities of events are fuzzified into the fuzzy numbers in the unit interval [0,1]. However, we note that these works are based on probist FTA (i.e., conventional FTA). More precisely, we can say that these works fall within the scope of fuzzy probist fault tree analysis.

Furuta and Shiraishi [35] proposed a kind of importance measure using fuzzy integrals assuming that the basic events in a fault tree are fuzzy. Feng and Wu [36] developed a model of profust fault tree analysis based on the theory of "probability of fuzzy events" and provided partial quantitative analysis when the state space is discrete. Their model is based on two assumptions: (1) the failure behavior of components is defined in a fuzzy way and (2) the probability assumption is used.

Based on the foregoing overview, we can itemize the main categories of the methods of FTA to date:

(1) Probist FTA (conventional FTA),

(2) Fuzzy probist FTA (or fuzzy FTA), and

(3) Profust FTA (corresponding to profust reliability theory).

Furthermore, we can find that the study of fuzzy probist FTA is confined to the algorithm itself, the cited engineering applications are overly simplified, and the obtained results lack comparability. On the other hand, the study of profust FTA has appeared only recently and the study of posbist FTA is not reported at all.

11.4.1 Posbist Fault Tree Analysis of Coherent Systems

11.4.1.1 Basic Definitions of Coherent Systems

Here we give several basic definitions of coherent systems that are indispensable to the model of posbist fault tree analysis that we will construct. Refer to [37] for a more detailed treatment of coherent systems.

Definition 6: In the context of posbist reliability theory, the i-th component is irrelevant to the structure ϕ if ϕ is constant in z_i, that is,

$$\phi(1_i, Z) = \phi(0_i, Z)$$

for all $(\bullet_i, Z)$. Otherwise the i-th component is relevant to the structure. Here we employ notations

$$(1_i, Z) = (z_1, \cdots, z_{i-1}, 1, z_{i+1}, \cdots, z_n)$$

$$(0_i, Z) = (z_1, \cdots, z_{i-1}, 0, z_{i+1}, \cdots, z_n)$$

$$(\bullet_i, Z) = (z_1, \cdots, z_{i-1}, \bullet, z_{i+1}, \cdots, z_n)$$

Definition 7: A system of components is coherent in the context of posbist reliability theory if (1) every component is relevant to the system and (2) its structure function ϕ is increasing in every component. We can denote a coherent system by ϕ, or more precisely by (C, ϕ), where the set C is a set of integers designating the components.

To be brief, coherent systems are monotone systems wherein no unit irrelevant to the system exists since the units irrelevant to the system are removed by Boolean calculation after their reliability behavior is analyzed. We will use possibility measures rather than probability measures to characterize the failure behavior of coherent systems. We note that the definition of coherent systems here is the same as that of coherent systems in conventional reliability theory. This is because the binary-state assumption is also valid in posbist reliability theory [8].

Definition 8: A path set, denoted by P, of a coherent system (C, ϕ) in the context of posbist reliability theory is a subset of C that makes ϕ functioning. P is minimal if any real subset of it will not make ϕ functioning. A cut set, denoted by K, of a coherent system (C, ϕ) is a subset of C that makes ϕ failed. K is minimal if any real subset of it will not make ϕ failed.

Suppose a coherent system ϕ in the context of posbist reliability theory with p minimal path sets $(P_1, P_2, \cdots, P_p)$ and k minimal cut sets $(K_1, K_2, \cdots, K_k)$. Define $P_j(X) = \bigcap_{i \in P_j} x_i$ and $K_j(X) = \bigcup_{i \in K_j} x_i$.

Then the structure function ϕ can be expressed as

$$\phi(X) = \bigcup_{j=1}^{p} P_j(X) = \max_{1 \le j \le p} \min_{i \in P_j} x_i \tag{25}$$

or

$$\phi(X) = \bigcap_{j=1}^{k} K_j(X) = \min_{1 \le j \le k} \max_{i \in K_j} x_i . \tag{26}$$

11.4.1.2 Basic Assumptions

It is necessary for the construction of the model of posbist FTA to make the following assumptions:

(1) The states of events are crisp: occurrence or nonoccurrence. However, the event state is uncertain at a given future instant.

(2) The failure behaviors of events are characterized in the context of possibility measures. Furthermore, the possibility distribution functions of events have been obtained by adopting a certain technique (or several techniques) for estimating possibility distributions.

(3) The events are mutually unrelated.

11.4.2 Construction of the Model of Posbist Fault Tree Analysis

According to the equivalent conversion of special logic gates [38], we can convert an arbitrary fault tree of coherent systems into a basic fault tree that consists only of AND gates, OR gates and basic events.

11.4.2.1 The Structure Function of Posbist Fault Tree

Consider a coherent system S of n components. The failure of the system is the top event and the failures of the components are basic events. Since the system and its components demonstrate only two crisp states, i.e., fully functioning or completely failed, we can use 0 and 1 to represent the states of the top event and basic events. Thus, we assume

$$X_i = \begin{cases} 1, & \text{if the basic event } i \text{ occurs} \\ 0, & \text{if the basic event } i \text{ dosen't occur} \end{cases} \quad i = 1,2,\cdots,n$$

$$\varphi(X) = \begin{cases} 1, & \text{if the top event occurs} \\ 0, & \text{if the top event dosen't occur} \end{cases}$$

$$X = (X_1, X_2, \cdots, X_n).$$

Then the function $\varphi(X)$ is called the structure function of a posbist fault tree. We can call it a posbist fault tree $\varphi(X)$, or more precisely, a posbist fault tree $(C, \varphi(X))$, where the set C is a set of integers designating basic events.

Analogous to the conventional fault tree, we can easily obtain the following results:

For a posbist fault tree consisting of AND gates, we have

$$\varphi(X) = \prod_{i=1}^{n} X_i = \min(X_1, X_2, \cdots, X_n). \tag{27}$$

For a posbist fault tree consisting of OR gates, we have

$$\varphi(X) = 1 - \prod_{i=1}^{n} (1 - X_i) = \max(X_1, X_2, \cdots, X_n). \tag{28}$$

Definition 9: A path set, denoted by P_a, of a posbist fault tree $\varphi(X)$ is a subset of C that will not make the top event occur. P_a is minimal if any real subset of it will make the top event occur. A cut set, denoted by K_u, of a posbist fault tree $\varphi(X)$ is a subset of C that makes the top event occur. K_u is minimal if any real subset of it will not make the top event occur.

Then, suppose a fault tree $\varphi(X)$ of a coherent system with p minimal path sets $(P_{a1}, P_{a2}, \cdots, P_{ap})$ and k minimal cut sets $(K_{u1}, K_{u2}, \cdots, K_{uk})$. Define $P_{aj}(X) = \bigcap\limits_{i \in P_{aj}} x_i$ and $K_{uj}(X) = \bigcup\limits_{i \in K_{uj}} x_i$.

Thus, the structure function $\varphi(X)$ of the posbist fault tree can be expressed as

$$\varphi(X) = \bigcap_{j=1}^{p} P_{aj}(X) = \min_{1 \le j \le p} \max_{i \in P_{aj}} x_i \tag{29}$$

or

$$\varphi(X) = \bigcup_{j=1}^{k} K_{uj}(X) = \max_{1 \le j \le k} \min_{i \in K_{uj}} x_i. \tag{30}$$

11.4.2.2 Quantitative Analysis

According to posbist reliability theory based on state variables (i.e., system states) and the basic assumptions in section 4.1.2, we have:
The failure possibility of the basic event i is

$$P_{oss_i} = P_{oss}(X_i = 1). \tag{31}$$

The failure possibility of the top event is

$$P_{oss_T} = P_{oss}(\varphi = 1). \tag{32}$$

Theorem 3: For the AND gate, the operator is

$$P_{oss_T}^{AND} = \min(P_{oss_1}, P_{oss_2}, \cdots, P_{oss_n}). \tag{33}$$

Proof

$$P_{oss_T}^{AND} = P_{oss}(\varphi = 1) = P_{oss}\left(\prod_{i=1}^{n} X_i = 1\right)$$

$$= P_{oss}(\min(X_1, X_2, \cdots, X_n) = 1) = P_{oss}(X_1 = 1, X_2 = 1, \cdots, X_n = 1).$$

Since the basic events are mutually unrelated, we have

$$P_{oss_T}^{AND} = \min(P_{oss}(X_1 = 1), P_{oss}(X_2 = 1), \cdots, P_{oss}(X_n = 1))$$

$$= \min(P_{oss_1}, P_{oss_2}, \cdots, P_{oss_n})$$

Q.E.D.

Theorem 4: For the OR gate, the operator is

$$P^{OR}_{oss_T} = \max\left(P_{oss_1}, P_{oss_2}, \cdots, P_{oss_n}\right). \tag{34}$$

Proof

$$P^{OR}_{oss_T} = P_{oss}\left(\varphi = 1\right) = P_{oss}\left(\max(X_1, X_2, \cdots, X_n) = 1\right)$$

$$= P_{oss}\left(\{X_1 = 1\} \cup (X_2 = 1) \cup \cdots \cup \{X_n = 1\}\right)$$

$$= \max\left(P_{oss}(X_1 = 1), P_{oss}(X_2 = 1), \cdots, P_{oss}(X_n = 1)\right)$$

$$= \max\left(P_{oss_1}, P_{oss_2}, \cdots, P_{oss_n}\right)$$

Q.E.D.

Theorem 5: For a posbist fault tree $\varphi(X)$ of a coherent system with mutually unrelated basic events, suppose there are k minimal cut sets $(K_{u1}, K_{u2}, \cdots, K_{uk})$. Let P_{oss_i} be the failure possibility of the basic event i. Then

$$P_{oss_T} = \max_{1 \leq j \leq k}\left(\min_{i \in K_{uj}} P_{oss_i}\right). \tag{35}$$

Proof

From Eq. (30), we have $\varphi(X) = \bigcup_{j=1}^{k} K_{uj}(X) = \max_{1 \leq j \leq k} \min_{i \in K_{uj}} x_i$.

Since all the basic events of the posbist fault tree are mutually unrelated, we may arrive at

$$P_{oss}\left(K_{uj}(X) = 1\right) = P_{oss}\left(\min_{i \in K_{uj}} x_i = 1\right) = \min_{i \in K_{uj}}\left(P_{ossi}\right).$$

The above equation is due to Theorem 3. Then according to Theorem 4, we have

$$P_{oss_T} = P_{oss}\left(\varphi(X) = 1\right) = P_{oss}\left(\max_{1 \leq j \leq k} K_{uj}(X) = 1\right) = \max_{1 \leq j \leq k}\left(\min_{i \in K_{uj}} P_{oss_i}\right)$$

where we can find that the unrelatedness of $\{K_{uj}(X), j = 1, 2, \cdots, k\}$ is not required.

Q.E.D.

Thus, as long as we know the failure possibility of every basic event, we can use the above-mentioned operators to obtain the failure possibility of the top event.

11.5 The Methods for Developing Possibility Distributions

In the theory of possibilistic reliability, the concept of a possibility distribution plays a role that is analogous—though not completely—to that of a probability distribution in the theory of probabilistic reliability. Because the concept of membership functions bears a close relation to the concept of possibility distributions [39], in this section, we believe that all the methods for generating membership functions can be used to construct the relevant possibility distributions in principle. However, we should realize that it might be difficult, if not impossible, to come up with a general possibility distribution method which will work for all applications. Much future work is yet to be done on this subject.

Here, we present several techniques for estimating possibility distributions from probability distributions.

11.5.1 Possibility Distributions Based on Membership Functions

As Zadeh [24] pointed out, a possibility distribution can be viewed as a fuzzy set which serves as an elastic constraint on the values that may be assigned to a variable. Therefore, the possibility distribution numerically equals to the corresponding membership function, i.e.,

$$\pi_X(x) = \mu_A(x), \tag{36}$$

where X is a fuzzy variable and A is the fuzzy set induced by X.

Note that although a possibility distribution and a fuzzy set have a common mathematical expression, the underlying concepts are different. The fuzzy set A is a fuzzy value that can be assigned to a certain variable. However, the possibility constraint A is a fuzzy set of nonfuzzy values that can possibly be assigned to X.

According to the above-mentioned viewpoint, we can use the methods for constructing membership functions to generate the corresponding possibility distributions. That is to say, from Eq. (36), if the membership function of a fuzzy set has been obtained, the possibility distribution of the fuzzy variable of the fuzzy set is obtained too. In the following, we present a few commonly used methods for generating membership functions.

11.5.1.1 Fuzzy Statistics [40]

Fuzzy statistics are analogous to probability statistics in form and they all use certainty approaches to deal with uncertainty problems in real-life systems or structures. When fuzzy statistics are used, a definite judgment must be made on whether a fixed element u_0 in the universe of discourse belongs to an alterable crisp set A^* or not. In other words, based on n observations, we have

$$\text{the grade of membership of } u_0 \text{ in } A = \frac{\text{the number of times of } "u_0 \in A^*"}{n}.$$

The following principles must be observed in evaluation of fuzzy statistics:

(1) The user should be familiar with concepts of fuzzy sets and capable of quantifying the entity being observed. In other words, the user should be an expert in the field of application.

(2) A preliminary analysis of the raw data should be conducted so that abnormal data may be removed.

For further details and examples of fuzzy statistics, readers are referred to [40].

11.5.1.2 Transformation of Probability Distributions to Possibility Distributions

According to Kosko's argument that "fuzziness contains probability as a special case" [41], if we have obtained estimates of the probability density function (pdf) or other statistical properties of an entity being measured, we can construct its corresponding membership function following the approach outlined in [42]. Based on the technique in [42], we summarize the following simple method for constructing the membership function from the probability density function of a Gaussian random variable, i.e.,

$$\mu(x) = \lambda p(x), \tag{37}$$

$$\lambda = \frac{1}{\max(p(x))}, \tag{38}$$

where $p(x)$ is the pdf of a Gaussian random variable and $\mu(x)$ is the corresponding membership function based on $p(x)$. For example, the pdf and the corresponding membership function of the yield strength of an aluminum alloy developed with this approach are shown in Fig. 3.

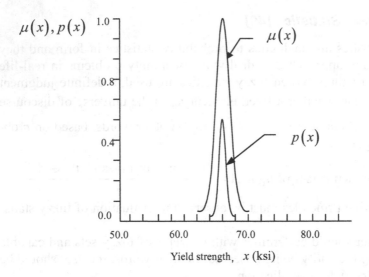

Fig. 3. The pdf $p(x)$ and the corresponding membership function $\mu(x)$ for the yield strength of an aluminum alloy.

11.5.1.3 Heuristic Methods [43]

With heuristic methods, we first select a predefined shape of the membership function to be developed. The specific parameters of the membership function with the selected shape are determined from the data collected. In most real-life problems, the universe of discourse of the membership functions is the real number line R. The commonly used membership function shapes are the piecewise linear function and the piecewise monotonic function. Linear and piecewise linear membership functions have the advantages of reasonably smooth transitions and easy manipulation through fuzzy operations. However, the shapes of many heuristic membership functions are not flexible enough to model all kinds of data. Moreover, the parameters of the membership functions must be provided by experts. In many applications, the parameters need to be adjusted extensively to achieve a certain performance level.

A few commonly used piecewise linear functions are given below:

(1) $\mu(x) = 1 - \dfrac{x}{a}$ or $\mu(x) = \dfrac{x}{a}$, where $x = [0, a]$;

(2) $\mu(x) = \begin{cases} 1 - \dfrac{|a-x|}{\alpha}, & \alpha - a \leq x \leq \alpha + a \\ 0, & \text{otherwise} \end{cases}$;

(3) $\mu(x) = \begin{cases} 0, & x \leq a \\ w_1 \dfrac{x-a}{b-a}, & a \leq x \leq b \\ 1, & b \leq x \leq c \\ w_2 \dfrac{d-x}{d-c}, & c \leq x \leq d \\ 0, & x > d \end{cases}$;

(4) $\mu(x) = \begin{cases} 0, & x < a_1 \\ \dfrac{x}{a_1 - a_2} + \dfrac{a_1}{a_1 - a_2}, & a_1 \leq x \leq a_2 \\ 1, & x > a_2 \end{cases}$.

Some commonly used piecewise monotonic functions are as follows:

(1) $s(x;a,b,c) = \begin{cases} 0, & x \leq a \\ 2\left(\dfrac{x-a}{c-a}\right)^2, & a < x \leq \dfrac{a+c}{2} \\ 1 - 2\left(\dfrac{x-a}{c-a}\right)^2, & \dfrac{a+c}{2} < x \leq c \\ 1, & x > c \end{cases}$;

(2) $\mu(x;a,b,c) = \begin{cases} s\left(x; c-b, c-\dfrac{b}{2}, c\right), & x \leq c \\ 1 - s\left(x; c, c+\dfrac{b}{2}, c+b\right), & x > c \end{cases}$;

(3) $\mu(x) = e^{-b(x-a)^2}$, $-\infty < x < +\infty$;

(4) $\mu(x) = \dfrac{1}{2} - \dfrac{1}{2}\sin\left(\dfrac{\pi}{b-a}\left(x - \dfrac{a+b}{2}\right)\right)$, $x \in [a,b]$.

In practical applications, we often combine fuzzy statistics with heuristic methods. First, the shape of the membership function is suggested by statistical data. Then, the suggested shape is compared with the predefined shape and the more appropriate ones are selected. Finally, the most suitable membership function is determined through practical tests.

11.5.1.4 Expert Opinions

Sometimes, the opinions of experts are used to construct the membership functions. In these situations, the universes of discourse are usually discrete.

In addition to the methods reviewed above, trichotomy [40], multiphase fuzzy statistics [40], and neural network based methods [43, 44] have been used in construction of membership functions. It should be pointed out that developing new methods for constructing membership functions is still a hot research topic. The reported methods for constructing membership functions are not as mature as those for constructing probability distribution functions. Constructing membership functions still depends on experience and feedback from actual use and continuous revisions have to be made to achieve satisfactory results. This situation results in the immaturity of the methods for constructing possibility distributions.

11.5.2 Transformation of Probability Distributions to Possibility Distributions

The methods for transforming probability distributions to possibility distributions are based on the possibility/probability consistency principle. The possibility/probability consistency principle states:

If a variable X can take the value $u_1, \cdots, u_n$ with respective possibilities $\pi = \left(\pi(u_1), \cdots, \pi(u_n) \right)$ and probabilities $p = \left(p(u_1), \cdots, p(u_n) \right)$, then the degree of consistency of the probability distribution p with possibility distribution π is expressed by

$$C_z(\pi, p) = \sum_{i=1}^{n} \pi(u_i) p(u_i).$$

For more details on this principle, readers are referred to [24].

11.5.2.1 The Bijective Transformation Method [45]

Let $X = \left\{ x_i \middle| i = 1, 2, \cdots, n \right\}$ be the universe of discourse. If the histograms (or the probability density function) of the variable X has a decreasing trend, that is,

$$p(x_1) \geq p(x_2) \geq \cdots \geq p(x_n),\qquad(39)$$

then, the corresponding possibility distribution can be constructed as follows:

$$\pi_X(x_i) = \sum_{j=1}^{n} \min(p(x_i), p(x_j)) = ip(x_i) + \sum_{j=i+1}^{n} p(x_j).\qquad(40)$$

Generally, the histograms can be normalized by setting the maximal value to 1, i.e.,

$$\pi'_X(x_i) = \frac{\pi_X(x_i)}{\max_{j=1}^{n} \pi_X(x_j)} = \frac{p(x_i)}{\max_{j=1}^{n} p(x_j)}.\qquad(41)$$

Fig. 4 illustrates the construction of a possibility distribution by use of a general histogram.

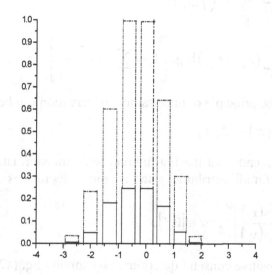

Fig. 4. Construction of a possibility distribution from a histogram

11.5.2.2 The Conservation of Uncertainty Method

Klir [46] presented a method for constructing possibility distributions based on the principle of uncertainty conservation. When uncertainty is transformed from one theory T_1 to another T_2, the following requirements must be met:

(1) The amount of inherent uncertainty should be preserved and

(2) All relevant numerical values in T_1 must be converted to their counterparts in T_2 by an appropriate scale.

The probabilistic measure of uncertainty is the well known Shannon entropy and is given by

$$H(p) = -\sum_{i=1}^{n} p_i \log_2 p_i$$

In the possibility theory, there are two types of uncertainties, nonspecificity $N(\pi)$, and discord $D(\pi)$, and they are given by

$$N(\pi) = -\sum_{i=2}^{n} \pi_i \log_2 \left(\frac{i}{i-1} \right),$$

and

$$D(\pi) = -\sum_{i=1}^{n-1} (\pi_i - \pi_{i+1}) \log_2 \left[1 - i \sum_{j=i+1}^{n} \frac{\pi_i}{j(j-1)} \right].$$

Therefore, the principle of uncertainty conservation can be expressed as

$$H(p) = N(\pi) + D(\pi).$$

Klir [32] contends that the log-interval scale transformation is the only one that exists for all distributions and is unique. Its form is

$$\pi_X(x_i) = \left[\frac{p(x_i)}{p(x_1)} \right]^{\alpha}, \quad \alpha \in [0,1]. \tag{42}$$

where α is a positive constant determined by solving Eq. (42); Klir conjectures that α lies in the interval [0,1].

Clearly the obtained possibilistic information is less precise than the original probabilistic information after the transformation of the probability distribution to the possibility distribution, because only upper bounds on the probability values are derived. However, fuzzy arithmetic operations are generally much easier to handle than operations with random variables [47]. Thus, such a transformation method is intended for situations where the manipulation of randomness is hard. However, this transformation method has the disadvantage of requiring a large amount of data to estimate the probability distribution first.

11.5.3 Subjective Manipulations of Fatigue Data

For products requiring high reliability, manufacturers often conduct laboratory tests to obtain a certain quantity of lifetime data. Due to budget limitations, it is usually difficult or impossible to obtain sufficient statistical data. Although the number of data points available may be too small for us to perform a statistical analysis, it may be sufficient for subjective estimation of the possibility distribution. If we have constructed a model of the possibilistic reliability of the device under study and derived the needed possibility distribution, we can perform a quantitative analysis of the possibilistic reliability of the device.

Assume that we have obtained fatigue life data of a device, denoted by $\left(n_i^j\right)_{1 \le j \le N}, 1 \le i \le M$, where M is the number of stress levels, N is the number of data points at each stress level. Then the mean fatigue life at stress level I can be expressed as

$$m_{n_i} = \frac{1}{N} \sum_{j=1}^{N} n_i^j, \quad 1 \le i \le M. \tag{43}$$

The lifetime data at each stress level can be divided into two groups, that is,

$$G_1 = \left\{ n_i^j, j = 1, 2, \cdots, N \mid n_i^j < m_{n_i} \right\}, \tag{44}$$

$$G_2 = \left\{ n_i^j, j = 1, 2, \cdots, N \mid n_i^j > m_{n_i} \right\}. \tag{45}$$

The mean value m_{n_i} is assigned a possibility degree of 1 and the possibility degree of 0.5 is assigned to the means of the lifetime data in the two groups G_1 and G_2, that is,

$$m_{l_{n_i}} = \frac{1}{\#(G_1)} \sum_{n_i^j \in G_1} n_i^j, \quad \pi_{n_i}\left(m_{l_{n_i}}\right) = 0.5, \quad i = 1, 2, \cdots, M \tag{46}$$

$$m_{r_{n_i}} = \frac{1}{\#(G_2)} \sum_{n_i^j \in G_2} n_i^j, \quad \pi_{n_i}\left(m_{r_{n_i}}\right) = 0.5, \quad i = 1, 2, \cdots, M \tag{47}$$

where $\#(\cdot)$ denotes the number of data points in a set.

By use of the above-mentioned analysis, we can express the $L-R$ type possibility distribution of fatigue lifetime as follows:

$$\pi_{n_i}\left(n_i^j\right) = \begin{cases} L\left(\dfrac{m_{n_i} - n_i^j}{\alpha_{n_i}}\right), & n_i^j \leq m_{n_i} \\[3mm] R\left(\dfrac{n_i^j - m_{n_i}}{\beta_{n_i}}\right), & n_i^j > m_{n_i} \end{cases} \tag{48}$$

where $\alpha_{n_i} = \dfrac{m_{n_i} - m_{l_{n_i}}}{L^{-1}(0.5)}$ and $\beta_{n_i} = \dfrac{m_{r_{n_i}} - m_{n_i}}{L^{-1}(0.5)}$.

Considering the various types of $L-R$ type possibility distributions mentioned earlier in this chapter, we can use Eq. (48) to get specific possibility distributions to represent fatigue lifetime data. For example, the following triangular possibility distribution may be used to represent fatigue lifetime data:

$$\pi_{n_i}\left(n_i^j\right) = \begin{cases} 0, & n_i^j \leq m_{n_i} - \alpha_{n_i} \\[3mm] 1 - \dfrac{m_{n_i} - n_i^j}{\alpha_{n_i}}, & m_{n_i} - \alpha_{n_i} \leq n_i^j \leq m_{n_i} \\[3mm] 1 - \dfrac{n_i^j - m_{n_i}}{\beta_{n_i}}, & m_{n_i} \leq n_i^j \leq m_{n_i} + \beta_{n_i} \\[3mm] 0, & m_{n_i} + \beta_{n_i} \leq n_i^j \end{cases} \tag{49}$$

where $\alpha_{n_i} = 2\left(m_{n_i} - m_{l_{n_i}}\right)$ and $\beta_{n_i} = 2\left(-m_{n_i} + m_{r_{n_i}}\right)$.

Similarly, we may use the following Gaussian possibility distribution to represent fatigue lifetime data:

$$\pi_{n_i}\left(n_i^j\right) = \begin{cases} \exp\left[-\left(\dfrac{m_{n_i} - n_i^j}{\alpha_{n_i}}\right)^2\right], & n_i^j \le m_{n_i} \\[4mm] \exp\left[-\left(\dfrac{n_i^j - m_{n_i}}{\beta_{n_i}}\right)^2\right], & n_i^j > m_{n_i} \end{cases},$$

where $\alpha_{n_i} = \dfrac{m_{n_i} - m_{l_{n_i}}}{\sqrt{\ln 0.5}}$ and $\beta_{n_i} = \dfrac{m_{r_{n_i}} - m_{n_i}}{\sqrt{\ln 0.5}}$.

11.6 Examples

11.6.1 Example 1

Calculate the posbist reliability of a series system, a parallel system, and a cold standby system with perfect sensing and switching mechanism. It is assumed that each system has two components and they are mutually unrelated. We also assume that the lifetime of each component is a Gaussian fuzzy variable, i.e.,

$$\mu_{X_1}(t) = \begin{cases} \exp\left(-\left(\dfrac{120-t}{40}\right)^2\right), & t \le 120 \\[4mm] \exp\left(-\left(\dfrac{t-120}{40}\right)^2\right), & t > 120 \end{cases}$$

$$\mu_{X_2}(t) = \begin{cases} \exp\left(-\left(\dfrac{130-t}{50}\right)^2\right), & t \le 130 \\[4mm] \exp\left(-\left(\dfrac{t-130}{50}\right)^2\right), & t > 130 \end{cases}$$

11.6.1.1 The Series System

Using Eq. (8), we can express the posbist reliability of the series system as:

$$R_s(t) = \sup_{u>t} \mu_X(u)$$

$$= \begin{cases} 1, & t \le m_1 \\ \exp\left(-\left(\dfrac{m_1-t}{b_1}\right)^2\right), & m_1 < t \le m_2 \\ \min\left(\exp\left(-\left(\dfrac{t-m_1}{b_1}\right)^2\right), \exp\left(-\left(\dfrac{t-m_2}{b_2}\right)^2\right)\right), & t > m_2 \end{cases}$$

$$= \begin{cases} 1, & t \le 120 \\ \exp\left(-\left(\dfrac{120-t}{40}\right)^2\right), & 120 < t \le 130. \\ \min\left(\exp\left(-\left(\dfrac{t-120}{40}\right)^2\right), \exp\left(-\left(\dfrac{t-130}{50}\right)^2\right)\right), & t > 130 \end{cases}$$

For example, when $t = 140$, we have:

$$R_s(140) = \exp\left(-\left(\frac{140-120}{40}\right)^2\right) = 0.7788.$$

11.6.1.2 The Parallel System

Using Eq. (13), we can express the posbist reliability of the parallel system as:

$$R_p(t) = \sup_{u>t} \mu_X(u) = \begin{cases} 1, & t \le a_2 \\ \max(\mu_{X_1}(t), \mu_{X_2}(t)), & t > a_2 \end{cases}$$

$$= \begin{cases} 1, & t \le m_2 \\ \max\left(\exp\left(-\left(\dfrac{t-m_1}{b_1} \right)^2 \right), \exp\left(-\left(\dfrac{t-m_2}{b_2} \right)^2 \right) \right), & t > m_2 \end{cases}$$

$$= \begin{cases} 1, & t \le 130 \\ \max\left(\exp\left(-\left(\dfrac{t-120}{40} \right)^2 \right), \exp\left(-\left(\dfrac{t-130}{50} \right)^2 \right) \right), & t > 130 \end{cases}.$$

For example, $t = 140$, we have:

$$R_p(140) = \exp\left(-\left(\frac{140-130}{50} \right)^2 \right) = 0.9608.$$

11.6.1.3 The Cold Standby System

Using Eq. (22), we can express the posbist reliability of the standby system as

$$R_{cr}(t) = \sup_{u>t} \mu_X(u) = \sup_{u>t} \mu_{(X_1+X_2)}(u)$$

$$= \begin{cases} 1, & t \le m_1 + m_2 \\ \exp\left(-\left(\dfrac{t-(m_1+m_2)}{b_1+b_2} \right)^2 \right), & t > m_1 + m_2 \end{cases}$$

$$= \begin{cases} 1, & t \le 250 \\ \exp\left(-\left(\dfrac{t-250}{90} \right)^2 \right), & t > 250 \end{cases}.$$

When $t = 140$, we have $R_{cr}(140)=1$.

From this example, we can see that the posbist reliability of a parallel system is higher than that of a series system and the posbist reliability of a cold standby system is higher than that of a parallel system.

11.6.2 Example 2

Consider the problem of a failure caused by the break of the hoisting rope of a crane. For a failure analysis of a broken hoisting rope, we can refer to

[48]. In [48], the authors concluded that the main reasons for the failure of the crane's hoisting rope were fatigue and poor inspection. But they considered only the failures of the steel wires themselves. In fact, there are many factors (materials and/or human errors) that caused the break of the hoisting rope of the crane. It is not enough to consider the failure of the steel wires only.

The fault tree of a failure of the hoisting rope of a crane has been constructed in Fig. 5. By means of the technique for analyzing a fault tree, we can derive almost all the main reasons for the failure of the crane's hoisting rope. The events of the fault tree are illustrated in Table 1.

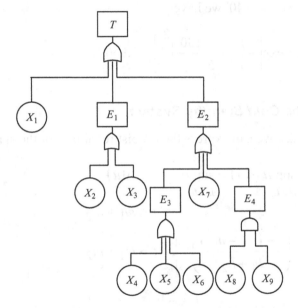

Fig. 5. The fault tree of a failure of the rope on a crane

For a failure such as that of the hoisting rope of crane, it is often very difficult to estimate precise failure rates or failure probabilities of individual components or failure events. This is because the failure events consist of not only the failure of components (e.g., drawback of materials) but also human factors (e.g., insufficient inspection). According to Zadeh's consistency principle [39], it may be feasible to use possibility measures as a rough estimate of probability measures.

Using the judgments of experts or technologists, we can obtain the failure possibility of every basic event illustrated in Table 1. Thus, we can deduce the failure possibility of the top event by use of the technique presented in this chapter.

Table 1. The events and the failure possibility of every basic event of the fault tree

Symbol	Event	Failure possibility
T	rope broken	
E_1	hoisting objects aslant	
E_2	inadequate strength	
E_3	manufacturing defects	
E_4	misuse	
X_1	overloading	0.050
X_2	dragging	0.030
X_3	obstacles present	0.004
X_4	materials defects	0.002
X_5	machining defects	0.001
X_6	poor inspection	0.003
X_7	inappropriate diameter size	0.005
X_8	poor maintenance	0.020
X_9	arriving at limit of failure	0.500

According to Eq. (34), we arrive at

$$P_{oss}(E_1) = \max(P_{oss}(X_2), P_{oss}(X_3)) = \max(0.03, 0.004) = 0.03.$$
$$P_{oss}(E_3) = \max(P_{oss}(X_4), P_{oss}(X_5), P_{oss}(X_6))$$
$$= \max(0.002, 0.001, 0.003) = 0.003.$$
$$P_{oss}(E_2) = \max(P_{oss}(E_3), P_{oss}(X_7), P_{oss}(E_4))$$
$$= \max(0.003, 0.005, 0.02) = 0.02.$$

Further, according to Eq. (33), we arrive at

$$P_{oss}(E_4) = \min(P_{oss}(X_8), P_{oss}(X_9)) = \min(0.02, 0.5) = 0.02.$$

In this way, we can arrive at the failure possibility of the top event according to Eq. (34)

$$P_{oss}(T) = \max(P_{oss}(X_1), P_{oss}(E_1), P_{oss}(E_2))$$
$$= \max(0.05, 0.03, 0.02) = 0.05.$$

11.6.3 Example 3

We illustrate the method presented in Section 5.3 for constructing the possibility distribution of the fatigue lifetime data given in [49]. An experimental investigation was conducted in [49] on the reliability of gear-tooth of hardened and tempered steel 40Cr with the failure mode of fatigue. Four loading stress levels were used. The collected data of bending fatigue lifetime are shown in Table 2. Only four data points at each stress level are given in Table 2 and they are sufficient for subjective estimation of the possibility distribution of fatigue lifetime.

Table 2. The bending fatigue lifetime data of gear-teeth made of hardened and tempered steel 40Cr(in units of 10^6 loading cycles)

Data points	Stress levels S (MPa)			
	S_1=467.2	S_2=424.3	S_3=381.6	S_4=339.0
1	0.1404	0.1573	0.2919	0.3879
2	0.1508	0.1723	0.3024	0.4890
3	0.1572	0.1857	0.3250	0.5657
4	0.1738	0.1872	0.3343	0.5738

First, we calculate the average fatigue lifetime at each stress level m_{n_i}.

We will illustrate the procedure for constructing the possibility distribution of the lifetime using the data collected at the first stress level. The same procedure should be followed for analysis of data at other stress levels. Using Eq. (43) and Table 2, we have

$$m_{n_1} = \frac{1}{N} \sum_{j=1}^{N} n_1^j = \frac{1}{4} \sum_{j=1}^{4} n_1^j = \frac{1}{4}(0.1404 + 0.1508 + 0.1572 + 0.1738)$$

$$= 0.15555,$$

$$\pi_{n_1}(m_{n_1} = 0.15555) = 1.$$

The data points at the first stress level is divided into two groups separated by the calculated mean value m_{n_1}, i.e.,

$$G_1 = \left\{ 0.1404, 0.1508 \mid n_1^j < m_{n_1} \right\},$$

$$G_2 = \left\{ 0.1572, 0.1738 \mid n_1^j > m_{n_1} \right\}.$$

Further, from Eqs. (46) and (47), we have

$$m_{l_{n_1}} = \frac{1}{\#(G_1)} \sum_{n_1^j \in G_1} n_1^j = \frac{1}{2}(0.1404 + 0.1508) = 0.1456,$$

$$\pi_{n_1}\left(m_{l_{n_1}} = 0.1456\right) = 0.5;$$

$$m_{r_{n_1}} = \frac{1}{\#(G_2)} \sum_{n_1^j \in G_2} n_1^j = \frac{1}{2}(0.1572 + 0.1738) = 0.1655,$$

$$\pi_{n_1}\left(m_{r_{n_1}} = 0.1655\right) = 0.5.$$

Finally, with these calculated results and Eq. (49), we can construct the triangular possibility distribution of the bending fatigue lifetime of gear-teeth made of hardened and tempered steel 40Cr under the stress level of 467.2MPa as follows:

$$\alpha_{n_1} = 2\left(m_{n_1} - m_{l_{n_1}}\right) = 2(0.15555 - 0.1456) = 0.0199,$$

$$\beta_{n_1} = 2\left(-m_{n_1} + m_{r_{n_1}}\right) = 2(-0.15555 + 0.1655) = 0.0199,$$

$$\pi_{n_1}\left(n_1^j\right) = \begin{cases} 0, & n_1^j \le m_{n_1} - \alpha_{n_1} \\ 1 - \dfrac{m_{n_1} - n_1^j}{\alpha_{n_1}}, & m_{n_1} - \alpha_{n_1} \le n_1^j \le m_{n_1} \\ 1 - \dfrac{n_1^j - m_{n_1}}{\beta_{n_1}}, & m_{n_1} \le n_1^j \le m_{n_1} + \beta_{n_1} \\ 0, & m_{n_1} + \beta_{n_1} \le n_1^j \end{cases}$$

$$= \begin{cases} 0, & n_1^j \le 0.1357 \\ 1 - \dfrac{0.15555 - n_1^j}{0.0199}, & 0.13565 \le n_1^j \le 0.15555 \\ 1 - \dfrac{n_1^j - 0.15555}{0.0199}, & 0.15555 \le n_1^j \le 0.17545 \\ 0, & 0.17545 \le n_1^j \end{cases}.$$

The shape of this obtained possibility distribution is shown in Fig. 6. Note that the procedure for constructing the possibility distributions at other stress levels is the same. If we are interested in constructing other $L - R$ types of possibility distributions such as the Gaussian possibility distribution for the fatigue lifetime data of gear-teeth, a similar procedure can be followed.

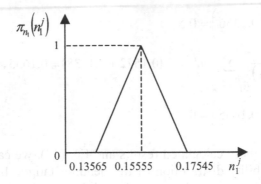

Fig. 6. The possibility distribution of the fatigue lifetime at the first stress level

After obtaining the possibility distribution of fatigue lifetime of the gear, we can derive the possibilistic reliability of bending fatigue strength of the gear at any time according to posbist reliability theory [8], e.g., under the stress level of 467.2MPa, we can figure out the possibilistic reliability of bending fatigue strength of the gear as follows:

$$R(t) = P_{oss}(X > t) = \sup_{x>t} P_{oss}(X = x) = \sup_{n_1^j > t} \pi_{n_1}\left(n_1^j\right)$$

$$= \begin{cases} 1, & t \leq 0.15555 \\ 1 - \dfrac{t - 0.15555}{0.0199}, & 0.15555 < t \leq 0.17545 \,. \\ 0, & t > 0.17545 \end{cases}$$

11.7 Conclusions

(1) Although the conventional reliability theory has been the dominant tool for evaluating system safety and analyzing failure uncertainty, the uncertainty within a system and its components cannot always be defined in the framework of probability. To analyze highly complex systems and deal with the vast variations of system characteristics, researchers have realized that the probability theory is not a panacea. In this chapter, based on the posbist reliability theory, the lifetime of a system is considered to be a Gaussian fuzzy variable. The posbist reliability of typical systems including series, parallel, series-parallel, parallel-series, and cold standby systems is derived.

(2) The universe of discourse on system lifetime defined in [8, 9] is expanded from $(0, +\infty)$ to $(-\infty, +\infty)$. We have illustrated in Section 3 that this expansion does not affect the nature of the problems to be solved. On the contrary, it makes the proofs in [8, 9] much more straightforward and the complexity of calculation is greatly reduced.

(3) In this chapter, we addressed the critical problem in the possibilistic reliability theory which is the construction of the possibility distribution and pointed out that all methods for generating membership functions can be used to construct the corresponding possibility distributions. We also presented a new method for constructing the possibility distribution with the possibilistic reliability analysis of fatigue lifetime of mechanical parts.

(4) The methods for constructing possibility distributions are not as mature as those for constructing probability distributions. The present chapter has provided a concise overview of the methods for constructing the possibility distributions in possibilistic reliability analysis. Further research is needed to develop a more general method for constructing possibility distributions.

(5) The model of posbist FTA constructed in this chapter can be used to evaluate the failure possibility of systems, in which the statistical data is scarce or the failure probability is extremely small (e.g., 10^{-7}). It is very difficult, however, to evaluate the safety and reliability of such systems using conventional fault tree analysis.

(6) As long as the failure possibilities of basic events can be obtained, the failure possibility of the top event can be derived according to the technique outlined in this chapter. Thus, it is crucial to estimate possibility distributions of basic events. In this chapter, we have pointed out that all the methods for generating membership functions can be used to construct the relevant possibility distributions in principle, and we have provided several methods for constructing possibility distributions. Nevertheless, further research is needed.

(7) We should note that the model of posbist FTA proposed in the present chapter, where the uncertainty is characterized in the context of possibility measures rather than probability measures, is different from the reported models of fuzzy FTA and the model of profust FTA.

Acknowledgements

This research was partially supported by the National Natural Science Foundation of China under the contract number 50175010, the Excellent Young Teachers Program of the Ministry of Education of China under the contract number 1766, the National Excellent Doctoral Dissertation Special Foundation of China under the contract number 200232, and the Natural Sciences and Engineering Research Council of Canada.

References

[1] Cai K Y, Wen C Y, Zhang M L. Fuzzy states as a basis for a theory of fuzzy reliability. Microelectronics and Reliability, 1993, 33: 2253-2263

[2] Tanaka H, Fan L T, Lai F S, Toguchi. Fault-tree analysis by fuzzy probability. IEEE Transactions on Reliability, 1983, 32: 453-457

[3] Singer D. A fuzzy set approach to fault tree and reliability analysis. Fuzzy Sets and Systems, 1990, 34: 145-155

[4] Onisawa T. An approach to human reliability in man-machine systems using error possibility. Fuzzy Sets and Systems, 1988, 27: 87-103

[5] Cappelle B, Kerre E E. On a possibilistic approach to reliability theory. In: Proceedings of the Second International symposium on Uncertainty Analysis, Maryland MD, 1993, 415-418

[6] Cremona C, Gao Y. The possibilistic reliability theory: theoretical aspects and applications. Structural Safety, 1997, 19(2): 173-201

[7] Utkin L V, Gurov S V. A general formal approach for fuzzy reliability analysis in the possibility context. Fuzzy Sets and Systems, 1996, 83: 203-213

[8] Cai K Y, Wen C Y, Zhang M L. Fuzzy variables as a basis for a theory of fuzzy reliability in the possibility context. Fuzzy Sets and Systems, 1991, 42:145-172

[9] Cai K Y, Wen C Y, Zhang M L. Posbist reliability behavior of fault-tolerant systems. Microelectronics and Reliability, 1995, 35(1): 49-56

[10] Huang H Z. Reliability analysis method in the presence of fuzziness attached to operating time. Microelectronics and Reliability, 1995, 35(12): 1483-1487

[11] Huang H Z. Reliability evaluation of a hydraulic truck crane using field data with fuzziness. Microelectronics and Reliability, 1996, 36(10): 1531-1536

[12] Huang H Z. Fuzzy multi-objective optimization decision-making of reliability of series system. Microelectronics and Reliability, 1997, 37(3): 447-449

[13] Huang H Z, Yuan X, Yao X S. Fuzzy fault tree analysis of railway traffic safety. Proceedings of the Conference on Traffic and Transportation Studies, ASCE, 2000, 107-112

[14] Huang H Z, Tong X, Zuo Ming J. Posbist fault tree analysis of coherent systems. Reliability Engineering and System Safety, 2004, 84(2): 141-148

[15] Xin Tong, Hong-Zhong Huang, Ming J Zuo. Construction of possibility distributions for reliability analysis based on possibility theory. Proceeding of the 2004 Asian International Workshop on Advanced Reliability Modeling, pp: 555-562

[16] Huang H Z, Sun Z Q, Zuo Ming J, Tian Z. Bayesian reliability assessment of gear lifetime under fuzzy environments. The 51st Annual Reliability & Maintainability Symposium (RAMS2005), Alexandria, Virginia, USA, 2005

[17] Huang H Z, Zuo Ming J, Fan X F, Tian Z. Quantitative analysis of influence of fuzzy factor on fuzzy reliability of structures. The European Safety and Reliability Conference (ESREL2005), 2005

[18] Huang H Z, Zuo Ming J, Sun Z Q. Bayesian reliability analysis for fuzzy lifetime data. Fuzzy Sets and Systems, in press.

[19] Cai K Y. System failure engineering and fuzzy methodology: An introductory overview. Fuzzy Sets and Systems, 1996, 83: 113-133

[20] Cooman G de. On modeling possibilistic uncertainty in two-state reliability theory. Fuzzy Sets and Systems, 1996, 83: 215-238

[21] Moller B, Beer M, Graf W, Hoffmann A. Possibility theory based safety assessment. Computer-Aided Civil and Infrastructure Engineering, 1999, 14: 81-91

[22] Savoia M. Structural reliability analysis through fuzzy number approach, with application to stability. Computers and Structures, 2002, 80: 1087-1102

[23] Guo S X, Lu Z Z, Feng L F. A fuzzy reliability approach for structures in the possibility context. Chinese Journal of Computational Mechanics, 2002, 19(1): 89-93

[24] Zadeh L A. Fuzzy sets as a basis for a theory of possibility. Fuzzy sets and Systems, 1978, 1(1): 3-28

[25] Dubois D, Prade H. Fuzzy real algebra: Some results, Fuzzy Sets and Systems, 1979, 2: 327-348

[26] Cai K Y. Introduction to Fuzzy Reliability. Boston: Kluwer Academic Publishers, 1996

[27] Soman K P, Misra K B. Fuzzy fault tree analysis using resolution identity and extension principle, Internat. J. Fuzzy Math. 1993, 1: 193~212

[28] Misra K B, Soman K P. Multistate fault tree analysis using fuzzy probability vectors and resolution identity. In: Onisawa T and Kacprzyk J Eds., Reliability and Safety Analysis under Fuzziness. Heidelberg: Physica-Verlag, 1995, 113~125

[29] Sawyer J P, Rao S S. Fault tree analysis of mechanical systems. Microelectronics and Reliability, 1994, 54(4): 653~667

[30] Pan H S, Yun W Y. Fault tree analysis with fuzzy gates. Computers and Industrial Engineering, 1997, 33(3~4): 569~572

[31] Lin C T, Wang M J J. Hybrid fault tree analysis using fuzzy sets. Reliability Engineering and System Safety, 1997, 58: 205~213

[32] Suresh P V, Babar A K, Raj V V. Uncertainty in fault tree analysis: A fuzzy approach. Fuzzy Sets and Systems, 1996, 83: 135~141

[33] Hua X Y, Hu Z W, Fan Z Y. A method to multistate fuzzy fault tree analysis. Journal of Mechanical Strength, 1998, 20(1): 35~40

[34] Huo Y, Hou C Z, Liu S X. Application of fuzzy set theory in fault tree analysis. Chinese Journal of Mechanical Engineering (English Edition), 1998, 11(4): 326~332

[35] Furuta H, Shiraishi N. Fuzzy importance in fault tree analysis. Fuzzy Sets and Systems, 1984, 12: 205~213

[36] Feng J, Wu M D. The profust fault tree and its analysis. Journal of National University of Defense Technology, 2001, 23(1): 85~88

[37] Barlow R E, Proschan F. Statistical Theory of Reliability and Life Testing: Probability Models. Holt, Rinehart and Winston, 1975

[38] Mei Q Z, Liao J S, Sun H Z. Introduction to system reliability engineering. Beijing: Science Press, 1987

[39] Zadeh L A. Fuzzy sets as a basis for a theory of possibility. Fuzzy sets and Systems, 1978, 1(1): 3~28

[40] Wang P Z. Fuzzy Sets and Their Applications. Shanghai: Shanghai Scientific & Technical Publishers (China), 1983

[41] Mcneill D, Freiberger P. Fuzzy Logic. New York: Simon and Schuster, 1993

[42] Civanlar M R, Trussell H J. Constructing membership functions using statistical data. Fuzzy sets and Systems, 1986, 18(1): 1-13

[43] Medasani S, Kim J, Krishnapuram R. An overview of membership function generation techniques for pattern recognition. International Journal of Approximate Reasoning, 1998, 19: 391-417

[44] Medaglia A L, Fang S C, Nuttle H L W, Wilson J R. An efficient and flexible mechanism for constructing membership functions. European Journal of Operational Research, 2002, 139: 84-95

[45] Dubois D, Prade H. Unfair coins and necessity measures: towards a possibilistic interpretation of histograms. Fuzzy Sets and Systems, 1983, 10:

[46] KLIR G. A principle of uncertainty and information invariance. International Journal of General Systems, 1990, 17(2/3): 249-275

[47] Dubois D, Prade H. Fuzzy Sets and Systems: Theory and Applications. New York: Academic Press, 1980

[48] Torkar M, Arzensek B. Failure of crane wire rope. Engineering Failure Analysis, 2002, 9: 227~233

[49] Tao J, Wang X Q, Tan J Z. Reliability of gear-tooth bending fatigue strength for through hardened and tempered steel 40Cr. Journal of University of Science and Technology Beijing (China), 1997, 19(5): 482-484

Analyzing Fuzzy System Reliability Based on the Vague Set Theory

Shyi-Ming Chen

Department of Computer Science and Information Engineering,
National Taiwan University of Science and Technology

12.1 Introduction

It is obvious that the reliability modeling is the most important discipline of reliable engineering (Kaufmann and Gupta, 1988). Traditionally, the reliability of a system's behavior is fully characterized in the context of probability measures. However, because of the inaccuracy and uncertainties of data, the estimation of precise values of probability becomes very difficult in many systems (Chen, 1996). In recent years, some researchers have used the fuzzy set theory (Zadeh, 1965) for fuzzy system reliability analysis (Cai *et al*, 1991a; Cai *et al*, 1991b; Cai *et al*, 1991c; Cai, 1996; Chen, 1994; Chen and Jong, 1996; Chen, 1996; Chen, 1997a; Cheng and Mon, 1993; Mon and Cheng, 1994; Singer, 1990; Wu, 2004).

Cai *et al*. (1991b) presented the following two assumptions for fuzzy system reliability analysis:

(1) Fuzzy-state assumption: At any time, the system may be either in the fuzzy success state or the fuzzy failure state.
(2) Possibility assumption: The system behavior can be fully characterized by possibility measures.

Cai (1996) presented an introduction to system failure engineering and its use of fuzzy methodology. Chen (1994) presented a method for fuzzy system reliability analysis using fuzzy number arithmetic operations. Chen and Jong (1996) presented a method for analyzing fuzzy system reliability using intervals of confidence. Chen (1996) presented a method for fuzzy system reliability analysis based on fuzzy time series and the α–cuts operations of fuzzy numbers. Cheng and Mon (1993) presented a method for fuzzy system reliability analysis by interval of confidence. Mon and Cheng (1994) presented a method for fuzzy system reliability analysis for components with different membership functions using non-linear programming techniques. Singer (1990) presented a fuzzy set approach for fault tree and

S.-M. Chen: *Analyzing Fuzzy System Reliability Based on the Vague Set Theory*, Computational Intelligence in Reliability Engineering (SCI) **40**, 347–362 (2007)
www.springerlink.com

reliability analysis. Suresh et al. (1996) presented a comparative study of probabilistic and fuzzy methodologies for uncertainty analysis using fault trees. Utkin and Gurov (1996) presented a general formal approach for fuzzy system reliability analysis in the possibility context. Wu (2004) presented a method for fuzzy reliability estimation using the Bayesian approach.

In this article, we present a method for analyzing fuzzy system reliability using the vague set theory (Chen, 1995; Gau and Buehrer, 1993), where the reliabilities of the components of a system are represented by vague sets defined in the universe of discourse [0, 1]. The grade of membership of an element x in a vague set is represented by a vague value $[t_x, 1 - f_x]$ in [0, 1], where t_x indicates the degree of truth, f_x indicates the degree of false, $1 - t_x - f_x$ indicates the unknown part, $0 \leq t_x \leq 1 - f_x \leq 1$, and $t_x + f_x \leq 1$. The notion of vague sets is similar to that of intuitionistic fuzzy sets (Atanassov, 1986). Both of them are generalizations of fuzzy sets (Zadeh, 1965). The proposed method can model and analyze fuzzy system reliability in a more flexible and convenient manner.

The rest of this article is organized as follows. In Section 2, we briefly review a method for fuzzy system reliability analysis from (Chen and Jong, 1996). In Section 3, we briefly review some definitions and arithmetic operations of vague sets from (Chen, 1995) and (Gau and Buehrer, 1993). In Section 4, we present a method for analyzing fuzzy system reliability based on the vague set theory. The conclusions are discussed in Section 5.

12.2 A Review of Chen and Jong's Fuzzy System Reliability Analysis Method

In this section, we briefly review a method for fuzzy system reliability analysis from (Chen and Jong, 1996).

In (Kaufmann and Gupta, 1988, pp. 184-208), the reliability $K(t)$ of a subsystem or system is represented by an interval of confidence $K(t) = [K_a(t), K_b(t)]$, where $K_a(t)$ and $K_b(t)$ are the lower and upper bounds of the survival function at time t ($t = 0, 1, 2, \ldots$), respectively, and $0 \leq K_a(t) \leq K_b(t) \leq 1$. For example, Fig. 1 shows the lower and upper bounds of the survival function given subjectively by an expert.

Chen and Jong (1996) considered the situation in which there are uncertainties associated with the survival interval of confidence $[K_a(t), K_b(t)]$ at time t ($t = 0, 1, 2, \ldots$). In such a situation, the reliability of a subsystem P_i can be represented by $[K_{ia}(t), K_{ib}(t)]/C_i(t)$, where $C_i(t)$ indicates the degree of certainty that the reliability of the subsystem P_i at time t lies in interval $[K_{ia}(t), K_{ib}(t)]$, $0 \leq K_{ia}(t) \leq K_{ib}(t) \leq 1$, $0 \leq C_i(t) \leq 1$, and $= 0, 1, 2, ...$ The values of $K_{ia}(t)$, $K_{ib}(t)$ and $C_i(t)$ at time t ($t = 0, 1, 2, ...$) are given by experts, respectively.

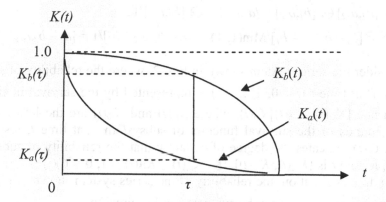

Fig. 1. Lower and upper bounds of the survival function

Chen and Jong (1996) presented a method for fuzzy system reliability analysis based on the interval of confidence, which is reviewed as follows. Let $[a_1, a_2]/c_1$ and $[b_1, b_2]/c_2$ be two survival intervals of confidence, where $0 \leq a_1 \leq a_2 \leq 1$, $0 \leq b_1 \leq b_2 \leq 1$, $0 \leq c_1 \leq 1$, and $0 \leq c_2 \leq 1$. The multiplication operation and the subtraction operation between the survival intervals of confidence $[a_1, a_2]/c_1$ and $[b_1, b_2]/c_2$ are defined as follows:

$$[a_1, a_2]/c_1 \otimes [b_1, b_2]/c_2 = [a_1 \times b_1, a_2 \times b_2]/Min(c_1, c_2), \tag{1}$$

$$[a_1, a_2]/c_1 \ominus [b_1, b_2]/c_2 = [a_1 - b_2, a_2 - b_1]/Min(c_1, c_2), \tag{2}$$

where $\otimes$ and $\ominus$ are the multiplication operator and subtraction operator between the survival intervals of confidence, respectively.

The complement of a survival interval of confidence $[b_1, b_2]/c_2$ is defined by

$$1 \ominus [b_1, b_2]/c_2 = [1, 1]/1 \ominus [b_1, b_2]/c_2$$
$$= [1 - b_2, 1 - b_1]/Min(1, c_2) = [1 - b_2, 1 - b_1]/c_2. \tag{3}$$

It is obvious that if $a_1 = a_2 = a$ and $b_1 = b_2 = b$, then

$$[a_1, a_2]/c_1 \otimes [b_1, b_2]/c_2 = [a, a]/c_1 \otimes [b, b]/c_2$$
$$= [a \times b, a \times b]/Min(c_1, c_2) = (a \times b)/Min(c_1, c_2), \tag{4}$$
$$[a_1, a_2]/c_1 \ominus [b_1, b_2]/c_2 = [a, a]/c_1 \ominus [b, b]/c_2$$
$$= [a - b, a - b]/Min(c_1, c_2) = (a - b)/Min(c_1, c_2). \tag{5}$$

Because $[x,y\,]$ can be written as $[x,y\,]/1$, where $0 \le x \le y \le 1$, we can get

$$[a_1, a_2] \otimes [b_1,b_2] = [a_1,a_2]/1 \otimes [b_1,b_2]/1 \tag{6}$$

$$= [a_1 \times b_1, a_2 \times b_2]/\text{Min}(1, 1) = [a_1 \times b_1,a_2 \times b_2]/1 = [a_1 \times b_1,a_2 \times b_2],$$

$$[a_1,a_2] \ominus [b_1,b_2] = [a_1,a_2]/1 \ominus [b_1,b_2]/1 \tag{7}$$

$$= [a_1 - b_2,a_2 - b_1]/\text{Min}(1, 1) = [a_1 - b_2,a_2 - b_1]/1 = [a_1 - b_2,a_2 - b_1].$$

Consider the series system shown in Fig. 2, where the reliability of sub-system P_i at time t ($t = 0, 1, 2, \ldots$) is represented by the survival interval confidence $[K_{ia}(t), K_{ib}(t)]/C_i(t)$, where $K_{ia}(t)$ and $K_{ib}(t)$ are the lower and upper bounds of the survival function of subsystem i at time t, respectively, $C_i(t)$ indicates the degree of certainty that the reliability of subsystem P_i at time t is $[K_{ia}(t), K_{ib}(t)]$, $0 \le K_{ia}(t) \le K_{ib}(t) \le 1$, $0 \le C_i(t) \le 1$, and $1 \le i \le n$. In this situation, the reliability of the series system shown in Fig. 2 at time t ($t = 0, 1, 2, \ldots$) can be evaluated and is equal to

$$[K_{1a}(t),K_{1b}(t)]/C_1(t) \otimes [K_{2a}(t),K_{2b}(t)]/C_2(t) \otimes \ldots \otimes [K_{na}(t),K_{nb}(t)]/C_n(t)$$

$$= [K_{1a}(t) \times K_{2a}(t) \times \ldots \times K_{na}(t), K_{1b}(t) \times K_{2b}(t)$$

$$\times \ldots \times K_{nb}(t)]/\text{Min}(C_1(t),C_2(t),\ldots,C_n(t)). \tag{8}$$

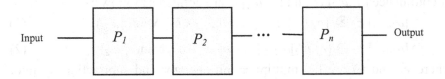

Fig. 2. A series system

Consider the parallel system shown in Fig. 3, where the reliability of subsystem P_i at time t ($t = 0, 1, 2, \ldots$) is $[K_{ia}(t), K_{ib}(t)]/C_i(t)$, where $0 \le K_{ia}(t) \le K_{ib}(t) \le 1$, $0 \le C_i(t) \le 1$, and $1 \le i \le n$. Then, the reliability of the parallel system shown in Fig. 3 at time t can be evaluated and is equal to $[K_a(t),K_b(t)]/C_p(t)$, where

(1) $[K_a(t),K_b(t)] = 1 \ominus (1\ominus [\,K_{1a}(t),K_{1b}(t)]) \otimes (1\ominus [\,K_{2a}(t),$

$K_{2b}(t)]) \otimes \ldots \otimes (1 \ominus [K_{2a}(t), K_{2b}(t)])] = 1 \ominus ([1 K_{1b}(t), 1$
$- K_{1a}(t)]) \otimes ([1 -K_{2b}(t), 1 K_{2a}(t)]) \otimes \ldots \otimes ([1 -K_{nb}(t), 1 K_{na}(t)]).$

(2) The value of $C_p(t)$ is evaluated as follows. Let X and Y be two real intervals in $[0, 1]$, where $X = [x_1, x_2]$, $Y = [y_1, y_2]$, $0 \le x_1 \le x_2 \le 1$, and $0 \le y_1 \le y_2 \le 1$. Based on the similarity function S presented in (Chen and Wang, 1995), we can calculate the degree of similarity between the intervals X and Y, where $S(X, Y) = 1 - (|x_1 - y_1| + |x_2 - y_2|)/2$ and $0 \le S(X, Y) \le 1$. The larger the value of $S(X, Y)$, the more the similarity between the intervals X and Y. Because the reliability of the subsystem P_i at time t ($t = 0, 1, 2, \ldots$) is $[K_{ia}(t), K_{ib}(t)]/C_i(t)$, where $0 \le K_{ia}(t) \le K_{ib}(t) \le 1$, $0 \le C_i(t) \le 1$, and $1 \le i \le n$. Based on the similarity function S, we can get

$S ([K_{1a}(t), K_{1b}(t)], [K_a(t), K_b(t)]) = s_1,$
$S ([K_{2a}(t), K_{2b}(t)], [K_a(t), K_b(t)]) = s_2,$

$$\ldots$$

$S ([K_{na}(t), K_{nb}(t)], [K_a(t), K_b(t)]) = s_n,$

where $0 \le s_i \le 1$ and $i = 1, 2, \ldots, n$. If s_j is the largest value among the values $s_1, s_2, \ldots$, and s_n, then let the value of $C_p(t)$ be equal to $C_j(t)$, where $1 \le j \le n$.

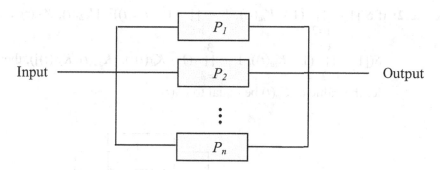

Fig. 3. A parallel system

Consider the series-parallel system shown in Fig. 4, where the reliability of the subsystem P_i at time t ($t = 0, 1, 2, \ldots$) is represented by $[K_{ia}(t), K_{ib}(t)]/C_i(t)$, where $0 \le K_{ia}(t) \le K_{ib}(t) \le 1$, $0 \le C_i(t) \le 1$, and $1 \le i \le 3$. Then, the reliability of the series-parallel system shown in Fig. 4 at time t ($t = 0, 1,$ 2, …) can be evaluated and is equal to

$$[K_{1a}(t), K_{1b}(t)]/C_1(t) \otimes [1 - \prod_{i=2}^{3} (1 - K_{ia}(t)), 1 - \prod_{i=2}^{3} (1 - K_{ib}(t))]/C_p(t)$$

$$= [K_{1a}(t), K_{1b}(t)]/C_1(t) \otimes [1 - (1 - K_{2a}(t))(1 - K_{3a}(t)), 1 - (1 - K_{2b}(t))$$

$$(1 - K_{3b}(t))]/C_p(t) = [K_{1a}(t)(1 - (1 - K_{2a}(t))(1 - K_{3a}(t)), K_{1b}(t)(1 - (1$$

$$- K_{2b}(t))(1 - K_{3b}(t))) /Min(C_1(t), C_p(t)) = [K_{1a}(t) - K_{1a}(t)(1 - K_{2a}(t))(1$$

$$- K_{3a}(t)), K_{1b}(t) - K_{1b}(t)(1 - K_{2b}(t))(1 - K_{3b}(t))] /Min(C_1(t), C_p(t)),$$

where the value of $C_p(t)$ is evaluated as follows:

Case 1: If $S([1 - \prod_{i=2}^{3} (1 - K_{ia}(t)), 1 - \prod_{i=2}^{3} (1 - K_{ib}(t))], [K_{2a}(t), K_{2b}(t)]) \geq$

$S([1 - \prod_{i=2}^{3} (1 - K_{ia}(t)), 1 - \prod_{i=2}^{3} (1 - K_{ib}(t))], [K_{3a}(t), K_{3b}(t)])$, then
let the value of $C_p(t)$ be equal to $C_2(t)$.

Case 2: If $S([1 - \prod_{i=2}^{3} (1 - K_{ia}(t)), 1 - \prod_{i=2}^{3} (1 - K_{ib}(t))], [K_{2a}(t), K_{2b}(t)]) <$

$S([1 - \prod_{i=2}^{3} (1 - K_{ia}(t)), 1 - \prod_{i=2}^{3} (1 - K_{ib}(t))], [K_{3a}(t), K_{3b}(t)])$, then
let the value of $C_p(t)$ be equal to $C_3(t)$.

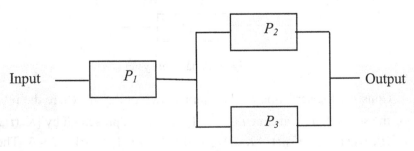

Fig. 4. A series-parallel system

The method presented in (Chen and Jong, 1996) is more flexible and more general than the one presented in (Kaufmann and Gupta, 1988, pp. 184-208) due to the fact that it allows the survival function of each subsystem at different times to be associated with different degrees of certainty between zero and one.

12.3 Basic Concepts of Vague Sets

In (Zadeh, 1965), Zadeh proposed the theory of fuzzy sets. Let U be the universe of discourse, $U = \{u_1, u_2, ..., u_n\}$. The grade of membership of an element u_i in a fuzzy set is represented by a real value between zero and one, where $u_i \in U$. However, Gau and Buehrer (1993) pointed out that this single value combines the evidence for $u_i \in U$ and the evidence against $u_i \in U$. They also pointed out that it does not indicate the evidence for $u_i \in U$ and the evidence against $u_i \in U$, respectively, and it does not indicate how much there is of each. Furthermore, Gau and Buehrer also pointed out that the single value tells us nothing about its accuracy. Therefore, Gau and Buehrer (1993) presented the concepts of vague sets. Chen (1995) have presented the arithmetic operations between vague sets.

Let U be the universe of discourse, $U = \{u_1, u_2, ..., u_n\}$, with a generic element of U denoted by u_i. A vague set $\tilde{A}$ in the universe of discourse U is characterized by a truth-membership function $_{\tilde{A}}$, $t_{\tilde{A}}: U \rightarrow [0, 1]$, and a false-membership function $f_{\tilde{A}}$, $f_{\tilde{A}}: U \rightarrow [0, 1]$, where $t_{\tilde{A}}(u_i)$ is a lower bound of the grade of membership of u_i derived from the evidence for u_i, $f_{\tilde{A}}(u_i)$ is a lower bound of the negation of u_i derived from the evidence against u_i, and $t_{\tilde{A}}(u_i) + f_{\tilde{A}}(u_i) \leq 1$. The grade of membership of u_i in the vague set $\tilde{A}$ is bounded by a subinterval $[t_{\tilde{A}}(u_i), 1 f_{\tilde{A}}(u_i)]$ of $[0, 1]$. The vague value $[t_{\tilde{A}}(u_i), 1 f_{\tilde{A}}(u_i)]$ indicates that the exact grade of membership $\mu_{\tilde{A}}(u_i)$ of u_i is bounded by $t_{\tilde{A}}(u_i) \leq \mu_{\tilde{A}}(u_i) \leq 1 f_{\tilde{A}}(u_i)$, where $t_{\tilde{A}}(u_i) + f_{\tilde{A}}(u_i) \leq 1$. For example, a vague set $\tilde{A}$ in the universe of discourse U is shown in Fig. 5.

If the universe of discourse U is a finite set, then a vague set $\tilde{A}$ of the universe of discourse U can be represented as follows:

$$\tilde{A} = \sum_{i=1}^{n} \left[t_{\tilde{A}}(u_i), 1 - f_{\tilde{A}}(u_i) \right] / u_i . \tag{9}$$

If the universe of discourse U is an infinite set, then a vague set $\tilde{A}$ of the universe of discourse can be represented as

$$\tilde{A} = \int_U \left[t_{\tilde{A}}(u_i), \ 1 - f_{\tilde{A}}(u_i) \right] / u_i \quad u_i \in U. \tag{10}$$

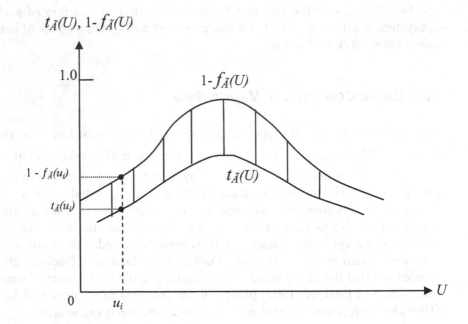

Fig. 5. A vague set

Definition 3.1: Let $\tilde{A}$ be a vague set of the universe of discourse U with the truth-membership function $t_{\tilde{A}}$ and the false-membership function $f_{\tilde{A}}$, respectively. The vague set $\tilde{A}$ is convex if and only if for all u_1, u_2 in U,

$$t_{\tilde{A}}(\lambda u_1 + (1 - \lambda)u_2) \geq \mathrm{Min}(t_{\tilde{A}}(u_1), \ t_{\tilde{A}}(u_2)), \tag{11}$$

$$1 - f_{\tilde{A}}(\lambda u_1 + (1 - \lambda)u_2) \geq \mathrm{Min}(1 - f_{\tilde{A}}(u_1), \ 1 - f_{\tilde{A}}(u_2)), \tag{12}$$

where $\lambda \in [0, 1]$.

Definition 3.2: A vague set $\tilde{A}$ of the universe of discourse U is called a normal vague set if $\exists \ u_i \in U$, such that $1 - f_{\tilde{A}}(u_i) = 1$. That is, $f_{\tilde{A}}(u_i) = 0$.

Definition 3.3: A vague number is a vague subset in the universe of discourse U that is both convex and normal.

In the following, we introduce some arithmetic operations of triangular vague sets (Chen, 1995). Let us consider the triangular vague set Ã shown in Fig. 6, where the triangular vague set Ã can be parameterized by a tuple

$<[(a, b, c); \mu_1], [(a, b, c); \mu_2]>$. For convenience, the tuple $<[(, b, c); \mu_1],$ $[(a, b, c); \mu_2]>$ can also be abbreviated into $<[(a, b, c); \mu_1; \mu_2]>$, where $0 \leq \mu_1 \leq \mu_2 \leq 1$.

Some arithmetic operations between triangular vague sets are as follows:

Case 1: Consider the triangular vague sets $\tilde{A}$ and $\tilde{B}$ shown in Fig. 7, where

$$\tilde{A} = <[(a_1, b_1, c_1); \mu_1], [(a_1, b_1, c_1); \mu_2]> = <[(a_1, b_1, c_1); \mu_1; \mu_2]>,$$

$$\tilde{B} = <[(a_2, b_2, c_2); \mu_1], [(a_2, b_2, c_2); \mu_2]> = <[(a_2, b_2, c_2); \mu_1; \mu_2]>,$$

and $0 \leq \mu_1 \leq \mu_2 \leq 1$. The arithmetic operations between the triangular vague sets $\tilde{A}$ and $\tilde{B}$ are defined as follows:

$$\tilde{A} \oplus \tilde{B} = <[(a_1, b_1, c_1); \mu_1], [(a_1, b_1, c_1); \mu_2]> \oplus <[(a_2, b_2, c_2); \mu_1],$$
$$[(a_2, b_2, c_2); \mu_2]>$$

$$= <[(a_1 + a_2, b_1 + b_2, c_1 + c_2); \mu_1], [(a_1 + a_2, b_1 + b_2, c_1 + c_2); \mu_2]>$$

$$= <[(a_1 + a_2, b_1 + b_2, c_1 + c_2); \mu_1; \mu_2]>, \tag{13}$$

$$\tilde{B} \ominus \tilde{A} = <[(a_2, b_2, c_2); \mu_1], [(a_2, b_2, c_2); \mu_2]> \ominus <[(a_1, b_1, c_1); \mu_1],$$
$$[(a_1, b_1, c_1); \mu_2]>$$

$$= <[(a_2 - c_1, b_2 - b_1, c_2 - a_1); \mu_1], [(a_2 - c_1, b_2 - b_1, c_2 - a_1); \mu_2]>$$

$$= <[(a_2 - c_1, b_2 - b_1, c_2 - a_1); \mu_1; \mu_2]>, \tag{14}$$

$$\tilde{A} \otimes \tilde{B} = <[(a_1, b_1, c_1); \mu_1], [(a_1, b_1, c_1); \mu_2]> \otimes <[(a_2, b_2, c_2); \mu_1],$$
$$[(a_2, b_2, c_2); \mu_2]>$$

$$= <[(a_1 \times a_2, b_1 \times b_2, c_1 \times c_2); \mu_1], [(a_1 \times a_2, b_1 \times b_2, c_1 \times c_2); \mu_2]>$$

$$= <[(a_1 \times a_2, b_1 \times b_2, c_1 \times c_2); \mu_1; \mu_2]>, \tag{15}$$

$$\tilde{B} \oslash \tilde{A} = <[(a_2, b_2, c_2); \mu_1], [(a_2, b_2, c_2); \mu_2]> \oslash <[(a_1, b_1, c_1); \mu_1],$$
$$[(a_1, b_1, c_1); \mu_2]>$$

$$= <[(a_2/c_1, b_2/b_1, c_2/a_1); \mu_1], [(a_2/c_1, b_2/b_1, c_2/a_1); \mu_2]>$$

$$= <[(a_2/c_1, b_2/b_1, c_2/a_1); \mu_1; \mu_2]>. \tag{16}$$

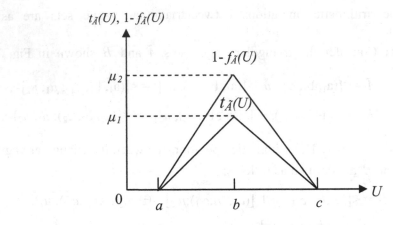

Fig. 6. A triangular vague set

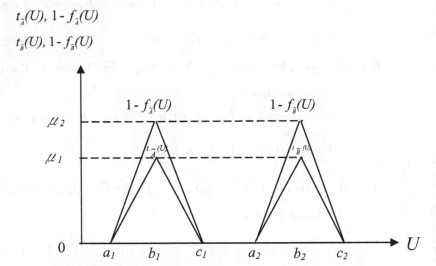

Fig. 7. Triangular vague sets $\tilde{A}$ and $\tilde{B}$ (Case 1)

Case 2: Consider the triangular vague sets $\tilde{A}$ and $\tilde{B}$ shown in Fig. 8, where

$$\tilde{A} = <[(a_1,b_1,c_1); \mu_1], [(a_1, b_1,c_1); \mu_2]>,$$
$$\tilde{B} = <[(a_2,b_2,c_2); \mu_3], [(a_2,b_2, c_2); \mu_4]>,$$

and $0 \le \mu_3 \le \mu_1 \le \mu_4 \le \mu_2 \le 1$.

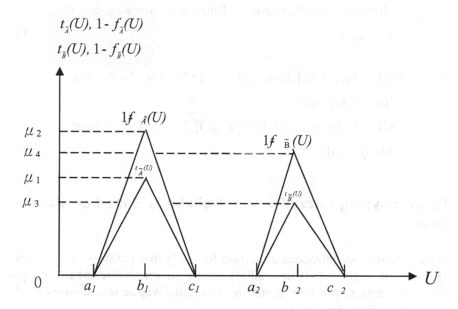

Fig. 8. Triangular vague sets $\tilde{A}$ and $\tilde{B}$ (Case 2)

The arithmetic operations between the triangular vague sets $\tilde{A}$ and $\tilde{B}$ are defined as follows:

$$\tilde{A} \oplus \tilde{B} = <[(a_1,b_1,c_1); \mu_1], [(a_1, b_1,c_1);\mu_2]> \oplus <[(a_2,b_2,c_2); \mu_3],$$
$$[(a_2,b_2,c_2); \mu_4]>$$
$$= <[(a_1+a_2,b_1+ b_2,c_1+ c_2); Min(\mu_1, \mu_3)], [(a_1+ a_2,b_1+ b_2,c_1+ c_2);$$
$$Min(\mu_2,\mu_4)]>, \tag{17}$$

$$\tilde{B} \ominus \tilde{A} = <[(a_2, b_2,c_2);\mu_3], [(a_2,b_2,c_2); \mu_4]> \ominus <[(a_1,b_1,c_1); \mu_1],$$

$$[(a_1, b_1, c_1); \mu_2]>$$

$$= <[(a_2 - c_1, b_2 - b_1, c_2 - a_1); \mathrm{Min}(\mu_1, \mu_3)], [(a_2 - c_1, b_2 - b_1, c_2 - a_1);$$

$$\mathrm{Min}(\mu_2, \mu_4)]>, \tag{18}$$

$$\tilde{A} \otimes \tilde{B} = <[(a_1, b_1, c_1); \mu_1], [(a_1, b_1, c_1); \mu_2]> \otimes <[(a_2, b_2, c_2); \mu_3],$$

$$[(a_2, b_2, c_2); \mu_4]>$$

$$= <[(a_1 \times a_2, b_1 \times b_2, c_1 \times c_2); \mathrm{Min}(\mu_1, \mu_3)], [(a_1 \times a_2, b_1 \times b_2, c_1 \times c_2);$$

$$\mathrm{Min}(\mu_2, \mu_4)]> \tag{19}$$

$$\tilde{B} \oslash \tilde{A} = <[(a_2, b_2, c_2); \mu_3], [(a_2, b_2, c_2); \mu_4]> \oslash <[(a_1, b_1, c_1); \mu_1],$$

$$[(a_1, b_1, c_1); \mu_2]>$$

$$= <[(a_2/c_1, b_2/b_1, c_2/a_1); \mathrm{Min}(\mu_1, \mu_3)], [(a_2/c_1, b_2/b_1, c_2/a_1);$$

$$\mathrm{Min}(\mu_2, \mu_4)]>. \tag{20}$$

12.4 Analyzing Fuzzy System Reliability Based on Vague Sets

In this section, we introduce a method for analyzing fuzzy system reliability based on vague sets (Chen, 2003), where the reliabilities of the components of a system are represented by triangular vague sets defined in the universe of discourse $[0, 1]$.

Consider a series system shown in Fig. 2, where the reliability $\tilde{R}_i$ of the subsystem P_i is represented by a triangular vague set $<[(a_i, b_i, c_i); \mu_{i1}; \mu_{i2}]>$, where $0 \le \mu_{i1} \le \mu_{i2} \le 1$, and $1 \le i \le n$. Then, the reliability $\tilde{R}$ of the series system shown in Fig. 2 can be evaluated as follows:

$$\tilde{R} = \tilde{R}_1 \otimes \tilde{R}_2 \otimes \ldots \otimes \tilde{R}_n = <[(a_1, b_1, c_1); \mu_{11}; \mu_{12}]> \otimes$$

$$<[(a_2, b_2, c_2); \mu_{21}; \mu_{22}]> \otimes \ldots \otimes <[(a_n, b_n, c_n); \mu_{n1}; \mu_{n2}]>$$

$$= <[(\prod_{i=1}^{n} a_i, \prod_{i=1}^{n} b_i, \prod_{i=1}^{n} c_i); \mathrm{Min}(\mu_{11}, \mu_{21}, \ldots, \mu_{n1});$$

$$\mathrm{Min}(\mu_{12}, \mu_{22}, \ldots, \mu_{n2})]>. \tag{21}$$

Furthermore, consider the parallel system shown in Fig. 3, where the reliability $\tilde{R}_i$ of the subsystem P_i is represented by a triangular vague set $<[(a_i, b_i, c_i); \mu_{i1}; \mu_{i2}]>$, where $0 \leq \mu_{i1} \leq \mu_{i2} \leq 1$, and $1 \leq i \leq n$. Then, the reliability $\tilde{R}$ of the parallel system shown in Fig. 3 can be evaluated as follows:

$$\tilde{R} = 1 \ominus \prod_{i=1}^{n} (1 \ominus \tilde{R}_i) = 1 \ominus (1 \ominus <[(a_1, b_1, c_1); \mu_{11}; \mu_{12}]>)$$

$$\otimes (1 \ominus <[(a_2, b_2, c_2); \mu_{21}; \mu_{22}]>) \otimes \ldots \otimes (1 \ominus <[(a_n, b_n, c_n); \mu_{n1}; \mu_{n2}]>)$$

$$= 1 \ominus <[(1 - c_1, 1 - b_1, 1 - a_1); \mu_{11}; \mu_{12}]> \otimes <[(1 - c_2, 1 - b_2, 1 - a_2);$$

$$\mu_{21}; \mu_{22}]> \otimes \ldots \otimes <[(1 - c_n, 1 - b_n, 1 - a_n); \mu_{n1}; \mu_{n2}]>$$

$$= 1 \ominus <[\prod_{i=1}^{n} (1 - c_i), \prod_{i=1}^{n} (1 - b_i), \prod_{i=1}^{n} (1 - a_i)); Min(\mu_{11}; \mu_{21}, \ldots, \mu_{n1});$$

$$Min(\mu_{12}; \mu_{22}, \ldots, \mu_{n2}))>$$

$$= <[(1 - \prod_{i=1}^{n} (1 - a_i), 1 - \prod_{i=1}^{n} (1 - b_i), 1 - \prod_{i=1}^{n} (1 - c_i));$$

$$Min(\mu_{11}; \mu_{21}, \ldots, \mu_{n1}); Min(\mu_{12}; \mu_{22}, \ldots, \mu_{n2}))> \tag{22}$$

In the following, we use an example to illustrate the fuzzy system reliability analysis process of the proposed method.

12.4.1 Example

Consider the system shown in Fig. 9, where the reliabilities of the subsystems P_1, P_2, P_3 and P_4 are represented by the triangular vague sets $\tilde{R}_1$, $\tilde{R}_2$, $\tilde{R}_3$ and $\tilde{R}_4$, respectively, where

$$\tilde{R}_1 = <[(a_1, b_1, c_1); \mu_{11}; \mu_{12}]>, \quad \tilde{R}_2 = <[(a_2, b_2, c_2); \mu_{21}; \mu_{22}]>,$$

$$\tilde{R}_3 = <[(a_3, b_3, c_3); \mu_{31}; \mu_{32}]>, \quad \tilde{R}_4 = <[(a_4, b_4, c_4); \mu_{41}; \mu_{42}]>,$$

$0 \leq \mu_{i1} \leq \mu_{i2} \leq 1$ and $1 \leq i \leq 4$. Based on the previous discussion, we can see that the reliability $\tilde{R}$ of the system shown in Fig. 9 can be evaluated as follows:

$$\tilde{R} = [1 \ominus (1 \ominus \tilde{R}_1) \otimes (1 \ominus \tilde{R}_2)] \otimes [1 \ominus (1 \ominus \tilde{R}_3) \otimes (1 \ominus \tilde{R}_4)]$$

$$= [1 \ominus (1 \ominus <[(a_1, b_1, c_1); \mu_{11}; \mu_{12}]>) \otimes (1 \ominus <[(a_2, b_2, c_2); \mu_{21}; \mu_{22}]>)$$
$$\otimes [1 \ominus (1 \ominus <[(a_3, b_3, c_3); \mu_{31}; \mu_{32}]>) \otimes (1 \ominus <[(a_4, b_4, c_4); \mu_{41}; \mu_{42}]>)]$$

$$= [1 \ominus <[(1 - c_1, 1 - b_1, 1 - a_1); \mu_{11}; \mu_{12}]> \otimes <[(1 - c_2, 1 - b_2, 1 - a_2);$$
$$\mu_{21}; \mu_{22}]>] \otimes [1 \ominus <[(1 - c_3, 1 - b_3, 1 - a_3); \mu_{31}; \mu_{32}]> \otimes$$
$$<[(1 - c_4, 1 - b_4, 1 - a_4); \mu_{41}; \mu_{42}]>]$$

$$= [1 \ominus <[((1 - c_1)(1 - c_2), (1 - b_1)(1 - b_2), (1 - a_1)(1 - a_2));$$
$$Min(\mu_{11}; \mu_{21}); Min(\mu_{12}, \mu_{22})]>] \otimes [1 \ominus <[((1 - c_3)(1 - c_4), (1 - b_3)$$
$$(1 - b_4), (1 - a_3)(1 - a_4)); Min(\mu_{31}; \mu_{41}); Min(\mu_{32}, \mu_{42})]>]$$

$$= <[(1 - (1 - a_1)(1 - a_2), 1 - (1 - b_1)(1 - b_2), 1 - (1 - c_1)(1 - c_2));$$
$$Min(\mu_{11}; \mu_{21}); Min(\mu_{12}, \mu_{22})]>] \otimes <[(1 - (1 - a_3)(1 - a_4), 1 -$$
$$(1 - b_3)(1 - b_4), 1 - (1 - c_3) \quad (1 - c_4)); Min(\mu_{31}; \mu_{41}); Min(\mu_{32}, \mu_{42})]>$$

$$= <[(a_1 + a_2 - a_1 a_2, b_1 + b_2 - b_1 b_2, c_1 + c_2 - c_1 c_2); Min(\mu_{11}; \mu_{21});$$
$$Min(\mu_{12}, \mu_{22})]> \otimes <[(a_3 + a_4 - a_3 a_4, b_3 + b_4 - b_3 b_4,$$
$$c_3 + c_4 - c_3 c_4); Min(\mu_{31}; \mu_{41}); Min(\mu_{32}, \mu_{42})]>$$

$$= <[((a_1 + a_2 - a_1 a_2)(a_3 + a_4 - a_3 a_4), (b_1 + b_2 - b_1 b_2)(b_3 + b_4 - b_3 b_4),$$
$$(c_1 + c_2 - c_1 c_2)(c_3 + c_4 - c_3 c_4)), Min(\mu_{11}, \mu_{21}, \mu_{31}, \mu_{41});$$
$$Min(\mu_{12}, \mu_{22}, \mu_{32}, \mu_{42})]>$$
$$= <[(a_1 a_3 + a_1 a_4 - a_1 a_3 a_4 + a_2 a_3 + a_2 a_4 - a_2 a_3 a_4 - a_1 a_2 a_3 - a_1 a_2 a_4$$
$$+ a_1 a_2 a_3 a_4, b_1 b_3 + b_1 b_4 - b_1 b_3 b_4 + b_2 b_3 + b_2 b_4 - b_2 b_3 b_4 - b_1 b_2 b_3 - b_1 b_2 b_4$$
$$+ b_1 b_2 b_3 b_4, c_1 c_3 + c_1 c_4 - c_1 c_3 c_4 + c_2 c_3 + c_2 c_4 - c_2 c_3 c_4 - c_1 c_2 c_3 - c_1 c_2 c_4$$
$$+ c_1 c_2 c_3 c_4); Min(\mu_{11}, \mu_{21}, \mu_{31}, \mu_{41}); Min(\mu_{12}, \mu_{22}, \mu_{32}, \mu_{42})]>. \qquad (23)$$

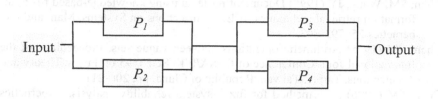

Fig. 9. A system with four subsystems P_1, P_2, P_3 and P_4

12.5 Conclusions

In this chapter, we have presented a method for analyzing fuzzy system reliability based on the vague set theory, where the components of a system are represented by triangular vague sets defined in the universe of discourse [0, 1]. The grade of membership of an element x in a vague set is represented by a vague value $[t_x, 1 - f_x]$ in [0, 1], where t_x indicates the degree of truth, f_x indicates the degree of false, $1 - t_x - f_x$ indicates the unknown part, $0 \leq t_x \leq 1 - f_x \leq 1$, and $t_x + f_x \leq 1$. The proposed method can model and analyze fuzzy system reliability in a more flexible and convenient manner.

References

Atanassov, K (1986) Intuitionistic fuzzy sets. Fuzzy Sets and Systems 20: 87-96.

Cai, KY, Wen, CY, Zhang, ML (1991a) Fuzzy variables as a basis for a theory of fuzzy reliability in possibility context. Fuzzy Sets and Systems 42: 145-172.

Cai, KY, Wen, CY, Zhang, ML (1991b) Posbist reliability behavior of typical systems with two types of failure. Fuzzy Sets and Systems 43: 17-32.

Cai, KY, Wen, CY, Zhang, ML (1991c) Fuzzy reliability modeling of gracefully degradable computing systems. Reliability Engineering and System Safety 33: 141-157.

Cai, KY (1996) System failure engineering and fuzzy methodology: An introductory overview. Fuzzy Sets and Systems 83: 113-133.

Chen, SM (2003) Analyzing fuzzy system reliability using vague set theory. Inter national Journal of Applied Science and Engineering 1: 82-88, 2003.

Chen, SM (1994) Fuzzy system reliability analysis using fuzzy number arithmetic operations. Fuzzy Sets and Systems 64: 31-38.

Chen, SM, Jong, WT (1996) Analyzing fuzzy system reliability using interval of confidence. International Journal of Information Management and Engineering 2: 16-23.

Chen, SM, Wang, JY (1995) Document retrieval using knowledge-based fuzzy information retrieval techniques. IEEE Transactions on Systems, Man, and Cybernetics 25: 793-803.

Chen, SM (1995) Arithmetic operations between vague sets. Proceedings of the International Joint Conference of CFSA/IFIS/SOFT'95 on Fuzzy Theory and Applications, Taipei, Taiwan, Republic of China, pp. 206-211.

Chen, SM (1996) New method for fuzzy system reliability analysis. Cybernetics and Systems: An International Journal 27: 385-401.

Chen, SM (1997a) Fuzzy system reliability analysis based on vague set theory. Proceedings of the 1997 IEEE International Conference on Systems, Man, and Cybernetics, Orlando, Florida, U. S. A., pp. 1650-1655.

Chen, SM (1997b) Similarity measures between vague sets and between elements. IEEE Transactions on Systems, Man, and Cybernetics-Part B: Cybernetics 27: 153-158.

Chen, SM, Shiau, YS (1998) Vague reasoning and knowledge representation using extended fuzzy Petri nets. Journal of Information Science and Engineering 14: 391-408.

Cheng, CH, Mon, DL (1993) Fuzzy system reliability analysis by interval of confidence. Fuzzy Sets and Systems 56: 29-35.

Gau, WL, Buehrer, DJ (1993) Vague sets. IEEE Transactions on Systems, Man, and Cybernetics 23: 610-614.

Kandel, A (1986) Fuzzy Mathematical Techniques with Applications. Addison-Wesley, Massachusetts, U. S. A.

Kaufmann, A, Gupla, MM (1988) Fuzzy Mathematical Models in Engineering and Management Science. North-Holland, Amsterdam, The Netherlands.

Mon, DL, Cheng, CH (1994) Fuzzy system reliability analysis for components with different membership functions. Fuzzy Sets and Systems 64: 147-157.

Singer, D (1990) A fuzzy set approach to fault tree and reliability analysis. Fuzzy Sets and Systems 34: 145-155.

Song, Q, Chissom, BS (1993a) Fuzzy time series and its models. Fuzzy Sets and Systems 54: 269-277.

Song, Q, Chissom, BS (1993b) Forecasting enrollments with fuzzy time series — Part I. Fuzzy Sets and Systems 54: 1-9.

Song, Q, Chissom, BS (1994) Forecasting enrollments with fuzzy time series — Part II. Fuzzy Sets and Systems 62: 1-8.

Suresh, PV, Babar, AK, Raj, VV (1996) Uncertainty in fault tree analysis: A fuzzy approach. Fuzzy Sets and Systems 83: 135-141.

Utkin, LV, Gurov, SV (1996) A general formal approach for fuzzy reliability analysis in the possibility context. Fuzzy Sets and Systems 83: 203-213.

Wu, HC (2004) Fuzzy reliability estimation using Bayesian approach. Computers & Industrial Engineering 46: 467-493.

Zadeh, LA (1965) Fuzzy sets. Information and Control 8: 338-353.

Zimmermann, HJ (1991) Fuzzy Set Theory and Its Applications. Kluwer Academic Publishers, Boston, U. S. A.

Fuzzy Sets in the Evaluation of Reliability

Olgierd Hryniewicz

Systems Research Institute, Warsaw, Poland

13.1 Introduction

Theory of reliability is more than fifty years old. Its basic concepts were established in the 1950s as useful tools for the analysis of complex technical systems. The rapid development of the theory of reliability was closely related to the importance of its main field of applications - military and space. For this reason the origins of the research in the area of reliability are still not well known. Ralph A. Evans, one of the founders of the IEEE Transactions on Reliability, wrote in an Editorial in this journal that all important theoretical results published in the 1960s and 1970s had been already obtained even in the 1950s, and for many years remained classified. The authors of the most important publications on reliability from those years belonged to the group of the most important scientists working in theory of probability, mathematical statistics, electronics and computer sciences.

When we look at the theory of reliability as the application of a basic mathematical theory, we could see without any doubt that it should be reagarded as one of the most important applications of the theory of probability. All important events which are of interest for the theory and practice of reliability have undoubtedly stochastic character, and all processes that lead to failures can be described by stochastic processes. Therefore, the theory of probability has been for many years used as the only tool for description, prediction and optimization of reliability. As the consequence of applying that approach, mathematical statistics has been used for the analysis of reliability data.

In its initial phase of development, statistical methods used in the area of reliability were based on a classical approach to statistics. Classical concepts of statistics, such as estimators, confidence intervals and tests of hypotheses, that have their interpretations in terms of frequencies, were widely used in the analysis of reliability data. However, together with a continuous improvement of reliability of components and systems these classical methods became not sufficient for practical applications. Therefore, new statistical methods that were based on the Bayesian paradigm found their applications both in theory

O. Hryniewicz: *Fuzzy Sets in the Evaluation of Reliability*, Computational Intelligence in Reliability Engineering (SCI) **40**, 363–386 (2007)
www.springerlink.com   © Springer-Verlag Berlin Heidelberg 2007

and practice of reliability. It is worthy noting that in that time the Bayesian approach to statistics was heavily attacked by the majority of the statistical community. However, practical successes of this approach have resulted nowadays with common acceptance of the Bayesian methodology in the area of reliability.

During the last fifteen years we have witnessed a similar situation in the case of the application of the theory of fuzzy sets in the area of reliability. First, in the early 1980s the quality of components used mainly in the aerospace industry became so high that the probabilities of their failures had the order of magnitude close to 10^{-7} and less. Classical statistical methods of estimation, based on the observation of a random sample, are not applicable in that case. On the other hand, the methods based on the Bayesian approach are usually too complicated to be used in practice. As the result of these difficulties researchers and practitioners working in the area of reliability were able to provide only imprecisely defined values of probabilities of failures. In order to describe those imprecise values of probabilities they proposed to use the theory of fuzzy sets introduced by Lotfi A. Zadeh in the 1960s. Moreover, this new methodology appeared to be very useful in all cases where the information related to reliability were based on imprecise expert opinions, imprecisely reported reliability data, etc. Another impulse for the development of the fuzzy reliability methodology was given in the investigation of complex man-machine systems, and complex multistate systems with imprecise definitions of failures. New methods for the reliability analysis that are based on the theory of fuzzy sets (and the related theory of possibility) and its mixture with the theory of probability have been proposed during last fifteen years, and are now ready for practical applications. An excellent overview of the problems mentioned above can be found in the paper by Cai [5].

The number of papers devoted to the applications of fuzzy sets in the analysis of reliability has become quite large, and it is rather impossible to present a comprehensive review of all of them in one paper. The readers who are interested in a broad introduction to the problem are encouraged to read collections of papers on that topic edited by Onisawa and Kacprzyk [43] and Misra [33]. Therefore, we have decided to give a rather general overview of the main results in this area. In the second section of the paper we consider problems related to the reliability analysis of systems with the usage of imprecise probabilities. In the third section of the paper we present the most important applications of the theory of possibility in the area of reliability. The fourth section is devoted to another very important from a practical point of view problem: statistical analysis of imprecise reliability data in both classical and Bayesian frameworks. Throughout the paper we present only main ideas and results that have been published in a few selected papers. The reader is encouraged, however, to find other related results that have been already published in the papers referenced by the papers that are listed in the bibliography to this paper.

13.2 Evaluation of Reliability in Case of Imprecise Probabilities

The methodology for the evaluation of reliability of a system characterized by binary states of its elements and binary states of the whole system was proposed in the early 1960s. Its detailed description can be found in fundamental books by Barlow and Proschan [1],[2]. We recall now only some basic notions of this theory.

Let $x = (x_1, x_2, ..., x_n)$ be a vector that describes the state of n elements of the system such that

$$x_i = \begin{cases} 1 & \text{if the element } i \text{ is functioning} \\ 0 & \text{if the element } i \text{ is failed} \end{cases}, i = 1, ..., n,$$

and ϕ describes a binary state of the whole system, i.e.

$$\phi = \begin{cases} 1 & \text{if the system is functionining} \\ 0 & \text{if the system is failed} \end{cases}$$

We assume that the state of the whole system is completely determined by the states of its elements, i.e. $\phi = \phi(x_1, x_2, ..., x_n)$. Function $\phi = \phi(x_1, x_2, ..., x_n)$ is called the structure function of the system, or, simply, the structure. It is possible to show that every structure can be expressed by the following general formula:

$$\phi(x) = \sum_y \prod_{j=1}^{n} x_j^{y_j} (1 - x_j)^{1-y_j} \phi(y), \tag{1}$$

where the summation is taken over all n-dimensional binary vectors y $(0^0 \equiv 1)$. Hence, every structure can be expressed as a polynomial of binary functions x_i that describe elements of the system.

Now, let us introduce the following notation:

$$(1_i, x) \equiv (x_1, ..., x_{i-1}, 1, x_{i+1}, ..., x_n)$$
$$(0_i, x) \equiv (x_1, ..., x_{i-1}, 0, x_{i+1}, ..., x_n)$$
$$(*_i, x) \equiv (x_1, ..., x_{i-1}, *, x_{i+1}, ..., x_n)$$

The i-th element of the system is irrelevant if $\phi(1_i, x) = \phi(0_i, x)$ for all $(*_i, x)$; otherwise such element is relevant. The system is called *coherent* if (a) its structure function ϕ is increasing in every component, and (b) all its elements are relevant. For the coherent systems there exist many algorithms for efficient calculations of their reliability defined as the probability that the system is functioning.

One of the fundamental concepts of reliability of systems is the notion of a minimal path. A minimal path is a subset of system's elements such that if all these elements work, the whole system works. A dual concept to the minimal path is that of a minimal cut. A minimal set of system's elements is called a minimal cut if the failures of all its elements cause the failure of the whole system. Suppose that the considered system has n_c minimal cuts, and n_p minimal paths. Denote by C_s, $s \in \{1,...,n_c\}$ a minimal cut of a system, and by P_r, $r \in \{1,...,n_p\}$ its minimal path. According to the fundamental result of Birnbaum et al. [3] the structure of any binary system can be decomposed using either minimal paths or minimal cuts, and the following formula holds:

$$\phi(x_1,...,x_n) = \bigvee_{1 \leq r \leq n_p} \bigwedge_{i \in P_r} x_i = \bigwedge_{1 \leq s \leq n_c} \bigvee_{i \in C_s} x_i . \qquad (2)$$

Therefore, the knowledge of all minimal cuts and/or minimal paths is sufficient for the full reliability description of a system.

Let us now recall basic results that are used in the calculation of reliability of a system. The reliability state of a system X_s and of each of its elements $(X_i, i = 1,...,n)$ is a random variable distributed according to a two-point probability distribution. Let $q_i, i = 1,...,n$ be the reliability of the i-th element of a system, and q_s the reliability of the whole system. Then, the following general expression holds:

$$q_s = E(X_s) = h(q_1, q_2,...,q_n). \qquad (3)$$

When the system is coherent and failures of its elements are statistically independent, then $h(q_1, q_2, ..., q_n)$, technically, is constructed by replacing $x_1, x_2,...,x_n$ in (2) with $q_1, q_2,...,q_n$; next by changing $\wedge$ to a product operator on [0, 1], and $\vee$ to a probabilistic sum on [0, 1], and finally replacing the powers like $q_k^m, m \geq 2$ (if exist) with respective values of q_k. Thus, the knowledge of $\phi(x)$ and the values of $q_1, q_2,...,q_n$, in case of coherent binary structures and independent failures of elements, is fully sufficient for the calculation of the reliability of the whole system.

Reliability analysis of complex systems can be divided into two phases: determination of the structure function and evaluation of the reliabilities of system's elements. The sets of minimal cuts and minimal paths can be obtained using different methods. However, the most efficient, and thus the most frequently used, method is the fault tree analysis. This method was introduced more than forty years ago, and since that time has been successfully used in many areas, such as aerospace industry, nuclear power plants, etc. The method consists in defining a structure of physical events related to failures of system's elements. There exist methods for the extraction of minimal paths and minimal cuts from the information contained in a fault tree when its events are precisely defined. However, it is much more difficult to evaluate probabilities of specific failures, and thus the reliabilities of systems components. In a classical approach to a fault tree analysis it is assumed that all these probabilities are precisely known. However, in many practical situations, especially in case of reliable components, the knowledge of probabilities of failures (or reliabilities) is hardly precise. Even if we use statistical data for the evaluation of those probabilities, we cannot be sure that these data have been obtained in exactly same conditions. Usually, we use data from reliability tests of similar objects conducted in similar conditions, but very often our data come from tests conducted in completely different conditions, e.g. from accelerated life tests. In all these cases there is a need to recalculate the results of reliability tests to the case of the considered system. Such recalculation very often needs opinions of experts, and these opinions are usually expressed in a natural language using vague and imprecise expressions. The formal description of this lack of precision is one of the most important practical problems of reliability analysis. Some researchers claim that the language of the probability theory is the only one that can be used for the description of uncertainty. However, there exist multitude counterexamples that indicate a necessity to apply other approaches. Moreover, the application of the theory of probability for the description of all imprecise information in the case of the reliability analysis of complex systems will make this analysis impossible to do due to an extremely high complexity of necessary computations. Therefore, the theory of fuzzy sets introduced by Lotfi A. Zadeh seems to be much better suited for this purpose.

In this paper we assume that the theory of fuzzy sets gives us tools appropriate for modeling and handling vague data such as imprecisely defined probabilities of failures. In the theory of fuzzy sets all objects of interest (events, numbers, etc.) have associated values of the so called membership function μ. The value of the membership function can be interpreted in different ways depending on the context. In the context of the evaluation of imprecise probabilities the value of the membership function $\mu(p)$ can be interpreted a possibility that the unknown probability adopts the value of p.

Let us now recall some basic notions of the theory of fuzzy sets that will be used in this paper. We start with the definition of a *fuzzy number*.

Definition 1. *The fuzzy subset A of the real line* R, *with the membership function* $\mu_A : R \rightarrow [0,1]$, *is a fuzzy number iff*

(a) *A is normal, i.e. there exists an element x_0, such that $\mu_A(x_0)=1$;*

(b) *A is fuzzy convex, i.e. $\mu_A(\lambda x_1 + (1-\lambda)x_2) \geq \mu_A(x_1) \wedge \mu_A(x_2)$,*

$\forall x_1, x_2 \in R, \quad \forall \lambda \in [0,1]$;

(c) *μ_A is upper semi-continuous;*

(*d*) *supp*(*A*) *is bounded.*

This definition is due to Dubois and Prade (see [13]). It is easily seen from this definition that if *A* is a fuzzy number then its membership function has the following general form:

$$\mu_A(x) = \begin{cases} 0 & \text{for } x < a_1 \\ r_l(x) & \text{for } a_1 \le x < a_2 \\ 1 & \text{for } a_2 \le x \le a_3 \\ r_u(x) & \text{for } a_3 < x \le a_4 \\ 0 & \text{for } x > a_4 \end{cases} \tag{4}$$

where $a_1, a_2, a_3, a_4 \in R$, $a_1 \le a_2 \le a_3 \le a_4$, $r_l : [a_1, a_2] \to [0,1]$ is a non-decreasing upper semi-continuous and $r_u : [a_3, a_4] \to [0,1]$ is a non-increasing upper semi-continuous function. Functions r_l and r_u are called sometimes the left and the right arms (or sides) of the fuzzy number, respectively.

By analogy to classical arithmetic we can add, subtract, multiply and divide fuzzy numbers (for more details we refer the reader to [13] or [35]). In a general case all these operations become rather complicated, especially if the sides of fuzzy numbers are not described by simple functions. Thus, only simple fuzzy numbers - e.g. with linear or piecewise linear sides - are preferred in practice. Such fuzzy numbers with simple membership functions have more natural interpretation. Therefore the most often used fuzzy numbers are trapezoidal fuzzy numbers, i.e. fuzzy numbers whose both sides are linear. Trapezoidal fuzzy numbers can be used for the representation of such expressions as, e.g., "more or less between 6 and 7", "approximately between 12 and 14", etc. Trapezoidal fuzzy numbers with $a_2 = a_3$ are called triangular fuzzy numbers and are often used for modeling such expressions as, e.g., "about 5", "more or less 8", etc. Triangular fuzzy numbers with only one side may be useful for the description of opinions like "just before 50" ($a_2 = a_3 = a_4$) or "just after 30" ($a_1 = a_2 = a_3$). If $a_1 = a_2$ and $a_3 = a_4$ then we get, so called, rectangular fuzzy numbers which may represent such expressions as, e.g., "between 20 and 25". It is easy to notice that rectangular fuzzy numbers are equivalent to well known interval numbers. In a special case of $a_1 = a_2 = a_3 = a_4 = a$ we get a crisp (non-fuzzy) number, i.e. a number which is no longer vague but represents precise value and can be identified with the proper real number *a*.

A useful tool for dealing with fuzzy numbers is the concept of α–cut or α–level set. The α–cut of a fuzzy number *A* is a non-fuzzy set defined as

$$A_\alpha = \{x \in R : \mu(x) \ge \alpha\}. \tag{5}$$

A family $\{A_\alpha : \alpha \in [0,1]\}$ is a set representation of the fuzzy number *A*. Basing on the resolution identity introduced by L. Zadeh, we get:

$$\mu_A(x) = \sup\{\alpha I_{A_\alpha}(x) : \alpha \in [0,1]\},$$
(6)

where $I_{A_\alpha}(x)$ denotes the characteristic function of A_α. From Definition 1 we can see that every α-cut of a fuzzy number is a closed interval. Hence we have $A_\alpha = [A_\alpha^L, A_\alpha^U]$, where

$$A_\alpha^L = \inf\{x \in R : \mu(x) \ge \alpha,\}$$
$$A_\alpha^R = \sup\{x \in R : \mu(x) \ge \alpha.\}$$
(7)

Hence, by (4) we get $A_\alpha^L = r_l^{-1}(\alpha)$, $A_\alpha^U = r_u^{-1}(\alpha)$.

In the analysis of fuzzy numbers and their functions we use the *extension principle* introduced by Zadeh [61], and described by Dubois and Prade [15] as follows:

Definition 2. *Let X be a Cartesian product of universes, $X_1 \times X_2 \times \cdots \times X_r$, and $A_1, A_2,..., A_r$ be r fuzzy sets in $X_1, X_2,..., X_r$, respectively. Let f be a mapping from $X = X_1 \times X_2 \times \cdots \times X_r$ to a universe Y such that $y = f(x_1, x_2,..., x_r)$. The extension principle allows us to induce from r fuzzy sets A_i a fuzzy set B on Y through f such that*

$$\mu_B(y) = \sup_{x_1,x_2,...,x_r:y=f(x_1,x_2,...,x_r)} \min\left[\mu_{A_1}(x_1), \mu_{A_2}(x_2),..., \mu_{A_r}(x_r)\right],$$
(8)

$$\mu_B(y) = 0, \ f^{-1}(y) = \varnothing.$$
(9)

Using the extension principle we can calculate membership functions of fuzzy sets that are defined as functions of other fuzzy sets. In their pioneering work Tanaka et al. [51] used the concept of fuzzy numbers for the description of imprecise probabilities in the context of fault tree analysis. They assumed that probabilities of events of a fault tree are described by the mentioned above trapezoidal fuzzy numbers. In such a case it is easy to show that the fuzzy probability of the failure (or fuzzy reliability) of a whole system is also a fuzzy number, but its membership function does not preserve trapezoidal shape. However, we can use the concept of α-cuts for relatively simple computations.

Let us assume that the reliabilities of systems components are described by fuzzy numbers defined by their α-cuts: $\left(q_{i,L}^\alpha, q_{i,U}^\alpha\right), i = 1,..., n$. Then, the α-cut $\left(q_{s,L}^\alpha, q_{s,U}^\alpha\right)$ for a coherent system can be calculated from (3) as follows:

$$q_{s,L}^\alpha = h\left(q_{1,L}^\alpha, q_{2,L}^\alpha,..., q_{n,L}^\alpha\right),$$
(10)

$$q_{s,U}^\alpha = h\left(q_{1,U}^\alpha, q_{2,U}^\alpha,..., q_{n,U}^\alpha\right).$$
(11)

This relatively simple way of calculations can be used only in the case of a known function $h(*,...,*)$. Formal description of the general procedure for the calculation of fuzzy system reliability can be also found in Wu [58]. However, when the calculations have to be made using directly the information from a fault tree, the methodology proposed in [51] has some drawbacks as it cannot be used for the fault trees with repeated events, and fault trees that contain events and their complementary events at the same tree. These drawbacks have been resolved by Misra and Soman who in [34] proposed a more general methodology for dealing with multistate systems and vectors of dependent fuzzy probabilities.

The general methodology described above is valid for any fuzzy description of fuzzy reliabilities $\tilde{q}_1, \tilde{q}_2, ..., \tilde{q}_n$. However, for practical calculations it is recommended to select several values of α, and to calculate α-cuts of the fuzzy reliability of the system $\tilde{q}_s$ for these values of α. Then the membership function of $\tilde{q}_s$ may be approximated by a piecewise linear function which connects the ends of consecutive α-cuts. More precise results can be obtained if for the description of imprecise probabilities we use the so called L-R fuzzy numbers introduced by Dubois and Prade [14]. For this case Singer [49] has presented recursive formulae that can be used for the calculation of the fuzzy reliability of a system.

Interesting application of fuzzy sets in the analysis of fault trees can be found in the paper by Lin and Wang [31], who considered the problem of estimating fuzzy probabilities of events using imprecise linguistic assessments for human performance and vague events. Fuzzy measures of importance of the elements of a fault tree described by fuzzy probabilities were considered in the paper by Suresh et al. [50]. Practical example of the fault tree analysis with fuzzy failure rates can be found in the paper by Huang et al. [24].

The general approach presented in this section can be used for solving any well defined problem of reliability analysis with imprecisely defined parameters. For example, Cheng [9] used fuzzy sets to describe reliability of repairable systems using a fuzzy GERT methodology. In all such cases the extension principle and the concept of α-cuts is quite sufficient for making necessary computations. However, if in these computations non-monotonic functions are involved, then it may be necessary to solve non-linear programming problems in order to arrive at required solutions.

13.3 Possibilistic Approach to the Evaluation of Reliability

In the previous section we have described the results of research in the area of system reliability for the case of imprecise (linguistic) description of probabilities of failures (or probabilities of survival, i.e. reliabilities). In all these papers life times were assumed to have probabilistic nature, but their distributions were imprecisely defined, resulting with imprecise probabilities of failures. Imprecise values in these models were described by fuzzy sets, and this description was often interpreted in terms of the theory of possibility introduced by L. A. Zadeh [62].

Zadeh [62] introduced the notion of possibility for the description of vaguely defined events whose interpretation in terms of probabilities is at least questionable. He introduced the notion of the possibility distribution, and showed that it can be formally described by fuzzy sets. This theory was further developed by many authors in the framework of the theory of fuzzy sets, and in the late 1980s found its applications in the area of reliability. The distinctive feature of the theory of possibility is not the way it describes vaguely defined concepts, but how it is used for merging uncertainties of possibilistic nature. In this respect it is basically different from the theory of probability, as it is not additive, and is governed by fuzzy logic.

For the readers who are not familiar with fuzzy logic we recall now two its most important features. Suppose we have two fuzzy sets $\tilde{A}$ and $\tilde{B}$ described by the membership functions $\mu_A(x)$ and $\mu_B(x)$, respectively. Then, the membership function of the logical sum (union) of $\tilde{A}$ and $\tilde{B}$ is given by

$$\mu_{A \cup B}(x) = \max[\mu_A(x), \mu_B(x)], \tag{12}$$

and the membership function of the logical product (intersection) of $\tilde{A}$ and $\tilde{B}$ is given by

$$\mu_{A \cap B}(x) = \min[\mu_A(x), \mu_B(x)]. \tag{13}$$

Thus, possibility measures are rather "maxitive" in contrast to the "additivity" of their probabilistic counterparts.

Possibilistic approach to reliability was introduced in works of Cai and his collaborators (for references see [4], [6], [5]) and Onisawa(see [41], [42]). Cai in his papers has given practical examples which let him conclude that in many cases life times have no probabilistic meaning but should be described by possibilistic (fuzzy) variables. The rationale behind that reasoning was the following: in many cases failures such as, e.g. software failures, cannot happen more than once. In such cases, Cai claims, probabilistic approach with its interpretation in terms of frequencies is not appropriate. Thus, times to such singular failures should be rather described by possibility distributions than by probability distributions. Introduction of possibilistic models of reliability from a purely mathematical point of view can be found in [8] and [12].

The agreement to possibilistic assumptions has many far reaching consequences for the analysis of system reliability. Let us define the system (or its component) life time X as a *fuzzy variable* [6]:

$$X = u : \mathsf{U}_X = u : \pi_X(u), u \in \mathsf{R}^+ = [0, +\infty), \tag{14}$$

where $\pi_X(u)$ is the possibility distribution of X. In such a case the possibilistic reliability ("posbist " reliability in Cai's terminology) is defined as the possibility that for given conditions the system performs its assigned functions, and is calculated from the following formula [6]:

$$R(t) = \sigma(X > t) = \sup_{u > t} \mathsf{U}_X(u), \tag{15}$$

where σ is a possibility measure.

Now, let us present two important theorems (formal definitions of some concepts used in these theorems are given in [6]).

Theorem 1. *(Cai et al. [6]) Suppose a series system has two components. Let X_1, X_2 be the component lifetimes, respectively. Further we assume X_1, X_2 are both normalized unrelated fuzzy variables, defined on (Γ, G, σ), with continuous possibility distribution functions, and induce strictly convex fuzzy sets, $X_1 = u : \mathsf{U}_{X_1}(u)$, $X_2 = u : \mathsf{U}_{X_2}(u)$. Let X be the system lifetime. Then there exists a unique pair (a_1, a_2), $a_1, a_2 \in \mathsf{R}^+$, such that the possibility distribution function of X, denoted by $\mathsf{U}_X(x)$, is given by*

$$\mathsf{U}_X(x) = \begin{cases} \max\left(\mathsf{U}_{X_1}(u), X_2 = u : \mathsf{U}_{X_2}(u)\right) & \text{if } x \leq a_1 \leq a_2 \\ \mathsf{U}_{X_1}(u) & \text{if } a_1 < x \leq a_2 \\ \min\left(\mathsf{U}_{X_1}(u), X_2 = u : \mathsf{U}_{X_2}(u)\right) & \text{if } a_1 \leq a_2 < x \end{cases} \tag{16}$$

Theorem 2. *(Cai et al. [6]) Suppose a parallel system has two components. Let X_1, X_2 be the component lifetimes, respectively. Further we assume X_1, X_2 are both normalized unrelated fuzzy variables, defined on (Γ, G, σ), with continuous possibility distribution functions, and induce strictly convex fuzzy sets, $X_1 = u : \mathsf{U}_{X_1}(u)$, $X_2 = u : \mathsf{U}_{X_2}(u)$. Let X be the system lifetime. Then there exists a unique pair (a_1, a_2), $a_1, a_2 \in \mathsf{R}^+$, such that the possibility distribution function of X, denoted by $\mathsf{U}_X(x)$, is given by*

$$U_X(x) = \begin{cases} \min\left(U_{X_1}(u), X_2 = u : U_{X_2}(u)\right) & \text{if } x \le a_1 \le a_2 \\ U_{X_1}(u) & \text{if } a_1 < x \le a_2 \\ \max\left(U_{X_1}(u), X_2 = u : U_{X_2}(u)\right) & \text{if } a_1 \le a_2 < x \end{cases} \qquad (17)$$

Similar results have been also given in [6] for other reliability systems like a k-out-of-n system, and for the most general case of a binary coherent system.

The consequences of both theorems (and their extensions) are somewhat strange. Cai et al. [6] already noticed: "the reliability of a parallel system with an arbitrary number of unrelated components coincides with the reliability of a series system with another arbitrary number of unrelated components, provided that all of the components contained in the systems are identical". This feature, in our opinion, indicates that the notion of the possibilistic reliability of systems should be used very cautiously.

In the possibilistic model described above it has been assumed that reliability states of the system and its components are binary. However, in many real cases, especially for large and complex systems, this assumption is not true. In the classical (probabilistic) theory of reliability the notion of "multistate systems" is used in order to cope with this problem. Unfortunately, the existing reliability data is usually not sufficient for the proper identification of such systems. Moreover, for multistate components and systems it is usually very difficult to define precisely the failures, especially in the case of failures made by human (operator) errors. Therefore, some researchers proposed to use fuzzy sets for the description of vaguely defined failures.

The importance of the problem of vaguely defined failures was recognized for the first time in the papers by Nowakowski [39], Nishiwaki [38], Nishiwaki and Onisawa [44], and Onisawa [40], [42] devoted to the problem of reliability analysis of man-machine systems. Interesting approach to that problem, both from probabilistic and fuzzy point of view, was also proposed by Rotshtein [48]. Similar problems have been also noticed in the analysis of fault trees constructed for complex systems. Fault trees, or more general event trees, are used for the description of the relationships between physical states of a system and its reliability states. In the classical case of binary systems this relationship is well defined, and described using logical gates AND, OR, and NOT. However, in many practical cases we do not have enough information to establish sure links between particular physical states of a system and its particular failures.

Different approaches have been used to model imprecise relationships between physical and reliability states of a system. Pan and Yun [45] proposed to use fuzzy gates with outputs described by triangular fuzzy numbers instead of crisp values 0 or 1. Another generalization of fault tree gates was proposed by Onisawa (see [42]) who considered parametric operations called Dombi t-norm and Dombi t-conorm instead of AND and OR operators, respectively.

Full application of the theory of possibility in the analysis of fault trees has been proposed by Nahman [37] and Huang et al. [25] who used possibility measures for the description of transition between states of a fault tree, and fuzzy logic for the description of its gates.

One of the most challenging problems of the reliability of complex systems is the multistate nature of their behaviour. Structure function describing the behaviour of systems composed of multistate elements could be extremely difficult to find and very often even impossible to be precisely identified. An attempt to describe such complex situation with the usage of fuzzy sets has been proposed by Montero et al. [36] and Cutello et al. [11].

Possibilistic approach to reliability has been also used for the analysis of repairable systems. Utkin and Gurov [52], [53] presented a mathematical model for the description of exploitation processes of systems using functional equations that describe transition processes between different states of a system. In a probability context these equations describe a stochastic process of the random behaviour of the system. However, the same equations can be used for that description in the possibilistic context. The resulting formulae look very awkwardly, but rather surprisingly they are easier to solve.

13.4 Statistical Inference with Imprecise Reliability Data

13.4.1 Fuzzy Estimation of Reliability Characteristics

In the previous sections we have assumed that all probabilities, crisp or fuzzy, that are necessary for the computations of reliability are known. However, in practice they have to be estimated from statistical data. One of the most important problems of reliability analysis is the estimation of the *mean life time* of the item under study (system or component). In technical applications this parameter is also called *mean time to failure* (*MTTF*) and is often included in a technical specification of a product. For example, producers are interested in whether this time is sufficiently large, as large *MTTF* allows them to extend a warranty time. Classical estimators require precise data obtained from strictly controlled reliability tests (for example, those performed by a producer at his laboratory). In such a case a failure should be precisely defined, and all tested items should be continuously monitored. However, in real situation these requirements might not be fulfilled. In the extreme case, the reliability data come from users whose reports are expressed in a vague way. The vagueness of the data has many different sources: it might be caused by subjective and imprecise perception of failures by a user, by imprecise records of reliability data, by imprecise records of the rate of usage, etc. The discussion concerning different sources of vagueness of reliability data can be found in Grzegor-zewski and Hryniewicz [18]. Therefore we require different tools appropriate for modeling vague data and suitable statistical methodology to handle these data as well.

To cope with the formal description of data that are both random and im-precise (fuzzy) it is convenient to use the notion of a fuzzy random variable. It was introduced by Kwakernaak [30]. There exist also definitions of fuzzy random variables that have been proposed by other authors, for example by Kruse [27] or by Puri and Ralescu [47]. The definition, we present below, was proposed in [19], and is similar to those of Kwakernaak and Kruse (see [17]). Suppose that a random experiment is described as usual by a probability space (Ω, A, P), where Ω is a set of all possible outcomes of the experiment, A is a σ-algebra of subsets of Ω (the set of all possible events) and P is a probability measure.

Definition 3. *A mapping* $X : \Omega \to \mathsf{FN}$ *is called a fuzzy random variable if it satisfies the following properties:*

a) $\{X_\alpha(\omega) : \alpha \in [0,1]\}$ *is a set representation of* $X(\omega)$ *for all* $\omega \in \Omega$,

b) *for each* $\alpha \in [0,1]$ *both* $X_\alpha^L = X_\alpha^L(\omega) = \inf X_\alpha(\omega)$ *and*

$X_\alpha^U = X_\alpha^U(\omega) = \sup X_\alpha(\omega)$, *are usual real-valued random variables on* (Ω, A, P).

Thus a fuzzy random variable X is considered as a perception of an un-known usual random variable $V : \Omega \to \mathsf{R}$, called an original of X. Let V de-note a set of all possible originals of X. If only vague data are available, it is of course impossible to show which of the possible originals is the true one. Therefore, we can define a fuzzy set on V, with a membership function $\iota : V \to [0,1]$ given as follows:

$$\iota(V) = \inf\{\mu_{X(\omega)}(V(\omega)) : \omega \in \Omega\}, \tag{18}$$

which corresponds to the grade of acceptability that a fixed random variable V is the original of the fuzzy random variable in question (see Kruse and Meyer [28]).

Similarly n-dimensional fuzzy random sample $X_1, X_2, ..., X_n$ may be con-sidered as a fuzzy perception of the usual random sample $V_1, V_2, ..., V_n$ (where $V_1, V_2, ..., V_n$ are independent and identically distributed crisp random vari-ables). A set V^n of all possible originals of that fuzzy random sample is, in fact, a fuzzy set with a membership function

$$\iota(V_1, ..., V_n) = \min_{i=1,...,n} \inf\{\mu_{X_i(\omega)}(V_i(\omega)) : \omega \in \Omega\}. \tag{19}$$

Random variables are completely characterized by their probability distributions. However, in many practical cases we are interested only in some parameters of a probability distribution, such as expected value or standard deviation. Let $\theta = \theta(V)$ be a parameter of a random variable V. This parameter may be viewed as an image of a mapping $\Gamma : P \to R$, which assigns each random variable V having distribution $P_\theta \in P$ the considered parameter θ, where $P = \{P_\theta : \theta \in \Theta\}$ is a family of distributions. However, in case of fuzzy random variables we cannot observe parameter θ but only its vague image. Using this reasoning together with Zadeh's extension principle Kruse and Meyer [28] introduced the notion of a fuzzy parameter of a fuzzy random variable which may be considered as a fuzzy perception of the unknown parameter θ. It is defined as a fuzzy set with the following membership function:

$$\mu_{\Lambda(\theta)}(t) = \sup\{\iota(V): V \in \mathsf{V}, \theta(V) = t\}, \quad t \in \mathsf{R}, \qquad (20)$$

where $\iota(V)$ is given by (18). This notion is well defined because if our data are crisp, i.e. $X = V$, we get $\Lambda(\theta) = \theta$. Similarly, for a random sample of size n we get

$$\mu_{\Lambda(\theta)}(t) = \sup\{\iota(V_1,...,V_n): (V_1,...,V_n) \in \mathsf{V}^n, \theta(V_1) = t\} \quad t \in \mathsf{R}. \qquad (21)$$

One can easily obtain α-cuts of $\Lambda(\theta)$:

$$\Lambda_\alpha(\theta) = \{t \in R : \exists (V_1,...,V_n) \in \mathsf{V}^n, \theta(V_1) = t, \text{ such that} \qquad (22)$$
$$V_i(\omega) \in (X_i(\omega))_\alpha \text{ for } \omega \in \Omega \text{ and for } i = 1,...,n \}.$$

For more information we refer the reader to Kruse, Meyer [28].

First papers devoted to the analysis of fuzzy reliability data did not use explicitly the concept of a fuzzy random variable. Pioneering works in this field can be attributed to Viertl [54],[55], who found appropriate formulae for important reliability characteristics by fuzzifying formulae well known from classical statistics of reliability data. The results of those and other works have been presented in the paper by Viertl and Gurker [56], who considered such problems as estimation of the mean life-time, estimation of the reliability function, and estimation in the accelerated life testing (with a fuzzy acceleration factor). Original approach has been proposed in Hryniewicz [21] who did not model fuzzy time to failures, but fuzzy survival times. In his models only the right-hand side of the fuzzy numbers has been considered, but this approach let him consider in a one mathematical model such phenomena like censored life times and partial failures.

One of the first attempts to propose a comprehensive mathematical model of fuzzy life times as fuzzy random variables was given in Grzegorzewski and Hryniewicz [18]. Grzegorzewski and Hryniewicz considered the case of exponentially distributed fuzzy life time data, and proposed the methodology for point estimation, interval estimation, and statistical hypothesis testing for the fuzzy mean life time. These results have been further extended in [19] where they also considered the case of vague censoring times and vague failures. In the case of vague failures the number of failures observed during the life time test is also fuzzy. The methodology for the description of a fuzzy number of failures in the context of the life time estimation was considered in [16].

Let us now present a summary of the results given in [19]. To begin with, let us recall some basic results from a classical theory of the statistical analysis of life time data. The mean lifetime may be efficiently estimated by the sample average from the sample of the times to failure $W_1, \ldots, W_n$ of n tested items, i.e.

$$MTTF = (W_1 + \cdots + W_n)/n.$$
(23)

However, in the majority of practical cases the lifetimes of all tested items are not known, as the test is usually terminated before the failure of all items. It means that exact lifetimes are known for only a portion of the items under study, while remaining life times are known only to exceed certain values. This feature of lifetime data is called censoring. More formally, a fixed censoring time $Z_i > 0$, $i = 1, \ldots, n$ is associated with each item. We observe W_i only if $W_i \geq Z_i$. Therefore our lifetime data consist of pairs $(T_1, Y_1), \ldots, (T_n, Y_n)$, where

$$T_i = \min\{W_i, Z_i\},$$
(24)

$$Y_i = \begin{cases} 1 & \text{if } W_i = T_i \\ 0 & \text{if } W_i = Z_i \end{cases}.$$
(25)

There are many probability distributions that are used in the lifetime data analysis. In [19] the exponential distribution has been used for modeling the lifetime T. The probability density function in this case is given by

$$f(t) = \begin{cases} \dfrac{1}{\theta} e^{-t/\theta} & \text{if } t > 0 \\ 0 & \text{if } t \leq 0 \end{cases}.$$
(26)

where $\theta > 0$ is the mean lifetime. Let

$$T = \sum_{i=1}^{n} T_i = \sum_{i \in D} W_i + \sum_{i \in C} Z_i$$
(27)

be the total survival time (sometimes called a total time on test), where D and C denote the sets of items for whom exact life times and censoring times are observed, respectively. Moreover, let

$$r = \sum_{i=1}^{n} Y_i \qquad (28)$$

be the number of observed failures. In the considered exponential model the statistic (r, T) is minimally sufficient statistic for θ and the maximum likelihood estimator of the mean lifetime θ is (assuming $r > 0$)

$$\hat{\theta} = T / r. \qquad (29)$$

Now suppose that the life times (times to failure) and censoring times may be imprecisely reported. In case of precisely known failures we assume that the values of the indicators $Y_1, Y_2, ..., Y_n$ defined above are either equal to 0 or equal to 1, i.e. in every case we know if the test has been terminated by censoring or as a result of failure. In order to describe the vagueness of life data we use the previously defined notion of a fuzzy number.

Now we consider fuzzy life times $\tilde{T}_1, \tilde{T}_2, ..., \tilde{T}_n$ described by their membership functions $\mu_1(t), ..., \mu_n(t) \in$ NFN . Thus applying the extension principle to (27) we get a fuzzy total survival lifetime $\hat{T}$ (which is also a fuzzy number)

$$\tilde{T} = \sum_{i=1}^{n} \tilde{T}_i \qquad (30)$$

with the membership function

$$\mu_{\tilde{T}}(t) = \sup_{t_1, ..., t_n \in R^+ : t_1 + \cdots + t_n = t} \{ \mu_1(t_1) \wedge ... \wedge \mu_n(t_n) \} . \qquad (31)$$

Using operations on α–cuts we may find a set representation of $\tilde{T}$ given as follows

$$\tilde{T}_\alpha = (T_1)_\alpha + \cdots + (T_n)_\alpha \qquad (32)$$
$$= \left\{ t \in R^+ : t = t_1 + \cdots + t_n, \text{where } t_i \in (T_i)_\alpha, i = 1, ..., n \right\}, \quad \alpha \in [0,1].$$

In the special case of trapezoidal fuzzy numbers that describe both life times and censoring times the total time on test calculated according to (30) is also trapezoidal.

If the number of observed failures r is known we can use the extension principle once more, and define a fuzzy estimator of the mean lifetime $\tilde{\theta}$ in the presence of vague life times as

$$\tilde{\theta} = \tilde{T} / r. \qquad (33)$$

Since $r \in$ N we can easily find the following set representation of $\tilde{\theta}$:

$$\tilde{\theta}_\alpha = \left\{ t \in R^+ : t = \frac{x}{r}, \text{where } x \in \tilde{T}_\alpha \right\}. \tag{34}$$

For more details and the discussion on fuzzy confidence intervals we refer the reader to [18].

However, in many practical situations the number of failures r cannot be precisely defined. Especially in case of non-critical failures the lifetime data may not be reported in a precise way. In order to take into account such non-critical failures Grzegorzewski and Hryniewicz [19] consider the state of each observed item at the time Z_i. Let G denote a set of all items which are functioning at their censoring times Z_i. Therefore we can assign to each item $i = 1,...,n$ its degree of belongingness $g_i = \mu_G(i)$ to G, where $g_i \in [0,1]$. When the item hasn't failed before the censoring time Z_i, i.e. it works perfectly at Z_i, we set $g_i = 1$. On the other hand, if a precisely defined failure has occurred before or exactly at time moment Z_i, we set $g_i = 0$. If $g_i \in (0,1)$ then the item under study neither works perfectly nor is completely failed. This situation we may consider as a partial failure of the considered item. Let us notice that in the described above case G can be considered as a fuzzy set with a finite support.

There are different ways to define the values of g_i depending upon considered applications. However, in the majority of practical situations we may describe partial failures linguistically using such notions as, e.g. "slightly possible", "highly possible", "nearly sure", etc. In such a case we may assign arbitrary weights $g_i \in (0,1)$ to such imprecise expressions. Alternatively, one can consider a set D of faulty items and, in the simplest case, the degree of belongingness to D is equal $d_i = \mu_D(i) = 1 - g_i$. Further on we'll call g_i and d_i as degrees of the up state and down state, respectively. Having observed the degrees of down states it is possible to count the number of failures with the degrees of down states exceeding certain rejection limit. Hence, we get a following (fuzzy) number of failures:

$$\tilde{r}_{opt}^f = |D|_f, \tag{35}$$

where $|D|_f$ denotes fuzzy cardinality of fuzzy set D. We may also start from up states. Therefore

$$\tilde{r}_{pes}^f = n - |G|_f, \tag{36}$$

where $|G|_f$ denotes fuzzy cardinality of fuzzy set G. However, contrary to the crisp counting $|D|_f \neq n - |G|_f$. It is seen that such fuzzy number of observed failures is a finite fuzzy set. Moreover, if we assume that at least one crisp failure is observed it is also a normal fuzzy set.

Using the extension principle, we may define a fuzzy estimator of the mean life-time $\tilde{\tilde{\theta}}$ in the presence of fuzzy life times and vague number of failures. Namely, for crisp failure counting methods we get the following formula

$$\tilde{\tilde{\theta}} = \tilde{T} / \tilde{r}, \tag{37}$$

where $\tilde{T}$ is the fuzzy total survival time and $\tilde{r}$ denotes the number of vaguely defined failures. Actually (37) provides a family of estimators that depend on the choice of $\tilde{r}$. However, in the case of a fuzzy failure number we have

$$\tilde{\tilde{\theta}} = \tilde{T} / conv(\tilde{r}), \tag{38}$$

where $conv(\tilde{r})$ is the convex hull of the fuzzy set $\tilde{r}$, and is defined as follows

$$conv(\tilde{r}) = \inf\{A \in \mathsf{NFN} : \tilde{r} \subseteq A\}. \tag{39}$$

Since now the denominator of (38) is a fuzzy number, our estimator of the mean life time is a fuzzy number whose membership function can be calculated using the extension principle.

First look at the results presented above which gives impression that even in the simplest case of the estimation of the mean life time for the exponential distribution the analysis of fuzzy data is not simple. It becomes much more complicated in the case of other life time distributions, such as the Weibull distribution, and in the case of such characteristics like the reliability function. Further complications will be encountered if we have to evaluate the reliability of a system using fuzzy data obtained for its components. In these and similar cases there is an urgent need to find approximate solutions that will be useful for practitioners. An example of such attempt can be found in the paper by Hryniewicz [23]. In this paper Hryniewicz considers the problem of the estimation of reliability of a coherent system $R_s(t)$ consisted of independent components having exponentially distributed life times when available observed life times for components are fuzzy. He assumes that observed life times (and censoring times) of individual components are described by trapezoidal fuzzy numbers. Then, he finds the membership function for the probability of failure

$$P(t) = 1 - e^{-t/\theta}, t > 0. \tag{40}$$

The obtained formulae are too complicated for the further usage in the calculation of the reliability of complex systems. Therefore, Hryniewicz [23] proposes to approximate fuzzy total time on test by shadowed sets introduced by

Pedrycz [46] who proposes to approximate a fuzzy number by a set defined by four parameters: a_1, a_2, a_3, a_4. The interpretation of the shadowed set is the following: for values of the fuzzy number that are smaller than a_1 and greater than a_4 the value of the membership function is reduced to zero, in the interval (a_2, a_3) this value is elevated to 1, and in the remaining intervals, i.e. (a_1, a_2) and (a_3, a_4) the value of the membership function is not defined. It is easy to see that all arithmetic operations on so defined shadowed sets are simple operations on intervals, and their result is also a shadowed set. Thus, calculation of imprecise reliability of a system using (3) is quite straightforward.

13.4.2 Fuzzy Bayes Estimation of Reliability Characteristics

In statistical analysis of reliability data the amount of information from life tests and field data is usually not sufficient for precise evaluation of reliability. Therefore, there is a need to merge existing information from different sources in order to obtain plausible results. Bayesian methods, such as Bayes estimators and Bayes statistical tests, provide a mathematical framework for processing information of a different kind. Thus, they are frequently used in the reliability analysis, especially in such fields as reliability and safety analysis of nuclear power plants and reliability evaluation of the products of an aerospace industry. There are two main sources of imprecise information in the Bayesian approach to reliability. First source is related to imprecise reliability data, and second is connected with imprecise formulation of prior information. First papers on the application of fuzzy methodology in the Bayesian analysis of reliability can be traced to the middle of 1980s. For example, Hryniewicz [20] used the concept of a fuzzy set to model the prior distribution of the failure risk in the Bayes estimation of reliability characteristics in the exponential model. He proposed a method for building a membership function using experts' opinions. However, in his model the membership function is interpreted as a kind of an improper prior probability distribution. Thus he finally arrived at non-fuzzy Bayes point estimators. At the same time Viertl (see [56] and [57]) used fuzzy numbers in order to model imprecise life times in the context of Bayes estimators.

Despite the significant progress in the development of fuzzy Bayesian methodology important practical results in reliability applications have been published only recently. Wu [59] considered Bayes estimators of different reliability characteristics. For example, he found Bayes estimators of the survival probability (reliability) using the results of binomial sampling and Pascal sampling experiments. In the binomial sampling experiment n items are tested, and the number of survivors x is recorded. Re-parameterized beta distribution is then used for the description of the prior information about the estimated survival probability (reliability) q. The parameters of the prior distribution have the following interpretation: n_0 is a "pseudo" sample size, and x_0 is a "pseudo" number of survivors in an imaginary experiment whose results subsume our prior information about q. Then, the Bayes point estimator of q is given by

$$\hat{q}_B = (x + x_0)/(n + n_0).$$
(41)

Wu [59] considers now the situation when the parameter x_0 is known imprecisely, and is described by a fuzzy number. Straightforward application of the extension principle leads to formulae for the limits of α-cuts of $\hat{q}_B$:

$$\hat{q}_{B,L}^{\alpha} = (x + x_{0,L}^{\alpha})/(n + n_0)$$
(42)

and

$$\hat{q}_{B,U}^{\alpha} = (x + x_{0,U}^{\alpha})/(n + n_0).$$
(43)

In the case of Pascal sampling the number of failures s is fixed, and the number of tested items N is a random variable. The parameters of the prior distribution of q have the same interpretation as in the case of binomial sampling. Then, the Bayes point estimator of q is given by

$$\hat{q}_B = (n + x_0 - s)/(n + n_0),$$
(44)

where n is the observed value of N. When x_0 is known imprecisely, and described by a fuzzy number, the limits of α-cuts of $\hat{q}_B$ are given by [59]:

$$\hat{q}_{B,L}^{\alpha} = (n - s + x_{0,L}^{\alpha})/(n + n_0)$$
(45)

and

$$\hat{q}_{B,U}^{\alpha} = (n - s + x_{0,U}^{\alpha})/(n + n_0).$$
(46)

Wu [59] presents also Bayes estimators for the failure rate λ and reliability function $e^{-\lambda/t}$ in the exponential model. In his paper Wu [59] also proposes an algorithm for the calculation of the membership value $\mu(q)$ of $\hat{q}$.

The Bayes estimator of λ has been independently investigated by Hryniewicz [22] who considered the case of the crisp number of observed failures d, the fuzzy total time on test $\tilde{T}$, and gamma prior distribution of λ re-parameterized in such a way that one of its parameters (scale) had the interpretation either of the expected value of λ, denoted by E_λ, or its mode, denoted by D_λ. He also assumed that the shape parameter δ of the prior gamma distribution is known, but the values of $E_\lambda (D_\lambda)$ are fuzzy. Now the fuzzy Bayes estimators of λ are given by the following formulae [22]:

$$\hat{\lambda}_E = (d + \delta)/(\tilde{T} + \delta / \tilde{E}_\lambda),\tag{47}$$

and

$$\hat{\lambda}_D = (d + \delta)/(\tilde{T} + (\delta - 1)/\tilde{D}_\lambda), \delta > 1.\tag{48}$$

The α-cuts for these estimators can be calculated straightforwardly using the extension principle.

In his recent paper Wu [60] considers the case of Bayes estimators for the reliability of series, parallel, and k-out-of-n reliability systems in the case of the available results or reliability tests conducted according to the binomial sampling scheme. By applying the Mellin transform he finds the posterior distribution for the system reliability, and then fuzzifies its expected value arriving at the fuzzy Bayes point estimators.

13.5 Conclusions

Evaluation of reliability of complex systems seems to be much more difficult than it appeared to be even twenty years ago. At that time probabilistic models developed by mathematicians and statisticians were offered with the aim to solve all important problems. However, reliability practitioners asked questions that could have not been successfully answered using the probabilistic paradigm. The usage of fuzzy sets in the description of reliability of complex systems opened areas of research in that field. This work has not been completed yet. In this paper we have presented only some results that seem to be important both from a theoretical and practical point of view. We focused our attention on probabilistic—possibilistic models whose aim is to combine probabilistic uncertainty (risk) with possibilistic lack of precision (vagueness). We believe that this approach is the most promising for solving complex practical problems. It has to be stressed, however, that we have not presented all the applications of fuzzy sets to reliability. For example, we have not presented interesting applications of fuzzy sets for the strength—stress reliability analysis or for a more general problem of the reliability analysis of structural systems. The readers are encouraged to look for the references to papers devoted to these problems in the papers by Jiang and Chen [26] and Liu et al. [32]. Another important problem that has not been considered in this paper is the construction of possibility measures from the information given by experts. Interesting practical example of the application of fuzzy "IF-THEN" rules for the solution of this problem has been presented by Cizelj et al. [10]. To sum up the presentation of the application of fuzzy sets in reliability we have to conclude that the problem of the appropriate description and analysis of complex reliability systems is still far from being solved.

References

1. Barlow, R., Proschan, F.: Mathematical theory of reliability, J.Wiley, New York, 1965
2. Barlow, R., Proschan, F.: Statistical theory of reliability and life testing. Probability models, Holt, Rinehart and Winston, Inc., New York, 1975
3. Birnbaum, Z.W., Esary, J.D., Saunders, S.C.: Multicomponent systems and structures, and their reliability. Technometrics **3** (1961), 55—77.
4. Cai, K.Y.: Fuzzy reliability theories. Fuzzy Sets and Systems **40** (1991), 510—511
5. Cai, K.Y.: System failure engineering and fuzzy methodology. An introductory overview. Fuzzy Sets and Systems **83** (1996), 113—133
6. Cai, K-Y., Wen C.Y., Zhang, M.L.: Fuzzy variables as a basis for a theory of fuzzy reliability in the possibility context. Fuzzy Sets and Systems **42** (1991), 145—172
7. Cai, K-Y., Wen C.Y., Zhang, M.L.: Coherent systems in profust reliability theory. In: T. Onisawa, J. Kacprzyk (Eds.), Reliability and Safety Analyses under Fuzziness. Physica-Verlag, Heidelberg, 1995, 81—94.
8. Cappelle, B., Kerre, E.E.: Issues in possibilistic reliability theory. In: T. Onisawa, J. Kacprzyk (Eds.), Reliability and Safety Analyses under Fuzziness. Physica-Verlag, Heidelberg, 1995, 61-80.
9. Cheng, C.H.: Fuzzy repairable reliability based on fuzzy GERT. Microelectronics and Reliability **36** (1096), 1557—1563.
10. Cizelj, R.J., Mavko, B., Kljenak, I.: Component reliability assessment using quantitative and qualitative data. Reliability Engineering and System Safety **71** (2001), 81—95.
11. Cutello, V., Montero J., Yanez J.: Structure functions with fuzzy states. Fuzzy Sets and Systems **83** (1996), 189—202.
12. de Cooman, G.: On modeling possibilistic uncertainty in two-state reliability theory. Fuzzy Sets and Systems **83** (1996), 215—238
13. Dubois, D., Prade, H.: Operations on Fuzzy Numbers. Int. J. Syst. Sci. **9** (1978), 613—626.
14. Dubois, D., Prade, H.: Fuzzy real algebra. Some results. Fuzzy Sets and Systems **2** (1979), 327—348.
15. Dubois, D., Prade, H.: Fuzzy Sets and Systems. Theory and Applications. Academic Press, New York, 1980.
16. Grzegorzewski, P.: Estimation of the mean lifetime from vague data. Proceedings of the International Conference in Fuzzy Logic and Technology Eusflat 2001, Leicester, 2001, 348—351.
17. Grzegorzewski, P., Hryniewicz, O.: Testing hypotheses in fuzzy environment. Mathware and Soft Computing **4** (1997), 203—217.
18. Grzegorzewski, P., Hryniewicz, O.: Lifetime tests for vague data, In: L.A. Zadeh, J. Kacprzyk (Eds.) Computing with Words in Information/Intelligent Systems, Part 2. Physica-Verlag, Heidelberg, 1999, 176—193.
19. Grzegorzewski, P., Hryniewicz, O.: Computing with words and life data. Int. Journ. of Appl. Math. and Comp. Sci. **13** (2002), 337—345.

20. Hryniewicz, O.: Estimation of life-time with fuzzy prior information: application in reliability. In: J.Kacprzyk, M.Fedrizzi (Eds.), Combining Fuzzy Imprecision with Probabilistic Uncertainty in Decision Making. Lecture Notes in Economics and Mathematical Systems. Springer-Verlag, Berlin, 1988, 307—321.

21. Hryniewicz, O.: Lifetime tests for imprecise data and fuzzy reliability requirements, In: T. Onisawa, J. Kacprzyk (Eds.), Reliability and Safety Analyses under Fuzziness. Physica-Verlag, Heidelberg, 1995, 169-179.

22. Hryniewicz, O.: Bayes life-time tests with imprecise input information. Risk Decision and Policy **8** (2003), 1—10.

23. Hryniewicz, O.: Evaluation of reliability using shadowed sets and fuzzy lifetime data. In: K.Kolowrocki (Ed.), Advances in Safety and Reliability, vol.1. A.A.Balkema Publ., Leiden, 2005, 881—886.

24. Huang, D., Chen, T., Wang, M.J.: A fuzzy set approach for event tree analysis. Fuzzy Sets and Systems **118** (2001), 153—165.

25. Huang, H.Z., Tong, X., Zuo, M.J.: Posbist fault tree analysis of coherent systems. Reliability Engineering and System Safety **84** (2004), 141—148.

26. Jiang, Q., Chen, C.H.: A numerical algorithm of fuzzy reliability. Reliability Engineering and System Safety 80 (2003), 299—307.

27. Kruse, R.: The strong law of large numbers for fuzzy random variables. Inform. Sci. **28** (1982), 233-241.

28. Kruse, R., Meyer, K.D.: Statistics with Vague Data. D. Riedel Publishing Company, 1987.

29. Kruse, R., Meyer K.D.: Confidence intervals for the parameters of a linguistic random variable, In: J.Kacprzyk , M.Fedrizzi (Eds.), Combining Fuzzy Imprecision with Probabilistic Uncertainty in Decision Making. Springer-Verlag, Heidelberg, 1988, 113-123.

30. Kwakernaak, H.: Fuzzy random variables, Part I: Definitions and theorems, Inform. Sci. **15** (1978), 1-15; Part II: Algorithms and examples for the discrete case, Inform. Sci. **17** (1979), 253-278.

31. Lin, C.T., Wang, M.J: Hybrid fault tree analysis using fuzzy sets. Reliability Engineering and System Safety **58** (1997), 205—213.

32. Liu, Y., Qiao, Z., Wang, G.: Fuzzy random reliability of structures based on fuzzy random variables. Fuzzy Sets and Systems **86** (1997), 345-355.

33. Misra, K.B. (Ed.): New Trends in System Reliability Evaluation. Elsevier, Amsterdam, 1993.

34. Misra, K.B., Soman K.P: Multistate fault tree analysis using fuzzy probability vectors and resolution identity. In: T. Onisawa, J. Kacprzyk (Eds.), Reliability and Safety Analyses under Fuzziness. Physica-Verlag, Heidelberg, 1995, 113-125.

35. Mizumoto, M., Tanaka, K.: Some Properties of Fuzzy Numbers. In: M.M. Gupta, R.K. Ragade., R.R. Yager (Eds.), Advances in Fuzzy Theory and Applications, North–Holland, Publ., Amsterdam, 1979, 153-164.

36. Montero, J., Cappelle, B., Kerre, E.E.: The usefulness of complete lattices in reliability theory. In: T. Onisawa, J. Kacprzyk (Eds.), Reliability and Safety Analyses under Fuzziness. Physica-Verlag, Heidelberg, 1995, 95—110.

37. Nahman, J.: Fuzzy logic based network reliability evaluation. Microelectronics and Reliability **37** (1997), 1161—1164.

38. Nishiwaki, Y.: Human factors and fuzzy set theory for safety analysis. In: M.C. Cullingford, S.M. Shah, J.H. Gittus (Eds.) Implications of Probabilistic Risk Assessment. Elsevier Applied Science, Amsterdam, 1987, 253—274.

39. Nowakowski, M.: The human operator: reliability and language of actions analysis. In: W. Karwowski, A. Mital (Eds.), Applications of Fuzzy Sets Theory in Human Factors. Elsevier, Amsterdam, 1986, 165—177.

40. Onisawa, T.: An approach to human reliability in man-machine system using error possibility. Fuzzy Sets and Systems **27** (1988), 87—103.

41. Onisawa, T.: An Application of Fuzzy Concepts to Modelling of Reliability Analysis. Fuzzy Sets and Systems **37** (1990), 120—124.
42. Onisawa, T.: System reliability from the viewpoint of evaluation and fuzzy sets theory approach. In: T. Onisawa, J. Kacprzyk (Eds.), Reliability and Safety Analyses under Fuzziness. Physica-Verlag, Heidelberg, 1995, 43—60.
43. Onisawa, T., Kacprzyk, J. (Eds.): Reliability and Safety Analyses under Fuzziness. Physica-Verlag, Heidelberg, 1995.
44. Onisawa, T., Nishiwaki, Y.: Fuzzy human reliability analysis on the Chernobyl accident. Fuzzy Sets and Systems **28** (1988), 115—127.
45. Pan, H.S., Yun, W.Y.: Fault tree analysis with fuzzy gates. Computers and Industrial Engineering **33** (1997), 569—572.
46. Pedrycz, W.: Shadowed sets: Representing and processing fuzzy sets. IEEE Trans. on Syst., Man and Cybern. Part B. Cybern. **28** (1988), 103—109.
47. Puri, M.L., Ralescu, D.A.: Fuzzy random variables. Journ. Math. Anal. Appl. **114** (1986), 409-422.
48. Rotshtein, A.: Fuzzy reliability analysis of labour (man-machine) systems. In: T. Onisawa, J. Kacprzyk (Eds.),Reliability and Safety Analyses under Fuzziness. Physica-Verlag, Heidelberg, 1995, 245—269.
49. Singer, D.: A fuzzy set approach to fault tree and reliability analysis. Fuzzy Sets and Systems **34** (1990), 145—155.
50. Suresh, P.V., Babar, A.K., Venkat Raj, V.: Uncertainty in fault tree analysis: A fuzzy approach. Fuzzy Sets and Systems **83** (1996), 135—141.
51. Tanaka, H., Fan, L.T., Lai, F.S., Toguchi, K.: Fault-tree analysis by fuzzy probability. IEEE Transactions on Reliability **32** (1983), 453—457
52. Utkin, L.V., Gurov, S.V.: Reliability of Composite Software by Different Forms of Uncertainty. Microelectronics and Reliability **36** (1996), 1459—1473.
53. Utkin, L.V., Gurov, S.V.: A general formal approach for fuzzy reliability analysis in the possibility cotext. Fuzzy Sets and Systems **83** (1996), 203—213.
54. Viertl, R.: Estimation of the reliability function using fuzzy life time data. In: P.K. Bose, S.P.Mukherjee, K.G. Ramamurthy (Eds.), Quality for Progress and Development. Wiley Eastern, New Delhi, 1989.
55. Viertl, R.: Modelling for fuzzy measurements in reliability estimation. In: V. Colombari (Ed.), Reliability Data Collection and Use in Risk and Availability Assessment. Springer Verlag, Berlin, 1989.
56. Viertl, R., Gurker, W.: Reliability Estimation based on Fuzzy Life Time Data. In: T. Onisawa, J. Kacprzyk (Eds.), Reliability and Safety Analyses under Fuzziness. Physica-Verlag, Heidelberg, 1995, 153-168.
57. Viertl, R.: Statistical Methods for Non-Precise Data. CRC Press, Boca Raton, 1996.
58. Wu, H.C.: Fuzzy Reliability Analysis Based on Closed Fuzzy Numbers. Information Sciences **103** (1997), 135—159.
59. Wu, H.C.: Fuzzy reliability estimation using Bayesian approach. Computers and Industrial Engineering **46** (2004), 467—493.
60. Wu, H.C.: Bayesian system reliability assessment under fuzzy environments. Reliability Engineering and System Safety **83** (2004), 277—286.
61. Zadeh, L.A.: The concept of a linguistic variable and its application to approximate reasoning, Parts 1, 2, and 3. Information Sciences **8** (1975), 199-249, 301-357, **9** (1975), 43-80.
62. Zadeh, L.A.: Fuzzy sets as a basis for a theory of possibility. Fuzzy Sets and Systems **1** (1978), 3—28.

Grey Differential Equation GM(1,1) Models in Repairable System Modeling

Renkuan Guo

Department of Statistical Sciences
University of Cape Town, Cape Town, South Africa

14.1 Introduction

Theory and methodology of repairable system modeling is in nature a sto-chastic process modeling, particularly, point process modeling. Since Ascher and Feingold [2] foundational work in repairable system modeling, many works were contributing to this research field, for example, Ander-son et al [1], Cox [7], Dagpunar [9], Kijima [53], and Guo et al [27].

14.1.1 Small Sample Difficulties and Grey Thinking

In reliability engineering modeling, or more specifically, in repairable sys-tem modeling, most of the reliability engineers are using the maximum likelihood theory for facilitating empirical analysis, which is in nature a large-sample based asymptotic theory for (asymptotic) confidence inter-vals and hypothesis testing.

Researcher may argue that it is possible to use one data point for a point estimation of the single parameter under homogeneous Poisson process as-sumptions. However, that is no longer an exercise of standard maximum likelihood estimation but it is merely an approximation exercise hinted by maximum likelihood theory. Furthermore, there is no guarantee that the underlying process must follow homogeneous Poisson process assump-tions and thus one-parameter exponential distribution. Therefore, the thinking of collecting two or three data points and then fitting an assumed one-parameter exponential distribution or a two-parameter Weibull distri-bution is questionable. As long as we follow the classical statistical-reliability theory, the sample size issue will inevitably haunt the maximum likelihood based exercises. The paper "Does Size Matter? Exploring the Small Sample Properties of Maximum Likelihood Estimation" [51] ad-dressed this question clearly. The authors stated that their preliminary re-search showed that there are "the lack of Type I error problems in small

R. Guo: *Grey Differential Equation GM(1,1) Models in Repairable System Modeling*, Computational Intelligence in Reliability Engineering (SCI) **40**, 387–413 (2007)
www.springerlink.com © Springer-Verlag Berlin Heidelberg 2007

samples". They also pointed out that the results for Type II errors are much less comforting." They concluded that "with some reservation that is appears that scholars might need about 30 to 50 cases per independent variable to avoid Type II problems we observe." This research seems a good news to certain research fields, say, market research, political research and others. However, their research essentially claimed the death for today's reliability engineering modeling because the fast changing environments of the industrial and business with globalization trend. In other words, for a reliability engineer collecting 10 to 15 items in a sample is already expensive and time-consuming. Therefore, exploring system state evaluation techniques under very small sample size is an urgent task in reliability engineering.

In general, the small sample asymptotic theory developments obtained attention since 1954. Field and Ronchetti [16] systematically summarized the small sample asymptotics in their monograph. This area is still very active, for example, Field and Ronchetti [17], Beran and Ocker [3] and others. However, the efforts are mostly concentrating on certain statistics, say, mean, M-estimator and L-estimator and there is no direct application of small sample asymptotic theory into reliability engineering yet, except Guo [39] and Kolassa and Tanner [54].

Nevertheless, we should be aware that small sample asymptotics develops an approximation to the distribution of interested quantity that models the true state of system. This approximation is still on the route of the *probabilistic thinking* – once the distribution of the system state is available then the information about the state is fully available. *The real aim of modeling is actually to find the dynamic law of the system state.* Therefore, the probabilistic thinking route is one of the possible choices. There are other choices of thinking logics, for example, fuzzy thinking, rough sets thinking, grey thinking or other thinking logics rooted in modern approximation theory.

Grey thinking is an approximation methodology aiming at directly revealing a system state dynamic relation (without the priori assumptions). It roots from modern control theory, which classifies system dynamics into three categories: white, black and grey three systems based on the degree of information completeness. If the information of a system is available to modeler completely, it is a *white* system, for example, the Earth gravity system. On the other opposite, if the information of a system is totally unknown, it is a *black* system, for example, the social society system out of the Solar System. If the information is partially known and partially unknown, it is a *grey* system. A critical feature of the information incompleteness of a grey system is due to the sparse data. In other words, grey uncertainty, which is different from other form of uncertainty, for example, random uncertainty, fuzzy uncertainty, rough uncertainty, etc., is generated from too little information about the system under investigation. The task of establishing model under the guidance of grey theory is inevitably to seek model building based on data of small sample size. Its target is establishment of grey differential equation and emphasizes the exploration, utilization and processing dynamic information containing in data. However, it is necessary to point out here that grey differential equation modeling is not statistical modeling but it is an approximation modeling exercise, even though the least-square estimation engages. The usage of least-square approach in grey differential equation modeling is not a part of statistical inference or estimation. Rather, it is an optimization technique to help searching grey estimator of a particularly interested parameter.

Differential equation is a powerful mathematical tool for describing *continuous* system dynamics. Conventionally, it is impossible to establish differential equation model based on information from a discrete data sequence. Nevertheless, the innovative aspect of the grey differential equation modeling is to construct a differential equation like model on the discrete data sequence. Therefore, we will use the grey differential equation for repairable system analysis in the sparse data context. We will emphasize the necessity and advantage of grey differential equation modeling on repairable system from the following three aspects: sparse data availability and the repairable system research progress, i.e., the research awareness on repair effects modeling in the literature.

14.1.2 Repair Effect Models and Grey Approximation

Repair (or maintenance) effect is an important aspect in repairable system modeling. If we could evaluate the repair effects correctly, we would be more actively plan the maintenance, production and thus improve management decision-making. We can trace statistical repair effect modeling back to early 90's, for example, Stadje and Zuckerman [61]. Unfortunately, their framework with rich and large data structural assumptions is difficult to implement in practices. Therefore, an important focus point over the last decade in repairable system modeling and maintenance optimization has been the virtual age concept introduced by Kijima [53] since it provided an intuitive mechanism to describe imperfect repair of systems. The intrinsic weakness of Kijima's virtual age models, in that the system repair effects cannot be estimated statistically, was raised in Guo, Love and Bradley [24] and later expanded in Guo, Ascher and Love [27] unless more assumptions were imposed and therefore a more rich structured data are required.

The reliability community has since turned to exploring other (statistically estimable) repair effect models. We may classify these efforts into three categories. The first category is repair-regime-based. It assumes that repair effect links to failure or planned maintenance (abbreviated as PM) regimes or relates to covariates (e.g., normalized repair costs etc). For example, Bowles and Dobbins [4], Cheng and Chen [5], Cheng [6], Cui and Li [8], Doyen and Gaudoin [15], Finkelstein [19], Gasmi, Love and Kahle [21], Gaudoin, Yang and Xie [22], Shirmohammadi, Love and Zhang [60], Wang and Trivedi [64], Wang, Po, Hsu, and Liu, [65], Yang, Lin, and Cheng [67], Yun, Lee, Cho, and Nam [68], etc. The second category is fuzzy-repair-effect-based. It is assumes that repair effect evaluations could be facilitated by fuzzy sets concept and then the fuzzy repair effects are combined into a probabilistic structure for statistical analysis. Some important works are listed as following: Guo and Love [28], Huang and Cheng [52], Pillay and Wang [58], and Shen, Wang, and Huang [59], etc. Taheri [62] reviewed the trends in fuzzy statistics containing a few works in fuzzy reliability modeling. Two foundational books on fuzzy statistical inferences are Kruse and Meyer [55] and Viertl [63]. By noticing the complexity of fuzzy statistical analysis, Guo and Love [30] started their semi-statistical fuzzy modeling exercises on repairable system and have explored them in various ways, see Guo and Love [31-35, 38], Guo [36-38, 40]. The third category is (fuzzy) semi-martingale based stochastic age processes. For example, Guo and Love [29], Guo and Dunne [41] proposed a semi-martingale age process. The stochastic age process has a rich coherent internal structure so that it does not only allow system age diffusion during it functioning but also permits jumps via system shocks and maintenance impacts without any additional assumptions. Guo [44] introduces a non-diffusion version in the internal structure. However, in the semi-martingale age developments, there is a thorny issue remaining there – the intrinsic structure of the stochastic process model requires the *individual* repair improvement effect as data input. Therefore, an effective implementation of semi-martingale age model has to combine the developments in the fuzzy repair effect evaluations or other approach.

It is obvious that the statistical estimation of individual repair is impossible. What statistical methodology can offer is the distribution of repair effect or a few moments of the distribution. Facing the challenge of the individual repair effect evaluation problem it is better to look at it different angle and thus develop a small-sample based repairable system analysis methodology in terms of approximation theory and methodology. Guo began his efforts in applying grey theory to repair efforts modeling since 2004. The motivation of grey theory modeling lies on two aspects. On the one hands, it facilitates a structure for sparse data modeling. On the other hand, grey modeling aims to provide *individual estimated effect* for each repair or PM (not in statistical sense) directly. This is different from probabilistic or fuzzy modeling where the estimates are in average or cut-level sense. The grey estimate itself is still contains intrinsic uncertainty. Although the individual estimated effect of repair is imprecise and not unique (as a matter of facts, it is a whitenization of the grey interval number) the repair effect does not describe the underlying mechanism (system changes under repair or PM) in statistical or probabilistic sense. However, we can understand the grey individual repair effect information as "input" data into semi-martingale age model for further standard statistical analysis. In practices, the grey individual repair effect estimate provides reliability engineers or management the information for decision-making on production and maintenance planning. For more details, see Guo [36-38, 49], Guo and Love [40, 42, 43, 45, 46, 48], Guo and Guo [47], Guo and Dunne [50].

We work with grey approach is just because it let us directly work with system state dynamic changes without involving the distributional and independent sampling assumptions. The computation of grey model is also very simple and can carry on Excel spreadsheet. Nevertheless, we have to emphasize that grey approximation is definitely not the only approach to address the individual repair improvement problem. Other approximation approach may also work, for example, small sample asymptotic techniques even small sample asymptotic techniques involve very complex mathematical developments.

14.2 The Foundation of GM(1,1) Model

Grey differential equation models play the core role in grey theory for extracting the evolving law underlying the sparse data. In reliability engineering context, the basic one is the first-order grey differential equation with one-variable model (abbreviated as GM(1,1) model), which deals with positive discrete data sequence and possesses extreme predictive power. This section is drafted based on the work of Deng [10-14], Fu [20], Liu et al [56] and Wen [66].

14.2.1 Equal-Spaced GM(1,1) Model

Definition 1. Equation

$$x^{(0)}(k)+\alpha\, z^{(1)}(k)=\beta, k=2,\ldots,n \tag{1}$$

is called a one-variable first order grey differential equation (abbreviated as GM(1,1)) with respect to time series sequence $X^{(0)} = (x^{(0)}(1), x^{(0)}(2), ..., x^{(0)}(n))$, where $z^{(1)}(k)$ are generated by MEAN operator,

$$z^{(1)}(k)=\text{MEAN}(x^{(1)}(k))=0.5(x^{(1)}(k)+x^{(1)}(k-1)) \qquad (2)$$

and $x^{(1)}(k)$ are generated by accumulated generating operation (abbreviated as AGO) operator,

$$x^{(1)}(k) = \text{AGO}(X^{(0)})_k = \sum_{i=1}^{k} x^{(0)}(i), \ k = 1,2,\cdots,n \qquad (3)$$

In Eq. 1, α is called the grey developing coefficient, β is the grey input, $x^{(0)}(k)$ is a *grey derivative* and $z^{(1)}(k)$ are called the background values for the grey differential equation. As a matter of fact, $x^{(0)}(k)$ is the kth observation in the data sequence but the term derivative used comes from a treatment that term $(x^{(1)}(k)-x^{(1)}(k-1))/(k-(k-1))= x^{(0)}(k)$ is an approximation to the true derivative of function $x^{(1)}(t)$, i.e., $dx^{(1)}/dt$ at $t=k$. The adjective *grey* used here indicates the grey uncertainty associated with the derivative approximation. Because the data points are collected with every unit time-increment the model is called equal-spaced GM(1,1) model. Furthermore, the differential equation

$$dx^{(1)}/dt + \alpha x^{(1)} = \beta \qquad (4)$$

is called the whitenization differential equation or the *shadow* equation of the grey differential equation (Eq. 1). The relation between the grey differential equation and its corresponding whitenization differential equation will be explored later.

Using least-square approach, the parameters α and β can be estimated and denoted by a and b respectively. Rewriting Eq. 1. as

$$x^{(0)}(k)=\beta+\alpha(-z^{(1)}(k)), \ k=2,...,n \qquad (5)$$

then we obtain a standard matrix form of the Eq. 1. as following

$$Y = X \begin{bmatrix} \beta \\ \alpha \end{bmatrix} \qquad (6)$$

where

$$
Y = \begin{bmatrix} x^{(0)}(2) \\ x^{(0)}(3) \\ \vdots \\ x^{(0)}(n) \end{bmatrix} \quad \text{and } X = \begin{bmatrix} 1 & -z^{(1)}(2) \\ 1 & -z^{(1)}(3) \\ \vdots & \vdots \\ 1 & -z^{(1)}(n) \end{bmatrix} \tag{7}
$$

The least-square estimate for parameter (β, α) is

$$
\begin{bmatrix} b \\ a \end{bmatrix} = \left(X^T X \right)^{-1} X^T Y \tag{8}
$$

The least-square estimator of the parameter pair (β, α), (b, a), should carry some intrinsic information contained in the discrete data sequence $X^{(0)} = (x^{(0)}(1), x^{(0)}(2), ..., x^{(0)}(n))$ sampled from the system investigated. Based on this believe, the true dynamic law specified by the whitenization differential equation Eq. 4. is then replaced by its estimated version

$$
dx^{(1)}/dt + ax^{(1)} = b \tag{9}
$$

which assumes to catch the true dynamics of the system interested in some degree. The solution to Eq. 9 is easily to obtain because it is a first-order non-homogeneous ordinary differential equation with constant coefficients.

$$
\begin{cases} x^{(1)}(t) = e^{-\alpha t} \left(\int_1^t be^{\alpha t} dt + c \right) \\ x^{(1)}(1) = x^{(1)}(1) = x^{(0)}(1) \end{cases} \tag{10}
$$

The filtering-predictive equation takes the discrete version of solution (with GM(1,1) least-square estimated parameter-values).

$$
\hat{x}^{(1)}(k) = (x^{(0)}(1)b / a) \exp(-a(k-1)) + b/a \tag{11}
$$

As to the filtering grey derivative sequence (the estimated original data sequence), $\hat{X}^{(0)} = \{ \hat{x}^{(0)}(k), k = 1,2,...,n \}$, it can be obtained in terms of the inverse accumulative generating operation (abbreviated by IAGO), a difference operation in nature but it is called as a grey differentiation in grey theory context because of unit time difference in t.

$$
\hat{x}^{(0)}(k) = \hat{x}^{(1)}(k) - \hat{x}^{(1)}(k-1) , k = 2,...,n \tag{12}
$$

14.2.2 The Unequal-Spaced GM(1,1) Model

It is very clear that GM(1,1) model with equal-spaced data is established with respect to a sampled discrete data sequence. It should be noticed that the solution to the grey differential equation is obtained in terms of the corresponding shadow (differential) equation. The elegancy in grey differential equation models lies in the approximations of the derivatives and integrals of a given function. Deng [11-14] named them as inverse accumulated generating operation (I-AGO) and accumulated generating operation (AGO) respectively.

It should be further strongly emphasized that the sampled discrete data sequence itself implies that the data is naturally ordered in the sequence. The ordering index could be time or distance from a reference point. AGO roots in integration and is created with a smoothing role by accumulative additions which will partially iron out the fluctuations in original data. Therefore, in order to let AGO function effectively and correctly, the observations in the original data sequence must be strictly positive.

However, other approximations to derivatives or integrals in the numerical analysis can be also considered for data generation. The basic principle here should be what is the best approximation to them rather than just blindly follow the data generating schemes in grey theory literature.

In practical circumstances, there are chances process performance indices may be collected at unequal-spaced times (or spatial distances). Therefore, the unequal-spaced GM(1,1) model is of practical importance.

Deng [14] developed a GM(1,1) model for unequal-gapped data, denoted by $GM_u(1,1)$. Given $X^{(0)} = \left(x^{(0)}\left(t_1\right), x^{(0)}\left(t_2\right), \cdots, x^{(0)}\left(t_N\right) \right)$, the original data sequence, where $t_k - t_{k-1} \neq constant$, then grey differential equation is defined in the following manner:

$$\frac{\Delta x^{(1)}\left(t_k\right)}{\Delta t_k} + \alpha z^{(1)}\left(t_k\right) = \beta \tag{13}$$

where

$$x^{(1)}\left(t_k\right) = \sum_{i=1}^{k} x^{(0)}\left(t_i\right)$$

$$\Delta x^{(1)}\left(t_k\right) = x^{(1)}\left(t_k\right) - x^{(1)}\left(t_{k-1}\right) = x^{(0)}\left(t_k\right) \tag{14}$$

$$\Delta t_k = t_k - t_{k-1}$$

$$z^{(1)}\left(t_k\right) = \frac{1}{2}\left(x^{(1)}\left(t_k\right) + x^{(1)}\left(t_{k-1}\right)\right)$$

Therefore, the $GM_u(1,1)$ model can be written as

$$x^{(0)}\left(t_k\right)+\alpha z^{(1)}\left(t_k\right)\Delta t_k = \beta\Delta t_k, \ k=1,2,\cdots,n \tag{15}$$

Accordingly, estimators of (α,β), (a,b), and the intermediate parameters are

$$a = \frac{CD-(n-1)E}{(n-1)F-C^2}, \quad b = \frac{DF-CE}{(n-1)F-C^2}$$

$$C = \sum_{k=2}^{n} z^{(1)}\left(t_k\right)\Delta t_k, \quad D = \sum_{k=2}^{n} x^{(0)}\left(t_k\right) \tag{16}$$

$$E = \sum_{k=2}^{n} z^{(1)}\left(t_k\right)\Delta t_k x^{(0)}\left(t_k\right), \quad F = \sum_{k=2}^{n}\left[z^{(1)}\left(t_k\right)\Delta t_k\right]^2$$

It is worth to point out here that $x^{(1)}\left(t_k\right)$ is the approximate value of "integration" on interval $\left[t_1,t_k\right]$ with the "integrand" $x^{(0)}(t)$. Therefore, the formation of $x^{(1)}\left(t_k\right)$ given by Deng [14] may be oversimplified. An intuitive and better approximation formulation can be given by

$$x^{(1)}\left(t_k\right) = \sum_{i=2}^{k} \frac{x^{(0)}\left(t_i\right)+x^{(0)}\left(t_{i-1}\right)}{2}\Delta t_i, \ k=2,3,\cdots,n \tag{17}$$

then

$$\frac{dx^{(1)}\left(t_k\right)}{dt} \approx \frac{x^{(1)}\left(t_k\right)-x^{(1)}\left(t_{k-1}\right)}{\Delta t_k} = \frac{x^{(0)}\left(t_k\right)+x^{(0)}\left(t_{k-1}\right)}{2} \tag{18}$$

denoted as $z^{(0)}\left(t_k\right)$. Then the unequal-spaced differential equation is

$$z^{(0)}\left(t_k\right)+\alpha z^{(1)}\left(t_k\right) = \beta, \ k=1,2,\cdots,n \tag{19}$$

Such that the estimators for α and β and the intermediate parameters are

$$\hat{\alpha} = \frac{CD-(n-1)E}{(n-1)F-C^2}, \quad \hat{\beta} = \frac{DF-CE}{(n-1)F-C^2}$$

$$C = \sum_{k=3}^{n} z^{(1)}\left(t_k\right), \quad D = \sum_{k=3}^{n} z^{(0)}\left(t_k\right) \tag{20}$$

$$E = \sum_{k=3}^{n} z^{(1)}\left(t_k\right)z^{(0)}\left(t_k\right), \quad F = \sum_{k=3}^{n}\left[z^{(1)}\left(t_k\right)\right]^2$$

respectively.

14.2.3 A two-stage GM(1,1) Model for Continuous Data

Fu [20] proposed a two-stage GM(1,1) model for continuous data aiming at to handle the non equal-spaced recorded data sequence.. Let the s-coordinate sequence $Z_s^{(0)} = \left\{ Z_s^{(0)}(s_i), s_i \in \mathbf{R}^+, i = 1, 2, \cdots, N \right\}$ satisfying the following grey differential equation,

$$Z_s^{(0)}(s_i) + \alpha_s \overline{Z}_s^{(1)}(s_{i+1}) = \beta_s, i = 2, \cdots, N \tag{21}$$

where

$$\overline{Z}_s^{(1)}(s_{i+1}) = \frac{1}{2} \left(Z_s^{(0)}(s_{i+1}) + Z_s^{(0)}(s_i) \right), i = 2, \cdots, N \tag{22}$$

Accordingly, the first-stage least-square parameter estimators are

$$\left(\hat{\beta}_s, a_s \right) = \left(\mathbf{X}^T \mathbf{X} \right)^{-1} \mathbf{X}^T \mathbf{Y}_N^s \tag{23}$$

where $\hat{b}_S$ denotes the estimator for b_s,

$$\mathbf{X} = \begin{bmatrix} 1 & -\frac{1}{2} \left(Z_s^{(0)}(s_2) + Z_s^{(0)}(s_1) \right) \\ 1 & -\frac{1}{2} \left(Z_s^{(0)}(s_3) + Z_s^{(0)}(s_2) \right) \\ \vdots & \vdots \\ 1 & -\frac{1}{2} \left(Z_s^{(0)}(s_N) + Z_s^{(0)}(s_{N-1}) \right) \end{bmatrix} \tag{24}$$

and

$$\mathbf{Y}_N^s = \begin{bmatrix} \dfrac{Z_s^{(0)}(s_2) - Z_s^{(0)}(s_1)}{s_2 - s_1} \\ \dfrac{Z_s^{(0)}(s_3) - Z_s^{(0)}(s_2)}{s_3 - s_2} \\ \vdots \\ \dfrac{Z_s^{(0)}(s_N) - Z_s^{(0)}(s_{N-1})}{s_N - s_{N-1}} \end{bmatrix} \tag{25}$$

Thus the first-stage response equation can be rewritten as

$$Z_s^{(0)}(s) = b_s + c_s \exp(-a_s s) \tag{26}$$

Then we enter the second-stage lsqstere estimation in which parameter (b_s, c_s) will be re-estimated but parameter a_s will be kept as same as that from the first-stage estimation and treated as an input variable for the second-stage estimation.

$$(b_s, c_s)^T = (D_s^T D_s)^{-1} D_s^T z_N^s \tag{27}$$

where

$$D_x = \begin{bmatrix} 1 & \exp(-a_s s_1) \\ 1 & \exp(-a_s s_2) \\ \vdots & \vdots \\ 1 & \exp(-a_s s_N) \end{bmatrix} \tag{28}$$

and

$$z_N^S = \begin{bmatrix} Z_s^{(0)}(s_1) \\ Z_s^{(0)}(s_2) \\ \vdots \\ Z_s^{(0)}(s_N) \end{bmatrix} \tag{29}$$

Therefore after two-stage least-square fitting, the estimated response function is

$$Z_s^{(0)}(s) = b_s + c_s \exp(-a_s s) \tag{30}$$

14.2.4 The Weight Factor in GM(1,1) Model

Deng   [11-14]   argued   that $\{z^{(1)}(k), k = 2, 3, \cdots, n\}$ with $z^{(1)}(k) = 0.5 \times (x^{(1)}(k) + x^{(1)}(k-1))$ should be the only candidate at the integrated level, , while others, for example, Fan et al [18] argued that the weight factor , denoted by ω, between $x^{(1)}(k)$ and $x^{(1)}(k-1)$ should not be predetermined rather than determined by the optimization procedure. In other words, at the integral level, the candidate

$$z^{(1)}(k) = \omega x^{(1)}(k) + (1-\omega) x^{(1)}(k-1), \quad \omega \in [0,1] \tag{31}$$

should be considered and thus the least-square estimation will be:

$$\min_{\alpha,\beta,\gamma,\omega} \{J^{(1)}\} \tag{32}$$

where the objective function is defined by:

$$J^{(1)} = \sum_{i=2}^{n} \left(z^{(1)}(i) - x^{(1)}(i) \right)^2 \tag{33}$$

and the function $x^{(1)}(i)$ is given by:

$$x^{(1)}(t) = (\gamma - \beta/\alpha)\exp(-\alpha(t-1) + \beta/\alpha \tag{34}$$

The first order partial derivatives of $x^{(1)}(t)$ with respect to α, β, γ, and the derivative of $z^{(1)}(i)$ with respect to ω are

$$\frac{\partial x^{(1)}(t)}{\partial \alpha} = \frac{\beta}{\alpha^2} e^{-\alpha(t-1)} - \left(\gamma - \frac{\beta}{\alpha}\right)(t-1)e^{-\alpha(t-1)} - \frac{\beta}{\alpha^2}$$

$$\frac{\partial x^{(1)}(t)}{\partial \beta} = -\frac{1}{\alpha}\exp(-\alpha(t-1)) + \frac{1}{\alpha}$$

$$\frac{\partial x^{(1)}(t)}{\partial \gamma} = \exp(-\alpha(t-1)) \tag{35}$$

$$\frac{dz^{(1)}(i)}{d\omega} = x^{(0)}(i)$$

Then, it is required to solve the following nonlinear equation system for the parameters α, β, γ and ω:

$$\sum_{i=2}^{n} (i-1)e^{-\alpha(i-1)} \left[z^{(1)}(i) - x^{(1)}(i) \right] = 0$$

$$\sum_{i=2}^{n} \left[z^{(1)}(i) - x^{(1)}(i) \right] = 0$$

$$\sum_{i=2}^{n} e^{-\alpha(i-1)} \left[z^{(1)}(i) - x^{(1)}(i) \right] = 0 \tag{36}$$

$$\sum_{i=2}^{n} x^{(0)}(i) \left[z^{(1)}(i) - x^{(1)}(i) \right] = 0$$

Mathematically, it is obvious that the optimal choice of ω is not necessary to choose as 0.5.

Let us examine an example. Given a discrete data sequence $X^{(0)}$={2.874, 3.278, 3.337, 3.390, 3.679}. We perform the parameter searching for two cases: Model (1) with equal-spaced GM(1,1) and Model (2) based on Eq. 31 to Eq. 36.

In Model (2) a genetic algorithm-search procedure gives $\omega = 0.0486454$. This confirms our statement made early that w is not necessarily 0.5 as proposed by Deng [11-14] but it is a data-dependent parameter. Also the model (2) gives much small squared error at $x^{(1)}$-level, i.e.,

$$J^{(1)} = \sum_{i=2}^{n} \left(z^{(1)}(i) - x^{(1)}(i) \right)^2$$ is minimized in the Model (2). Table 1. lists the

estimated parameter a and b by assuming the same initial parameter value $(g=x^{(1)}(1)=x^{(0)}(1))$. Model (1) fixes the weight w at 0.5 while Model (2) w value is an optimally searched one. Model (2) gives almost one third of the squared error at $x^{(1)}$-level of that given by Model (1).

Table 1. The weight factor impact in GM(1,1) models

Model	α	β	$x^{(0)}(1)$	ω	Squared error at $x^{(1)}$-level
GM(1,1)	-0.03720	3.06536	2.874	0.5 (Deng)	0.006929752
$\min_{\alpha,\beta,\gamma,\omega} \{J^{(1)}\}$	-0.04870	2.91586	2.874	0.048645	0.002342187

However, if we think the computation convenience of classical GM(1,1) model which can carry on Excel, we would prefer to Model (1)-GM(1,1) modeling.

14.3 A Grey Analysis on Repairable System Data

14.3.1 Cement Roller Data

The data set presented in Table 2 was used for various analyses because it contains a quite rich structure. The analysis performed in this section is an exploration of the possible dynamics of the cement roller functioning times. Although the data set will be subdivided into small groups for the purpose to explore whether the grey approach could well reveal the under-lying mechanism behind the sub-data sets with small sample sizes.

Table 2. Cement Roller data [57].

Functioning	Failure	Covariate	Covariate	Covariate	Repair
54	pm	12	10	800	93
133	failure	13	16	1200	142
147	pm	15	12	1000	300
72	failure	12	15	1100	237
105	failure	13	16	1200	0
115	pm	11	13	900	525
141	pm	16	13	1000	493
59	failure	8	16	1100	427
107	pm	9	11	800	48
59	pm	8	10	900	1115
36	failure	11	13	1000	356
210	pm	8	10	800	382
45	failure	10	19	1300	37
69	pm	12	14	1100	128
55	failure	13	18	1200	37
74	pm	15	12	800	93
124	failure	12	17	1100	735
147	failure	13	16	1100	1983
171	pm	11	13	900	350
40	failure	13	16	1100	9
77	failure	14	17	1100	1262
98	failure	12	15	1100	142
108	failure	12	15	1100	167
110	pm	16	14	1100	457
85	failure	8	19	1300	166
100	failure	12	15	1000	144
115	failure	13	16	1200	24
217	pm	9	11	900	474
25	failure	15	18	1200	0
50	failure	11	13	1100	738
55	pm	8	10	800	119

14.3.2 An Interpolation-least-square Modeling

We only have the system functioning and failure (or planned maintenance) times, which will be called as system stopping times. Denote system stopping times as $\{T_1, T_2, ..., T_L\}$. It is immediately noticed that we have a situation that there is no direct or original sequence $\{x^{(0)}(1), x^{(0)}(2), ..., x^{(0)}(n)\}$ readily available for analysis. Then we first apply 1-AGO to $\{T_1, T_2, ..., T_L\}$ to obtain $\{t_1, t_2, ..., t_L\}$ where,

$$\{t_1, t_2, ..., t_L\} = AGO\{T_1, T_2, ..., T_L\} \tag{37}$$

It is obvious that

$$t_i = \sum_{j=1}^{i} T_j \quad j = 1, 2, \cdots, L \tag{38}$$

Now the "original" observation sequence will be

$$x^{(0)}(s_i) = t_i, \quad i = 1, 2, \cdots, L \tag{39}$$

Furthermore, it is noticed that $\left\{ x^{(0)}(s_1), x^{(0)}(s_2), \cdots, x^{(0)}(s_L) \right\}$ is not an equidistant spaced since $s_{i+1} - s_i \neq s_{j+1} - s_j$, $\forall i \neq j$, $i, j = 1, 2, \cdots, L$. Then in terms the following steps, we will create an equal-gap (i.e., equal-spaced) "original" sequence.

(1) Divide $\{T_1, T_2, \cdots, T_L\}$ by T_1 and obtain a new subscript (i.e., index) sequence,

$$\{s_1, s_2, \cdots, s_L\} = \{1, T_2/T_1, \cdots, T_L/T_1\} \tag{40}$$

It is obvious that the values in the sequence $\{s_1, s_2, \cdots, s_L\}$ are mostly non-integers, thus it is required to create a mixed real-valued indexed sequence $\{s_1, i_2, s_2, i_3, \cdots, i_L, s_L\}$ and the corresponding data sequence $X_i^{(0)} = \left\{ x^{(0)}(s_1), x^{(0)}(i_2), x^{(0)}(s_2), x^{(0)}(i_3), \cdots, x^{(0)}(i_L), x^{(0)}(s_L) \right\}$ respectively. It is necessary to point out that the symbol i_s is not necessarily representing a single integer and it should be these integers between s_{i-1} and s_i.

(2) Determine the integer(s) i_s such that $s_{i-1} < i_s < s_i$ such that the index sequence $\{s_1, i_2, s_2, i_3, \cdots, i_L, s_L\}$ is available.

(3) Determine $X_i^{(0)}$ in terms of interpolation method (calculating these $x^{(0)}(i_s)$ where $s_{i-1} < i_s < s_i$, i_s must be all the integers between s_{i-1} and s_i.

$$x^{(0)}(i_s) = x^{(0)}(s_{i-1}) + \frac{i_s - s_{i-1}}{s_i - s_{i-1}} \left(x^{(0)}(s_i) - x^{(0)}(s_{i-1}) \right) \tag{41}$$

(4) Apply $1 - AGO$ to $X_i^{(0)}$

$$x^{(1)}(r) = \begin{cases} \sum_{k=1}^{i_s} x^{(0)}(k), & \text{if } r = i_s \text{ (integer)} \\ x^{(1)}(i_s) + (s_i - i_s)\left[x^{(0)}(s_i) - x^{(0)}(i_s) \right] & \text{if } r = s_i \text{ (non-integer)} \end{cases} \tag{42}$$

(5) Define the grey derivative on $x^{(1)}(r)$ as

$$dL\left(X^{(1)}\left(s_i\right)\right) = \frac{x^{(1)}\left(s_i\right) - x^{(1)}\left(i_i\right)}{s_i - i_i} \tag{43}$$

(7) Using $\left(x^{(1)}\left(i_s\right) + x^{(1)}\left(s_i\right)\right)/2$ as the grey value at non-integer point s_i. Then the grey differential equation for non-integer point s_i.

$$x^{(0)}\left(s_i\right) + \frac{\alpha}{2}\left(x^{(1)}\left(s_i\right) + x^{(1)}\left(i_s\right)\right) = \beta \tag{44}$$

(8) Then in terms of augmented equation we obtain the estimate of parameter (α, β) and finally obtain the filtering-prediction equation.

What we need to emphasize here is that the difference between failure stopping times and planned maintenance time is no longer making too much sense because AGO applications will eventually weaken and even eliminate the random difference between them. Furthermore, it should be noticed that the grey differential equation of system functioning time is intrinsic to the system as well as repair impact, and thus it is called the *system characteristic time*.

For the Cement Roller data in Table 2, we performed the interpolation calculations and enlarged the 31 data points into 84 data points. Then we partition the 84 data sequence into 17 sub-data sequences. GM(1,1) modeling is carried on for each sub-sequence and all 17 GM(1,1) groups computation results are summarized in Table 3. For each group, the *starting time* listed in Table 3 is just the value of "$x^{(0)}(1)$". The partition of 17 groups is an illustrative attempt with an intention that each group contains a few data points and includes one or two "original" data points (either failure or PM times).

It is noticed that the interpolation-based GM(1,1) modeling can not be performed in computation toolbox offered in Liu et al [56] or Wen [66]. However, it can be done in Excel easily (although a bit tedious). For illustrative purpose, we tabularize the computations for the first two sub-sequences (sub-data 1-6 and sub-data 7-11) in Table 4.

We can easily catch up that in sub-data 1-6, only two original data points are included (listed in Column A where s_i are recorded, which are non-integers except $s_1=1$) and the remaining four data points (listed in Column B where i_s are recorded, which are integers).

Table 3. Summary of 17 group of GM(1,1)models for Cement Roller data

Group	Range	Starting time	$\hat{\alpha}$	$\hat{\beta}$
1	1-6	54	82.14264	-0.29351
2	7-11	324	301.1063	-0.12667
3	12-16	486	455.3279	-0.09761
4	17-22	648	642.9939	-0.0613
5	23-27	864	849.572	-0.04863
6	28-33	1026	988.895	-0.05118
7	34-38	1242	1219.862	-0.03575
8	39-43	1404	1358.791	-0.03942
9	44-47	1566	1508.428	-0.04202
10	48-51	1728	1673.738	-0.03651
11	52-55	1890	1864.493	-0.0234
12	56-60	1998	1962.813	-0.0262
13	61-64	2160	2153.215	-0.01853
14	65-69	2322	2291.204	-0.02127
15	70-73	2484	2453.476	-0.02093
16	74-78	2646	2597.978	-0.02252
17	79-84	2862	2809.663	-0.01993

Demonstration of step-wise computation details of the interpolation approach is given by using sub-data 1 (data point 1-6) and sub-data 2 (data point 7-11) shown in Table 4. The "unit" of time is 54, the first failure time. Column C records the system chronological times when system failure or PM occurred (T_i). Column A records the non-integer ratio $t_i = T_i/T_1$ (where $T_1 = 54$) and Column B records the integer-valued sequence of the interpolation points between T_i and T_{i+1}. Column D records the 1-*AGO* of Column C according to the following equation:

$$D(k) = \begin{cases} D(k-1) + (B(k) - B(k-1)) \cdot C(k) & \text{if } B(k), B(k-1) \text{ are integer} \\ D(k-1) + (A(k) - B(k-1)) \cdot C(k) & \text{if } A(k) \text{ is non-integer} \\ D(k-1) + (B(k) - B(k-2)) \cdot C(k) & \text{if } A(k-1) \text{ is non-integer} \end{cases} \tag{45}$$

Column E records the X matrix where

$$E(K) = -0.5(D(K) + D(K-1)) \tag{46}$$

Then use the Regression (Option) in Data Analysis within Tools Menu in Excel for calculating a and b. For example, for Group 1 (data point 1-7), in the regression input menu, we fill Input $\underline{Y}$ Range: $C$2:$C$7 and we fill the Input $\underline{X}$ Range: $E$2:$E$7. After regression, a and b are obtained, then we can calculate Column F (i.e., $X^{(1)}(k)$) according to Eq. 47

$$F(k+1) = (x^{(1)}(0) - b/a)e^{-ak} + b/a \tag{47}$$

Table 4. Computing Demonstration for data group 1-6 and 7-11 in Excel.

	A	B	C	D	E	F	G	H
1	1.000	1	54	54	54	54	54	54
2		2	108	162	-108	177	123	
3		3	162	324	-243	333	157	
4	3.463		187	411	-367	420	187	133
5		4	216	540	-475	533	114	
6		5	270	810	-675	790	256	
7		6	324	324	324	324	324	
8	6.185		334	386	-355	387	341	154
9		7	378	702	-544	686	299	
10	7.519		406	913	-807	895	403	61
11		8	432	1134	-1023	1103	208	

Column G ($\hat{X}^{(0)}(k)$ -the filtered values of system function times) is calculated by the following equation:

$$G(k+1) = \begin{cases} F(k+1)-F(k) & \text{if } B(k+1), B(k) \text{ are integer} \\ (F(k+1)-F(k))/(A(k+1)-B(k)) & \text{if } A(k+1) \text{ is non-integer} \end{cases} \qquad (48)$$

Column H, then only calculate the intrinsic functioning time for those row with A(k) being non-integer, the formula is H(k_i)=G(k_i)-G(k_{i-1}), where A(k_i) and A(k_{i-1}) are non-integers.

The results shown in Column H are the filtered (i.e., estimated) values corresponding to the system's failure or PM times which are called as intrinsic functioning times. The impression of the interpolation approach is that it is intuitive and easy to be interpreted. However, because of the interpolation the computation time is increased and also the statistical estimation error sometimes involve cross group fittings.

14.3.3 A Two-stage Least-square Modeling Approach

We performed more trial computations in order to explore the inside of GM(1,1) modeling on system function times. Our tentative results show that the system intrinsic functioning time (function) takes a form

$$IFT(t)=g \exp(-a(t-t_0))+b \qquad (4\ 9)$$

which depends upon four parameters: t_0 (initial time, observed system failure or PM (sojourn) time counting from last failure or PM chronological time), a (slope parameter from the first-stage regression), b , and g (intersection parameter and slope parameter respectively from the second-stage regression). Different from Equation previous treatments, Eq. 49

includes t_0 with the advantage for wider time coverage. It is obvious that the quality of estimated $IFT(t)$ function depends upon the number of data points being included in regression and the range of the data points. However, the following partition of a failure or PM time can be abstracted from repeated two-stage grey fitting where the (random) error term is

$$e_i = t_i - IFT(t_i) \tag{50}$$

and the system repair improvement is

$$r_i = IFT(t_i) - E[IFT(t_i)] \tag{51}$$

Basically, the partition shown in Table 5 is intend to be used for further analysis.

Table 5. Partition of a Functioning Time via 2-stage estimation.

W	t_i	$IFT(t_i)$	$E[IFT(t_i)]$	r_i	e_i
54	54	54.413	54.012	0.401	-0.413
187	133	132.767	134.513	-1.746	0.233
334	147	147.066	148.013	-0.947	-0.066
406	72	71.922	73.086	-1.164	0.078
511	105	104.546	106.883	-2.336	0.454
626	115	114.568	116.853	-2.285	0.432
767	141	140.922	142.251	-1.329	0.078
826	59	59.256	59.359	-0.103	-0.256
933	107	106.546	108.886	-2.341	0.454
992	59	59.256	59.359	-0.103	-0.256
1028	36	37.101	34.432	2.669	-1.101
1238	210	213.004	206.587	6.417	-3.004
1283	45	45.732	44.288	1.444	-0.732
1352	69	68.990	69.940	-0.950	0.010
1407	55	55.380	55.084	0.296	-0.380
1481	74	73.880	75.176	-1.296	0.120
1605	124	123.641	125.728	-2.086	0.359
1752	147	147.066	148.013	-0.947	-0.066
1923	171	171.876	170.717	1.159	-0.876
1963	40	40.931	38.829	2.102	-0.931
2040	77	76.822	78.301	-1.479	0.178
2138	98	97.569	99.832	-2.263	0.431
2246	108	107.546	109.886	-2.340	0.454
2356	110	109.549	111.883	-2.333	0.451
2441	85	84.693	86.572	-1.879	0.307
2541	100	99.559	101.852	-2.293	0.441
2656	115	114.568	116.853	-2.285	0.432
2873	217	220.494	212.908	7.586	-3.494
2898	25	26.619	22.199	4.420	-1.619
2948	50	50.549	49.706	0.843	-0.549
3003	55	55.380	55.084	0.296	-0.380

However, we have to provide some details of the two-stage estimation procedure for illustration purpose.

We select a trial computation where each group consists of 8 data points. Table 6 is Excel spreadsheet computing step by step for the first group of 8 data points (data point 1-8).

Column B records the functioning time since last failure or PM. Column C is the 1-AGO of Column B and Column D records the negative background value z. Column E records the value given in the following equation:

$$E(k)=[B(k)-B(k-1)]/[A(k)-A(k-1)] \qquad (52)$$

Then use Regression option in Tools Menu (Input Y Range $E$3:$E$9) and (Input X Range $D$3:$D$9) to perform the first-stage fitting for obtaining the value of $a = 0.001553574$. The next step is to calculate Column F according to the following equation

$$F(k)= \exp\left(-a*(B(k)-\$B\$2)\right) \qquad (53)$$

Now use Regression menu in Tools Bar (Input Y Range $B$2:$B$9) and (Input X Range $F$2:$F$9) to perform the second-stage fitting for obtaining the value of $b = 743.5890381$ and $c = -690.2399076$. Finally, Column G is calculated as following

$$G(k)= c*F(k)+ b \qquad (54)$$

which gives the estimated intrinsic functioning time $IFT(t_i)$.

Table 6. Two-stage fitting in Excel (data point 1-8).

	A W_i	B t_i	C 1-AGO	D $-Z(t_i)$	E $X^{(0)}(t_i)$	F $e^{-b(t_i-t_0)}$	G $IFT(t_i)$
1	54	54	54			1.0000	53.3491
2	187	133	187	-120.5	0.5940	0.8845	133.0716
3	334	147	334	-260.5	0.0952	0.8655	146.2070
4	406	72	406	-370	-1.0417	0.9724	72.3838
5	511	105	511	-458.5	0.3143	0.9238	105.9279
6	626	115	626	-568.5	0.0870	0.9096	115.7579
7	767	141	767	-696.5	0.1844	0.8736	140.6125
8	826	59	826	-796.5	-1.3898	0.9923	58.6901

As to the expected intrinsic functioning time $E[IFT(t_i)]$ in Table 5, we can evaluate it in various approaches. It can be noticed that the Cement Roller data set contains 31 data points, so that we divide them into four groups: Group 1 consists of data point 1 to 8, Group 2: 9 to 16, Group 3: 17 to 24 and Group 4: 25 to 31. (The division is arbitrary with the intention of creating small sample analysis.) Accordingly, we obtain four groups of parameters shown in Table 7:

Table 7. Estimated Parameters for the Four Groups.

Group	a	b	c
1 (data1-8)	0.001554	743.5890	-690.2399
2 (data9-16)	-0.0007593	-1191.744	1297.0959
3 (data17-24)	-0.0006332	-1473.038	1596.6794
4 (data25-31)	0.00450917	344.1091	-252.1844

A shocking fact is that all the four models give similar estimates of $IFT(t_i)$. This leads us to believe that we can calculate four estimated $IFT(t_i)$ and average them for obtaining an estimate for $E[IFT(t_i)]$.

14.3.4 Prediction of Next Failure Time

It can be noticed that the two-stage grey modeling of repairable system data generated delicate results although the two-stage grey approach does not make itself as intuitive as the interpolation approach. We use the term of prediction but in nature an *approximation*.

If we have n data points in a sample, for the estimated data values $\hat{x}^{(0)}(k)$, if $k \in \{2,3,\ldots,n\}$, i.e., $2 \le k \le n$ we call $\hat{x}^{(0)}(k)$ as *filtered* values, if $k \in \{n+1, n+2, \ldots\}$, i.e., $k > n$, we call $\hat{x}^{(0)}(k)$ as *predicted* values. It is obvious that filtering is interpolation while predicting is extrapolation.

However, an immediate interest is given the next PM time, say $W_{32} = 3103$ can we predict the next $E[IFT(t_{32})]$ (it is obvious that $t_{32} = 100$). Table 8 details the related grey prediction of expected intrinsic function time $E[IFT(t_i)]$, estimated intrinsic functioning time $IFT(t_i)$, and relative errors if the sojourn PM time is 100.

Table 8. Predictions given $t_{32} = 100$.

Group	$IFT(t_{32})$	Relative error
1	100.955	0.009554
2	98.477	-0.015230
3	99.559	-0.004410
4	108.418	0.084178
$E[IFT(t_{32})]$	101.852	0.018523

What we can predict that for a planning maintenance time at 100, the intrinsic function time falls in an interval $[98.477, 108.418]$ and its estimated expected intrinsic function time $E[IFT(t_i)] = 101.852$ with relative error less than 2%. The way we perform the so-call predicting next intrinsic system functioning time is actually a cautious step of model validation because we only use $n-1=30$ data points for GM(1,1) modeling but keep the 31st data point not participating modeling but reserved for a validation. If we allow 5% relative error in prediction, we can fit a GM(1,1) model with the 31st data point in and then perform the next intrinsic system functioning time i.e., the 31st stopping time.

In general, Deng [14] develop a class ratio test where the class ratio, denoted by $\sigma^{(0)}(k) = x^{(0)}(k)/x^{(0)}(k-1)$, $2 \le k \le n$ with respect to a discrete data sequence $X^{(0)} = \{ x^{(0)}(1), x^{(0)}(2),..., x^{(0)}(n) \}$ should fall in the range $\sigma^{(0)}(k) \in [\exp(2/(n+1)), \exp(2/(n+1))]$. For example, given the sample size $n=4$, if the class ratio fall in the range $[0.6703, 1.4918]$, i.e., $0.6703 \le \sigma^{(0)}(k) \le 1.4918$, for any $k=2,3,4,$, then the grey exponential law (i.e., a successful GM(1,1) model) can be guaranteed. Liu et al [56] gave more details Shown in Table 9 which relates the range of grey development coefficient α (the class ratio $\sigma^{(0)} = e^{\alpha}$) and predicting (i.e., extrapolation) steps with associated relative error (in terms of simulation).

Table 9. The relation between range of α and GM(1,1) prediction steps.

$-\alpha$	1-step error	2-step error	5-step error	10-step error
0.1	0.129%	0.137%	0.160%	0.855%
0.2	0.701%	0.768%	0..967%	1.301%
0.3	1.998%	2.226%	2.912%	4.067%
0.4	4.317%	4.865%	6.529%	9.362%
0.5	7.988%	9.091%	12.468%	18.330%
0.6	13.405%	15.392%	21.566%	32.599%
0.8	31.595%	36.979%	54.491%	88.790%
1.0	65.117%	78.113%	-	-
1.5	-	-	-	-
1.8	-	-	-	-

14.4 Concluding Remarks

In this chapter, based on the observations that grey methodologies are powerful in the circumstances of sparse data availability, we focus the discussions on the most useful grey differential equation model, GM(1,1) model, its basic theory, its variation – the unequal-spaced GM(1,1) model and the continuous-time GM(1,1) model. We use Cement Roller data for illustrations, particularly showing the computations in MicroSoft Excel.

It is necessary to point out that the GM(1,1) modeling is a *deterministic* approach in nature and it generates approximations to system dynamic behavior. Because of the sparse data availability, many traditional survival analysis or reliability terminologies are no longer meaningful in grey treatments. For example, stochastic processes and deterministic processes are not differentiating here and they are all called as grey processes. We can not judge grey modeling in terms of traditional statistical foundation since meaningful theoretical arguments need the support of adequate data information but in grey uncertainty circumstances censoring is meaningless.

Reference

[1] Andersen, P. K., Ø. Borgan, R. D. Gill, N. Keiding (1993) Statistical Models Based on Counting Processes, Springer-Verlag
[2] Ascher, H. E., H. Feingold (1984) Repairable System Reliability: Modelling, Inference, Misconceptions and Their Causes, Dekker, New York
[3] Beran, J., D. Ocker (2005) Small sample asymptotics for credit risk portfolios, American Statistical Association, Journal of Computational & Graphical Statistical, 14 (2): 339-351
[4] Bowles, J. B., J. G. Dobbins (2004) Approximate reliability and availability models for high availability and fault-tolerant systems with repair, Quality and Reliability Engineering International, 7: 679-697
[5] Cheng, C.Y., M. Chen (2003) The Periodic Preventive Maintenance Policy for Deteriorating Systems by Using Improvement Factor Model, International Journal of Applied Science and Engineering, 1: 114-122
[6] Cheng, C.Y. (2004) The Interval Estimation For the Imperfectly Maintained System With Age Reduction, Proceedings of the 9th International Conference on Industrial Engineering, Theory, Applications and Practice, 27-30 November, The University of Auckland, Auckland, New Zealand: 152-157
[7] Cox, D. R. (1972) Regression models and life tables, Journal of Royal Statistical Society, B26: 187-220
[8] Cui, L. R., J. L. Li (2004) Availability For a Repairable System with Finite Repairs, Proceedings of the 2004 Asian International Workshop (AIWARM 2004), 26-27 August, Hiroshima City, Japan: 97-100
[9] Dagpunar, J. S. (1998) Some properties and computational results for a general repair process, Naval Research Logistics, 45: 391-405
[10] Deng, J. L. (1982) Control Problems of Grey Systems, Systems and Control Letters, 1 (6), March
[11] Deng, J. L. (1985) Grey Systems (Social ' Economical), The Publishing House of Defense Industry, Beijing (in Chinese)

[12] Deng, J. L. (1993) Grey Control System, Huazhong University of Technology Press, Wuhan City, China (In Chinese)

[13] Deng, J. L. (2002[a]) The Foundation of Grey Theory, Huazhong University of Technology Press, Wuhan City, China (in Chinese)

[14] Deng, J. L. (2002[b]) Grey Prediction and Decision, Huazhong University of Technology Press, Wuhan City, China (in Chinese)

[15] Doyen, L., O. Gaudoin (2004) Classes of imperfect repair models based on reduction of failure intensity or virtual age, Reliability Engineering & System Safety, 84 (1): 45-56

[16] Field, C. A., E. Ronchetti (1990) Small Sample Asymptotics, Institute of Mathematical Statistics, Lecture Notes-Monographs Series 13, USA

[17] Field, C.A., E. Ronchetti (1991) An Overview of Small Sample Asymptotics, in: W. Stahel, S. Weisberg (eds.), Directions in Robust Statistics and Diagnostics, Part I, Springer-Verlag, New York

[18] Fan, X. H., Q. M. Miao, H. M. Wang (2003) The Improved Grey Prediction GM(1,1) Model and Its Application, Journal of Armored Force Engineering Institute, 17 (2): 21-23, (in Chinese)

[19] Finkelstein, M. S. (2003) On the performance quality of repairable systems, Quality and Reliability Engineering International, 19 (1): 67-72

[20] Fu, L. (1991) Grey System Theory and Applications, The Publishing House of Science and Technology, Beijing (in Chinese)

[21] Gasmi S., C. E. Love, W. Kahle (2003) A general repair, proportional-hazards, framework to model complex repairable systems, IEEE Transactions on Reliability, 52 (1): 26-32

[22] Gaudoin, O., B. Yang, M. Xie (2003) A simple goodness-of-fit test for the power-law process, based on the Duane plot, IEEE Transactions on Reliability, 52 (1): 69-74

[23] Guo R., C. E. Love (1992) Statistical Analysis of An Age Model for Imperfectly Repaired System, Quality and Reliability Engineering International, 8: 133-146

[24] Guo R., C. E. Love, E. Bradley (1995) Bad-As-Old Modelling of Complex Systems with Imperfectly Repaired Subsystems, Proceedings of the International Conference on Statistical Methods and Statistical Computing for Quality and Productivity Improvement, 17-19 August, Seoul, Korea: 131-140

[25] Guo R., C. E. Love (1996) A Linear Spline Approximation for Semi-Parametric Intensities, IEEE Transactions on Reliability, 45 (2): 261-266

[27] Guo R., H. Ascher, C. E. Love (2001) Towards Practical and Synthetical Modeling of Repairable Systems, Economic Quality Control, 16 (2): 147-182

[28] Guo R., C. E. Love (2002) Reliability Modeling with Fuzzy Covariates, Proceedings of 7th Annual International Conference on Industrial Engineering, Theory, Applications and Practice, 24-25 October, Busan, Korea: 5-8

[29] Guo R., C. E. Love (2003[a]) Statistical Modeling of System Ages, Proceedings of 8th International Conference on Industrial Engineering, Theory, Applications and Practice, 10-12 November, Las Vegas, Nevada, USA: 380-385

[30] Guo, R., C. E. Love (2003[b]) Reliability Modeling with Fuzzy Covariates, International Journal of Reliability, Quality and Safety Engineering, 10 (2): 131-157

[31] Guo, R., C. E. Love (2003^c) Reliability Modeling with Fuzzy Covariates, International Journal of Industrial Engineering – Theory, Applications, and Practice, Special Issue: IJIE 2002 Conference, December, 10 (4): 511-518

[32] Guo, R., C. E. Love (2004^a) Fuzzy Covariate Modeling of an Imperfectly Repaired System, Quality Assurance, 10 (37): 7-15

[33] Guo, R. (2004) Interval-Fuzzy Sets Modelling on Repairable System Data, Proceedings of the 2004 Asian International Workshop on Advanced Reliability Modelling, 26-27 August, Hiroshima City, Japan: 157-164

[34] Guo, R., C. E. Love (2004^b) Analysis of Repairable System Data via Fuzzy Set-Valued Statistical Methodology, Proceedings of the 2004 Asian International Workshop on Advanced Reliability Modelling, 26-27 August, Hiroshima City, Japan: 165-172

[35] Guo, R., C. E. Love (2004^d) A Virtual State Analysis of Repairable System via Fuzzy Logical Functions, Proceedings of 9th International Conference on Industrial Engineering, Theory, Applications and Practice, 26-28 November, Auckland, New Zealand: 140-145

[36] Guo, R. (2004^c) A Fuzzy System Age Analysis of Repairable System via Interval-Valued Fuzzy Sets Approach, Proceedings of 9th International Conference on Industrial Engineering, Theory, Applications and Practice, 26-28 November, Auckland, New Zealand: 146-151

[37] Guo, R. (2005^a) Repairable System Modeling Via Grey Differential Equations, Journal of Grey System, 8 (1): 69-91

[38] Guo, R. (2005^b) A Repairable System Modeling by Combining Grey System Theory with Interval-Valued Fuzzy Set Theory, International Journal of Reliability, Quality and Safety Engineering, 12 (3): 241-266

[39] Guo, R. (2006) Small Sample Asymptotic distribution of Cost-related Reliability Risk Measure, submitted for 2006 Asian International Workshop on Advanced Reliability Modeling, 24-25 August, Busan, South Korea

[40] Guo, R., C. E. Love (2005^a) Fuzzy Set-Valued and Grey Filtering Statistical Inferences on a System Operation Data, International Journal of Quality in Maintenance Engineering – Advanced reliability modeling, 11(3): 267-278

[41] Guo, R., T. Dunne (2005) Stochastic Age Processes. Advances in Safety and Reliability, in: Krzysztof Kolowrocki (ed.), Proceedings of the European Safety and Reliability Conference (ESREL 2005), 27-30 June, Tri City (Gdynia-Sopot-Gdansk), Poland: 745-752

[42] Guo, R., C. E. Love (2005^b) Grey Repairable System Analysis (Plenary Lecture). Advances in Safety and Reliability, in: Krzysztof Kolowrocki (ed.), Proceedings of the European Safety and Reliability Conference (ESREL 2005), 27-30 June, Tri City (Gdynia-Sopot-Gdansk), Poland: 753-766

[43] Guo, R. (2005^c) A Grey Semi-Statistical Fuzzy Modeling of an Imperfectly Repaired System, Proceedings of the 4th International Conference on Quality and Reliability (ICQR 2005), 9-11 August, Beijing, China: 391-399

[44] Guo, R. (2005) Fuzzy Stochastic Age processes, Proceedings of the 4th International Conference on Quality and Reliability (ICQR 2005), 9-11 August, Beijing, China: 471-476

[45] Guo, R. (2005^e) A Grey Modeling of Covariate Information in Repairable System, Proceedings of the 4th International Conference on Quality and Reliability (ICQR 2005), 9-11 August, Beijing, China: 667-674

[46] Guo, R., D. Guo, C. Thiart (2005) A New Very Small-Sample Based Non-Linear Statistical Estimation Method, Proceedings of International Workshop on Recent Advances Stochastic Operations Research, 25-26 August, Canmore, Alberta, Canada: 50-57

[47] Guo, R., D. Guo (2005) Grey Reliability Analysis of Complex System, Proceedings of 10th International Conference on Industrial Engineering, Theory, Applications and Practice, 4-7 December, Clearwater, Florida, USA: 432-437

[48] Guo, R., C. E. Love (2006) Grey repairable system analysis, International Journal of Automation and Computing (in press)

[49] Guo, R. (2006) Modeling Imperfectly Repaired System Data Via Grey Differential Equations with Unequal-Gapped Times, Reliability Engineering and Systems Safety (in press)

[50] Guo, R., T. Dunne (2006) Grey Predictive Control Charts, Communications in Statistics – Simulation and Computation (in press)

[51] Hart, Jr. R. A., D. H. Clark (1999) Does Size Matter? Exploring the Small Sample Properties of Maximum Likelihood Estimation. http:/www.polmeth.wustl.edu/retrieve.php?id=222

[52] Huang, Y. S., C. C. Chang (2004) A study of defuzzification with experts' knowledge for deteriorating repairable systems, European Journal of Operational Research, 157 (3): 658-670

[53] Kijima, M. (1989) Some results for repairable systems with general repair, Journal of Applied Probability, 26: 89-102

[54] Kolassa, J. E., M. A. Tanner (1999) Small Sample Confidence Regions in Exponential Families, Biometrics, 55 (4): 1291-1294

[55] Kruse, R., K. D. Meyer (1987) Statistics with Vague data, Volume 33, Reidel, Dordrecht

[56] Liu, S. F., Y. G. Dang, Z. G. Fang (2004) Grey System Theory and Applications, The Publishing House of Science, Beijing, China (in Chinese)

[57] Love, C. E., R. Guo (1991) Using Proportional Hazard Modelling in Plant Maintenance, Quality and Reliability Engineering International, 7: 7-17

[58] Pillay, A., J. Wang (2003) Modified failure mode and effects analysis using approximate reasoning, Reliability Engineering & System Safety, 79 (1): 69-85

[59] Shen, Z. P., Y. Wang, X. R. Huang (2003) A quantification algorithm for a repairable system in the GO methodology, Reliability Engineering & System Safety 80 (3): 293-298

[60] Shirmohammadi, A., C. E. Love, Z. G. Zhang (2003) An Optimal Fixed Maintenance Cycle Policy with Possible Skipping Imminent Preventive Replacements, Journal of Operational Research Society, 54: 40-47

[61] Stadje, W., D. Zuckerman (1991) Optimal Maintenance Strategies for Repairable Systems with General degree of Repair, Journal of Applied Probability, 28: 384-396

[62] Taheri, M. S. (2003) Trends in Fuzzy Statistics, Austrian Journal of Statistics, 32 (3): 239-257

[63] Viertl, R. (1996) Statistical Methods for Non-Precise Data, CRC Press, Boca Taton

[64] Wang, D. Z., K. S. Trivedi (2005) Computing steady-state mean time to failure for non-coherent repairable systems, IEEE Transactions on Reliability, 54 (3): 506-516

[65] Wang, K. S., H. J. Po, F. S. Hsu, C. S. Liu (2002) Analysis of equivalent dynamic reliability with repairs under partial information, Reliability Engineering & System Safety, 76 (1): 29-42

[66] Wen, K. L. (2004) Grey Systems: Modeling and Prediction, Yang's Scientific Research Institute, Taiwan

[67] Yang, C. S., C. C. Lin, C. Y. Cheng (2003) Periodic Preventive Maintenance Model For Deteriorating Equipment Using Improvement Factor Method, Proceedings of the 5th ROC Symposium on Reliability and Maintainability: 345-355

[68] Yun, W. Y., K. K. Lee, S. H. Cho, K. H. Nam (2004) Estimating Parameters of Failure Model for Repairable System with Different Maintenance Effects, Advanced Reliability Modelling, Proceedings of the 2004 Asian International Workshop (AIWARM 2004), 26-27 August, Hiroshima City, Japan: 157-164